THE WORLD OF
MARINE LIFE

THE WORLD OF
MARINE LIFE

From tropical reef fish to mighty sharks

MICHAEL WRIGHT
AND GILES SPARROW

First published in 2003 for Grange Books
An imprint of Grange Books plc
The Grange
Kingsnorth Industrial Estate
Hoo, Nr Rochester
Kent ME3 9ND
www.grangebooks.co.uk

ISBN: 1-84013-508-5

Editorial and design by
Amber Books Ltd
Bradley's Close
74–77 White Lion Street
London N1 9PF

Project Editor: James Bennett
Design: Neil Rigby at Stylus Design

Printed in Singapore

PICTURE CREDITS
David Shale/Naturepl.com: 7
Topham: 13

ARTWORK CREDITS
All artwork © Istituto Geografico De Agostini, Novara SPA except the following:
Mike Langman: 9, 10, 11, 12, 16, 17, 18, 20, 26, 27, 28, 29, 30, 31, 32, 33, 101, 111, 121, 130,
136, 137, 143, 148, 149, 150, 151, 162, 172, 205.
Marshall Editions: 173, 174, 175, 193, 196, 201, 202, 203, 204, 206, 207, 212, 214, 216, 224,
225, 227, 228, 235, 239, 245, 246, 254, 269, 279, 280, 281, 283, 285, 298, 299, 300, 301, 302,
303, 304, 305, 306, 307, 308, 309, 310, 311

CONTENTS

Introduction

The world beneath the sea's surface is fascinating in its variety, but the difficulty of directly observing it without special skills or equipment means that most aspects of marine life remain hidden to most people. And even marine biologists have an enormous amount still to discover; the underwater world is without doubt the natural world's 'final frontier'. This book takes you under the waves and introduces you to the vast range of marine animals that live there. It visits the brilliant, jewel-like life of the tropical coral reef, teeming with fish and other organisms; it visits the eternal gloom of the deepest oceans, where bizarre creatures never see any light except perhaps the eerie glow produced by luminescent organs on their own or their fellow creatures' bodies; and it visits every part of the marine world in between, from the tropics to the poles.

It is, of course, impossible in a book of this size – or even in a much bigger one – to cover every known type of fish, crustacean, mollusc and the myriad other types of marine animal that are known to exist. They number many, many thousands – with undoubtedly many more thousands still to be discovered. (Remember that the seas make up about 70 per cent of the Earth's surface.) So we have selected 300 representatives that illustrate the great diversity and geographical range of marine life on our planet. They include examples from every type of marine habitat and, as explained below, examples of virtually every important major division of the animal kingdom.

They are not always the most familiar types. For one thing, what is familiar in one part of the world – a particular fish on sale in the market, for example – may never be seen in another (although even marine produce is today traded internationally). But this helps to emphasize the great diversity of life in the seas. They include examples of the smallest microscopic animals and of the biggest ones that have probably ever existed on Earth. They include favourite food items and deadly poisonous, venomous or aggressively predatory types. They include brilliantly coloured creatures that are favourite aquarium species and ones so ugly that fishmongers need to cut them into ready-to-cook fillets so that customers are not repelled! (One quick note about what this book does *not* cover: it excludes marine plants, birds, and creatures that live only in fresh water – inland rivers and lakes – although it does include a number that inhabit both fresh and salt water, either exclusively or at different stages of their life.)

The book is organized in approximate order of evolutionary complexity (*see below*), from the simplest to the most highly evolved organisms. Each main species (distinct kind of creature) covered is illustrated in colour, while the 'data' panel at the foot of each page summarizes the important facts about that species in a standard format. The general text in between focuses on various special aspects of the species or its life – and in some cases that of related species or the group as a whole to which it belongs – for every marine creature is unique and fascinating in its own special way. We very much hope you enjoy finding out about them.

Above: Many bizarre and fierce-looking fish species live in the ocean depths, but despite its fearsome appearance this fangtooth (Anoplogaster cornuta) is only 15cm (6in) long. It is found worldwide down to about 5000m (16,500ft), and eats mainly small crustaceans and fish.

THE MOST PRIMITIVE ANIMALS

It is not surprising that the seas contain examples of almost all of the major life forms that exist on Earth, since scientists believe that life itself first arose in the primaeval ocean thousands of millions of years ago, and a major part of its evolution took place there. Even in many of those groups – such as the reptiles and mammals – that evolved after life emerged onto dry land, some types returned to make the seas their home. As a result, today's seas contain direct descendants illustrating most of the stages in life's evolutionary history. The major exceptions are the insects and amphibians. Insect species make up three-quarters of the animal kingdom, with an estimated 2 million species. Of these, some 30,000 live in water at some stage of their life history, yet only around 300 normally come into contact with seawater and even fewer are truly marine. Those few live only on the sea surface, not in it. Nor are there any truly marine amphibians – sea-frogs or sea-toads – for these would lose water from their bodies into the salty seawater through their porous skin, and could not survive.

The first and simplest animals to evolve were the single-celled protozoans, and present-day marine examples include the beautiful radiolarians and the chalky-shelled foraminiferans (see pp.14–15). These make up a large part of the zooplankton – the tiny floating animals that form part of the bottom-most link in the marine food-chain. Not a great deal more complex, zoologically speaking, are the sponges (see pp.16–17). They have a non-living skeleton made of horny, chalky or glassy material within which the jelly-like living part of the sponge is found. This consists of many cells, but they are only loosely organised and do not have specialized functions; if a sponge is broken into parts, the separate sections can continue to live without permanent disruption.

The next stage of complexity is shown by the jellyfishes, anemones, corals and their relatives – the group known as cnidarians (see pp.18–27) – and the rather similar but distinct ctenophores (see pp.28–29), which include the comb jellies. In these creatures, different types of cells have specialized functions – stinging, movement, digestion and so on. In some cases such as the Portuguese man o'war (see p.18) separate types of polyps – in effect, separate individual creatures – have separate functions and work in cooperation with each other.

To most people, a worm is simply a worm, but in fact worms differ greatly in their level of complexity. Marine worms include examples ranging from the primitive flatworms (see pp.30–31) to the much more highly evolved segmented or annelid worms (see pp.110–119). Strange forms include the

8

spoon worms (*see p.120*) and the deep-ocean rift worm (*see p.121*), which lives in extremely hot water at the edge of ocean-floor hot vents several thousand metres (more than a mile) below the surface.

ANIMALS WITH SHELLS

The most familiar marine invertebrates (animals without a backbone) belong to two major groups: the molluscs (*see pp.34–109*) and the crustaceans (*see pp.122–149*). Both generally have shells of one kind or another – although in some molluscs it is reduced to a small internal stiffening body called the pen (in the case of squids and cuttlefish) or is absent altogether (in the case of octopuses and some other species). In other molluscs the shell is very obvious, and hardly any of the soft, living part of the animal is ever seen – although, in the case of mussels, oysters, clams and their like, it is of course the part that is eaten. Oysters and mussels are examples of so-called bivalve molluscs; they have two hinged 'valves' or shells which can be held tightly closed when the creature is threatened or when it is uncovered by the falling tide.

MAIN EXTERNAL PARTS OF A SQUID

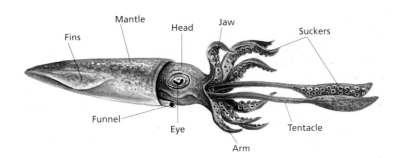

Fins • Mantle • Head • Jaw • Suckers • Funnel • Eye • Arm • Tentacle

The shells of crustaceans – which include shrimps, lobsters and crabs – are jointed in a much more complex way than those of bivalve molluscs. The main part of their body is made of many jointed segments, while their legs – five walking legs on each side in the major group that includes lobsters and crabs – have seven segments each. In many species, the first pair of legs have large pincers that are used for fighting and capturing food; the other legs also end in much smaller pincers. Crustaceans have complex sensory organs, including compound eyes made up of as many as 10,000 individual elements, touch-sensitive palps and antennae, and bristles on the antennae and mouth parts that detect tastes or smells.

MAIN EXTERNAL PARTS OF A LOBSTER

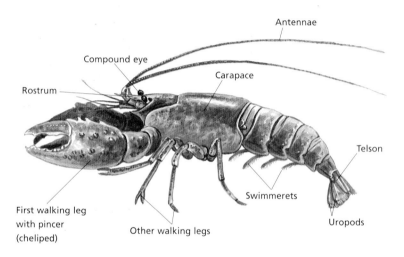

Antennae

Compound eye

Carapace

Rostrum

Telson

Swimmerets

First walking leg
with pincer
(cheliped)

Other walking legs

Uropods

Not every sea creature that looks like a crustacean is one, however. The horseshoe crabs (*see p.150*) belong to a group known as chelicerates, which also includes the arachnids (scorpions and spiders). Most of the latter live on land, but there are also marine spiders, including some remarkably large specimens that live in the ocean depths (*see p.151*). The final major group of marine invertebrates is the echinoderms, whose name means 'spiny-skins' (*see pp.152–171*). They do not have a true shell, but have a spiny skeleton just below the skin; they include such well-known marine creatures as sea urchins and starfish.

FISHES AND OTHER VERTEBRATES

The peak of animal evolution is represented by creatures that have a backbone – the vertebrates, which in the sea include fishes (*see pp.173–277*), reptiles (*see pp.278–283*) and mammals (*see pp.284–313*). Vertebrates in fact form the great majority of a larger group called chordates; non-vertebrate chordates include the sea squirts (*see p.172*), whose life history gives an insight into how vertebrates evolved.

Of all the vertebrates, it is of course the fishes that are most completely adapted to a marine environment – 'of course' because they first evolved there several hundred million years ago, and because they are virtually confined to the water. (There are a few fishes that can also breathe air, such as the mudskippers [*see p.264*].) In fact, however, the bony fishes that are now the dominant group first evolved in fresh water and later returned to the sea, where at the time fishes with a skeleton of cartilage (like today's sharks and rays) were far more numerous. Fishes have evolved into an enormous variety of distinct forms – some 24,000 known species in fresh and salt water, far more than of any other vertebrate group – and they fill by far the biggest single section of this book.

MAIN EXTERNAL PARTS OF A FISH

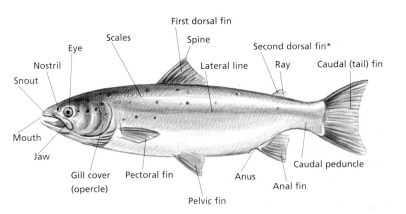

Snout · Nostril · Eye · Scales · First dorsal fin · Spine · Second dorsal fin* · Ray · Caudal (tail) fin · Lateral line · Mouth · Jaw · Gill cover (opercle) · Pectoral fin · Pelvic fin · Anus · Anal fin · Caudal peduncle

(*Many fish have only one dorsal fin)

11

In order to understand and explain the relationships between individual species of fishes (and of other creatures), ichthyologists (biologists who specialize in fishes) have organized them into various groupings. The basic category, already mentioned, is the *species*, which in some cases is subdivided into *subspecies* which show minor variations and often inhabit a certain part of the species's overall geographical range. A group of closely related species form a *genus* (plural *genera*), although sometimes a genus contains only one species. In the scientific (Latin) name, the genus name comes first, followed by the species name. For example, in the case of the cod, *Gadus morrhua*, *Gadus* is the genus name and *morrhua* the species name; you need to give both to be precise about which fish you are referring to. When an article refers to two or more species of the same genus, the genus name is abbreviated after the first mention.

MAIN INTERNAL PARTS OF A FISH

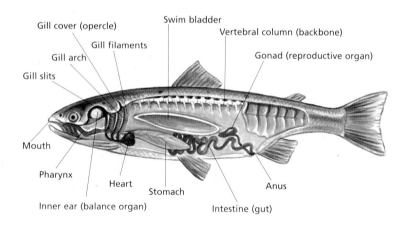

Gill cover (opercle)
Gill filaments
Gill arch
Gill slits
Swim bladder
Vertebral column (backbone)
Gonad (reproductive organ)
Mouth
Pharynx
Heart
Stomach
Anus
Inner ear (balance organ)
Intestine (gut)

There are also bigger classifications. One, two or more genera may be grouped together as a *family*, whose name usually ends in '...idae'; for example, Gadidae is the cod family, which includes haddock, pollack and other related fishes. A number of families are grouped to form an *order*, whose name ends in '...iformes'. Sometimes, orders are split into suborders, and families into subfamilies. The classification of each species is given in the 'data' panel at the foot of the page. The even larger groupings – such as Osteichthyes (bony fishes) and Chondrichthyes (cartilaginous fishes) – appear as 'signposts' at the very top of each page, above the common (ordinary English) name of the creature.

Above: Coral reefs are among the richest environments on Earth in terms of numbers of individuals and species. Many of the fish species found there are brightly coloured and defend small territories, but reefs are under threat from global warming, from alien species such as the crown-of-thorns starfish (Acanthaster planci; see p.167), and from illegal fishing with dynamite or cyanide.

Foraminiferans

The simplest animals living in the sea are the protozoa – single-celled and usually microscopic animals that are nevertheless capable of feeding and reproducing. The most widespread of all protozoans are foraminiferans, which have developed a calcium-rich shell around their bodies. These tiny creatures feed with finger-like appendages called pseudopods (false feet) that emerge through pores in the shell, capturing bacteria, algae, and smaller protozoa. Mud made of foraminiferan bodies forms around one third of the global sea floor. When compressed over millions of years, these crushed foraminiferan shells can form limestone and chalk.

Scientific name	*Nodosaria raphanus*
Classification	Kingdom Protista; Phylum Sarcodina;
	Order Foraminifera; Class Rotaliina
Size	Microscopic
Distribution	Global
Habitat	Seafloor
Diet	Bacteria; algae; protozoa
Reproduction	Binary fission – one individual splits to produce two

Radiolarians

Radiolarians are related to foraminiferans, but have far more elaborate and complex glassy skeletons. These skeletons, called tests, are made of silicate minerals the radiolarian extracts from seawater, and are often spherically symmetrical. Radiolarians frequently live in symbiosis (a mutually beneficial relationship) with algae. The radiolarian produces waste that feeds the algae, while the algae provide oxygen and food for the radiolarian. These strange creatures mostly float free in the sea, migrating through the water column daily, though no one is sure how or why they do this.

Scientific name	*Amphilonche elongata*
Classification	Kingdom Protista; Phylum Sarcodina; Superclass Actinopoda;
	Class Radiolaria; Order Entolithia; Family Acanthometrida
Size	Up to several millimetres (⅛–¼ inch)
Distribution	Global
Habitat	Free-floating
Diet	Algae; protozoa; copepods
Reproduction	Binary fission – one individual splits to produce two

Venus's flower-basket

Sponges are the simplest multicellular animals – loose collections of specialized cells combined to produce a more efficient feeding mechanism. They range from inconspicuous prickly layers growing on rocks to large, complex and beautiful structures like Venus's flower-basket. This conical 'glass sponge' is made of cells interlinked by six-rayed silicate spikes. Sponges feed by filtering water through large apertures called osculae, and into a network of internal passages, where feeding cells can filter out plankton and other food. Small prawns sometimes live inside Venus's flower-basket, unable to escape, but taking advantage of the regular food supply.

Scientific name	*Euplectella aspergillum*
Classification	Kingdom Metazoa; Phylum Porifera;
	Class Hexactinellida (glass sponges); Family Euplectellidae
Size	Up to 1.3m (4¼ft) high
Distribution	Japan; Philippines
Habitat	Deep seas
Diet	Plankton and organic debris
Reproduction	Probably hermaphroditic, distributing sperm on currents

Bath sponge

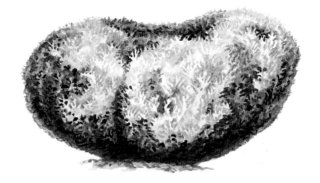

The largest and most widespread class of sponges are the horny sponges or demospongiae. They mostly build silica skeletons, but these lack the six-rayed symmetry of the glass sponges. Some, including the bath sponge, have frameworks built of spongin, a tough protein similar to those found in muscle tendons. Horny sponges include many encrusting sponges, but also larger, brightly coloured species. They often provide shelter for fish and other small sea creatures, but only a few animals will eat them, due to the arsenal of different toxins they have developed to defend themselves.

Scientific name	*Spongia officinalis*
Classification	Kingdom Metazoa; Phylum Porifera;
	Class Demospongiae (horny sponges)
Size	Up to 50cm (20in)
Distribution	Global
Habitat	Temperate and tropical waters
Diet	Plankton and organic matter
Reproduction	Can reproduce sexually or asexually depending on conditions

Portuguese man o'war

The Portuguese man o'war has a fearsome reputation as one of the most deadly creatures in the sea, though in fact its stings are rarely fatal to humans. Although it looks like a jellyfish, this strange creature is a colony of polyps closely working together. One of the polyps forms a large gas bladder that allows the colony to rise and sink in the waters. Others are specialized feeding cells, and stingers called nematocysts, forming long tendrils that spread out over a large area of water, stinging small fish that stray too close. This does not, however, prevent the small blue man o'war fish living among the tentacles for protection, or protect the man o'war from the sea turtles which like to eat it.

Scientific name	*Physalia physalis*
Classification	Phylum Cnidaria; Class Hydrozoa;
	Order Siphonophora; Family Physaliidae
Size	Tendrils stretch up to 20m (66ft) in all directions
Distribution	Worldwide in tropical and temperate waters
Habitat	Ocean surface and shallow waters
Diet	Small fish
Reproduction	Reproduces by 'budding' a group of cells to form a new colony

Purple jellyfish

Jellyfish are relatively simple animals with bodies that are more than 97 per cent water. They have no central nervous system and only rudimentary senses, but are nevertheless efficient hunters. The bell-like body of a jellyfish contains a substance called mesoglea, and it has a simple mouth on its underside, surrounded by 'arms' to grasp prey and trailing stingers to stun, paralyse, or even kill it outright. The purple jelly or purple stinger has a relatively small collection of tentacles, but makes up for it with stinging cells covering the surface of its bell. It sometimes glows at night, perhaps due to eating bioluminescent prey.

Scientific name	*Pelagia noctiluca*
Classification	Phylum Cnidaria; Class Scyphozoa; Order Semaeostomeae; Family Pelagiidae
Size	Bell up to 20cm (8in)
Distribution	Mediterranean and worldwide
Habitat	Warm temperate waters
Diet	Small fish; plankton
Reproduction	Adult jellyfish spawns polyps sexually, but polyps can split asexually

Sea wasp or Box jellyfish

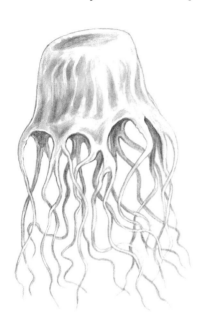

Infamous in Australian waters for their painful sting, box jellyfish have a unique shape that sets them apart from their relatives – their body is box-shaped, with tentacles spreading out from the corners. This feature may help to improve their swimming strength – unlike scyphozoans, which can only drift where the water takes them, cubozoans are strong swimmers. They also have unusually good eyesight for jellyfish, and pack some of the most vicious stings in the animal kingdom. All jellyfish have an unusual life-cycle. They mate with each other sexually to produce young called planulae. The planulae then settle on the sea bed and develop into polyps, which can in turn bud off new, free-swimming medusae.

Scientific name	*Chironex fleckeri*
Classification	Phylum Cnidaria; Class Cubozoa;
	Order Cubomedusae; Family Chirodropidae
Size	Body up to 20cm (8in), tentacles up to 3m (10ft)
Distribution	Australia; Indian Ocean
Habitat	Shallow coastal waters
Diet	Small fish, worms, and crustaceans
Reproduction	Sexual and asexual

Gem anemone

Sea anemones are cnidarians that have developed a generally stationary lifestyle – they attach themselves to rocks and use stinging tentacles to grab prey that passes too close, as well as waving to create currents that bring food to them. An anemone's body is called a polyp, and resembles the polyp form of a jellyfish. A muscular 'foot' on the base of the body, called the pedal disc, attaches to a suitable surface, and some anemones can even use this to move around, snail-fashion. Others relocate by simply inflating themselves and letting currents carry them to a new site.

Scientific name	*Bunodactis verrucosa*
Classification	Phylum Cnidaria; Class Anthozoap; Subclass Hexacorallia;
	Order Actinaria; Suborder Nyantheae; Family Actiniidae
Size	Up to 25cm (10in)
Distribution	Eastern Atlantic
Habitat	Rocky, battered coastlines
Diet	Small mussels
Reproduction	Apparently asexual, budding live young

Burrowing anemone

The tube or burrowing anemones are unusual because they create a hard tubular body case from sand particles mixed with their own mucus, and burrow backwards into soft sand and muddy sea floors. This means they look very like normal or 'true' anemones at first glance, but can have a hidden body up to a metre (40in) or more long. Normally, they lie on the sea floor with their tentacles outstretched, feeding on plankton and other small invertebrates. Many of the species have tentacles that are fluorescent, absorbing ultraviolet light and releasing it at visible wavelengths. The anemone can retreat into its tube if danger threatens.

Scientific name	*Cerianthus filiformis*
Classification	Phylum Cnidaria; Class Anthozoa; Subclass Hexacorallia; Order Ceriantharia; Suborder Spirularia; Family Cerianthidae
Size	Up to 25cm (10in) across
Distribution	Coasts of Japan
Habitat	Sand and mud on sea floor
Diet	Plankton; organic matter
Reproduction	Sexual and asexual

Sea fan

Despite their plant-like appearance, sea fans are in fact a form of soft coral or gorgonian (so-called because of their branching, snake-like appearance, similar to the hair of the gorgons in Greek myth). They are colonial animals made up of hundreds or even thousands of anemone-like creatures. These 'polyps' grow a broad, flat, semirigid skeleton from protein. Individual animals line the close-woven branches, with their tentacles spread out to catch passing plankton. Because they are filter-feeders, sea fans always keep their broad 'leaves' facing into the current. *Eunicella cavolinii* is the most common species in Mediterranean waters, found on steep rocky slopes.

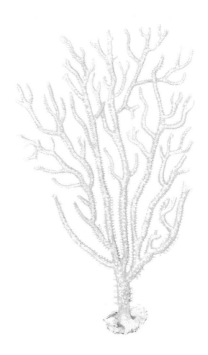

Scientific name	*Eunicella cavolinii*
Classification	Phylum Cnidaria; Class Anthozoa; Subclass Octocorallia;
	Order Alcyonacea; Family Gorgoniidae
Size	20–50cm (8–20in)
Distribution	Mediterranean and North Atlantic
Habitat	Steep rock shelves at moderate depths
Diet	Plankton
Reproduction	Eggs and sperm released from colonies fertilize in open water

Sea pen

Named because of their resemblance to old-fashioned feather quill pens, sea pens are colonial animals made from a large number of polyps working together. One large 'axial polyp,' called the rachis, buries itself deep into the seabed, while other secondary or lateral polyps form chains growing out from the axis to either side. The secondary polyps are specialized to pump water through the sea pen's arms, and to capture food. In the case of the deep-red *Pennatula phosphorea*, the polyps form large triangular 'leaves.' This is one of several seapens known to bioluminesce, emitting brilliant flashes and pulses of light passing in waves through the colony – though no one understands why.

Scientific name	*Pennatula phosphorea*
Classification	Phylum Cnidaria; Class Anthozoa; Subclass Octocorallia;
	Order Pennatulacea; Family Pennatulidae
Size	Up to 40cm (16in) long (25cm [10in] visible)
Distribution	North Atlantic; North Sea; Mediterranean
Habitat	Sandy and muddy sea floors of moderate depth
Diet	Plankton and organic matter
Reproduction	Colonies of a single sex release eggs or sperm into water

Devonshire cup coral

Hard corals are similar to anemones, sea fans and sea pens in their basic form, but different because they all secrete a hard shell of limestone called a corallum around their bodies for protection. Also, nearly all species are colonial, forming vast undersea reefs that are some of the world's richest marine habitats. The Devonshire cup coral is one of the few non-colonial corals, found on rocky coasts around western Europe. It is found in a wide variety of colours, and has a large number of tentacles to capture food and transfer it to the mouth. Although not considered colonial, cup corals do sometimes grow in clusters on suitable surfaces.

Scientific name	*Caryophyllia smithii*
Classification	Phylum Cnidaria; Class Anthozoa; Subclass Hexacorallia;
	Order Madreporaria; Family Dendrophilliidae
Size	Up to 25mm (1in)
Distribution	North Atlantic; North Sea; Mediterranean
Habitat	Rocky coasts down to 100m (330ft)
Diet	Plankton; organic matter
Reproduction	Sexual – males release sperm; eggs develop inside females

Brain coral

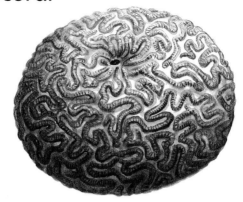

Brain corals are so-called because of the lumps and folds that bear a startling resemblance to a human brain. Being among the bulkiest and most solid corals, they can survive storms that damage and even destroy their delicate neighbours. Typically brain corals live on plankton, and nutrients from the algae that grows in their folds and grooves. Their structure gives them a large surface area and many passages through which water can flow. In fact brain corals are not a taxonomic group – the name is given to a wide variety of similar-looking creatures that are not necessarily close relations, of which *Oulophyllia*, illustrated here, is one.

Scientific name	*Oulophyllia* species
Classification	Phylum Cnidaria; Class Anthozoa; Subclass Hexacorallia;
	Order Madreporaria; Family Faviidae
Size	Up to 1m (40in) across, weighing over a tonne
Distribution	Pacific Ocean
Habitat	Coral reefs
Diet	Algal nutrients and plankton
Reproduction	Sexual – sperm and eggs fertilize in open water

Staghorn coral

While brain corals survive in coral reefs by sheer bulk, staghorn corals hold on to their environmental niche by rapid growth – although the price for this is that they are fragile and easily damaged. They may grow at a relatively breakneck speed of 10cm (4in) per year in the race to grab the best niches and collect as much sunlight as possible from above. Staghorn is used as a general name for corals of the *Acropora* genus, but not all the species have the forked appearance of antlers. Some grow to look more like bushes, while others develop flat tops that act as platforms to collect light for the symbiotic algae growing on them. They can also develop many different bright colours.

Scientific name	*Acropora* species
Classification	Phylum Cnidaria; Class Anthozoa; Subclass Hexacorallia;
	Order Madreporaria; Family Acroporidae
Size	Up to 2m (6½ft) tall
Distribution	Global
Habitat	Coral reefs
Diet	Algal nutrients and plankton
Reproduction	Sexual – sperm and eggs fertilize in open water

Sea gooseberry

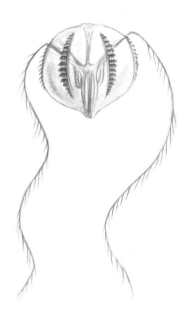

Looking superficially like a jellyfish, the sea gooseberry is in fact an animal from a completely different phylum – a comb jelly. The similarities are thought to be the result of similar adaptations to a floating lifestyle, but comb jellies lack the stinging nematocysts of jellyfish – instead they capture their food using specialized sticky cells called colloblasts. The comb jellies get their name from the eight rows of 'combs' along the sides of their bodies. In fact, these combs are fused plates called cilia which the jelly paddles back and forth to propel itself. Most comb jellies share the sea gooseberry's pair of long tentacles, which retract rapidly when stimulated.

Scientific name	*Pleurobrachia pileus*
Classification	Phylum Ctenophora; Class Tentaculata;
	Order Cydippida; Family Pleurobrachiidae
Size	Body up to 25mm (1in) across, tentacles up to 60cm (24in) long
Distribution	Atlantic Ocean; Mediterranean
Habitat	Open waters of varying depth
Diet	Plankton and organic matter
Reproduction	Animals are hermaphroditic, but release sperm and eggs into water

Venus's girdle

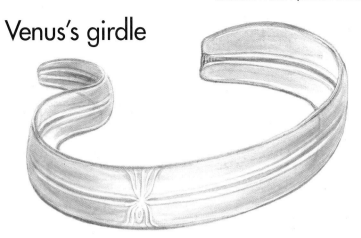

Although most comb jellies resemble the sea gooseberry, a couple of genera are startlingly different, with a long belt-like shape. The most beautiful of these is Venus's girdle, found around the world in tropical waters, and growing to 1m (40in) or more in length. In these animals the combs or cilia run along the edges of the body, allowing the animal to ripple through the water. Nearly all free-swimming comb jellies belong to the class Tentaculata – though in Venus's girdle the tentacles are not so obvious as in others. A second class of comb jellies, the Nuda, have no tentacles and generally dwell on the sea floor.

Scientific name	*Cestum veneris*
Classification	Phylum Ctenophora; Class Tentaculata;
	Order Cestida; Family Cestidae
Size	Up to 1m (40in) long
Distribution	Worldwide in tropical waters
Habitat	Open waters of varying depth
Diet	Plankton and organic matter
Reproduction	Hermaphroditic but gametes are released in water to mix with others

Polyclad flatworm

Flatworms are a simple type of worm with thin solid bodies and only a gut running down their centres. They have no blood and no lungs – instead they absorb oxygen directly from the waters around them through their paper-thin bodies. Many flatworms are parasitic, but turbellarians are mostly free-swimming predators. Many have brightly patterned skins and they move either by rippling through the water or by crawling like a land-based worm. Flatworms have a primitive head around which their sensory organs such as simple eyes and pseudotentacles (made from folds of the body wall) are clustered.

Scientific name	*Prostheceraeus vittatus*
Classification	Phylum Platyhelminthes; Class Turbellaria; Order Polycladida; Family Eryeptidae
Size	Up to 50mm (2in) long
Distribution	North Atlantic; North Sea
Habitat	Coastlines, generally below rocks
Diet	Plankton; smaller invertebrates
Reproduction	Hermaphrodites; also capable of asexual reproduction

Ribbon worm

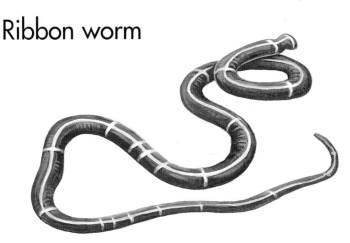

Up to 1,200 species of ribbon worms exist worldwide, and they form a unique phylum of the animal kingdom. Most are marine, and many have extremely long bodies – the group includes the world's longest animals, which can grow up to 55m (180ft) long. Although they are often brightly coloured with rings around their bodies, they are actually unsegmented. Ribbon worms are predators, often living wrapped around corals and other undersea outcrops. They hunt with a unique proboscis, a long tube pushed out from the mouth by hydraulic pressure, frequently equipped with thorn-like barbs that inject poisons into their prey.

Scientific name	*Tubulanus annulatus*
Classification	Phylum Nemertini; Class Anopla;
	Subclass Palaeonemertini; Family Tubulanidae
Size	Up to 0.5m (20in)
Distribution	Eastern North Atlantic; Mediterranean
Habitat	Coral reefs; rock outcrops; pilings
Diet	Small invertebrates
Reproduction	Sexual – worms have one sex only

Sea moss

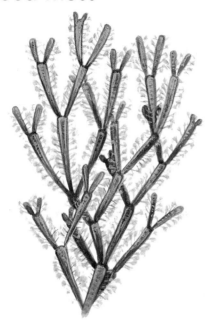

Bryozoans or moss animals look superficially plant-like, but are in fact colonial animals made of thousands of individual members called zooids. The zooids, less than 1mm (⅟₂₅in) long, each live inside a cubicle called a zooecium or house, which link together to form a leaf or mat-like structure depending on the species. They extend their tentacles to filter the water that flows around them for microscopic plankton. One species of bryozoan, *Bugula neritina*, produces a chemical called bryostatin that has proved to be a powerful anti-tumour agent, and is now being farmed off the coast of California.

Scientific name	*Bugula neritina*
Classification	Phylum Bryozoa; Class Gymnolaemata;
	Order Cheilostomata; Family Bugulidae
Size	25–50mm (1–2in)
Distribution	Temperate and tropical waters worldwide
Habitat	Sea floor, attached to rocks and other outcrops
Diet	Plankton
Reproduction	Asexual, producing larvae in ovicells on the corners of houses

Lamp shell

Lamp shells or brachiopods are externally similar to clams, living inside a bivalved shell with two halves, but they attach to a stable surface by a stalk called a pedicle (which either sticks to rock or burrows into sand). The animal inside the shell is also different, with small tentacles called lophophores filtering the water and supplying food to the mouth. Similarities to the tentacles of bryozoan zooids show that these two very different-looking groups of creatures are actually related. Brachiopods are very important in the fossil record, where they can help palaeontologists to date rock deposits, but only a few hundred species exist today.

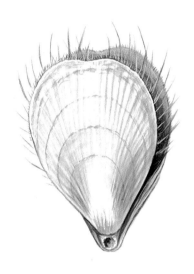

Scientific name	*Terebratulina retusa*
Classification	Phylum Brachiopoda; Class Articulata;
	Order Terebratulida
Size	Around 15mm (⅝in) across
Distribution	North Atlantic cold and temperate waters
Habitat	Sea floor to depths of 1500m (5000ft)
Diet	Plankton
Reproduction	Sexual – individuals of each sex release gametes into water

Abalone

Abalones are snail-like gastropods that graze on the sea bed and are widely farmed both for food and for their decorative shells, the pearly interiors of which are used to make jewellery. They are molluscs, with a muscular foot on the bottom of the body which they use to move around and anchor themselves to rocks. The body is protected by a low, bumpy, and characteristically ear-like shell, with a series of small holes that allows water to reach the internal gill and provides the animal with oxygen. This shell is surrounded by a frilly margin and may be one of many different colours, depending on the individual's diet.

Scientific name	*Haliotis lamellosa*
Classification	Phylum Mollusca; Class Gastropoda;
	Order Archaeogastropoda; Family Haliotidae
Size	Up to 5cm (2in) long
Distribution	Mediterranean
Habitat	Shallow coastal waters with plentiful plants
Diet	Phytoplankton (microscopic plants)
Reproduction	Separate sexes release sperm and eggs into the water

Emperor slit shell

One of the major problems for early gastropods was how to expel waste from the body. The anus is on top of the body between the gills, so either its waste must pass down the sides of the shell past the gills, or down the front, past the mouth. The slit shells were a group of abalone-like gastropods that evolved a unique solution to the problem – a slit-shaped extra opening in the shell, allowing exhaled water from the gills to pass out and carry the waste with it. At first, these primitive gastropods were only known from fossils, but then the first living one was discovered, and today some 16 species are known.

Scientific name	*Perotruchus hirasei*
Classification	Phylum Mollusca; Class Gastropoda;
	Order Archaeogastropoda; Family Pleurotomariidae
Size	Up to 10cm (4in)
Distribution	Japanese coast
Habitat	Deep water
Diet	Phytoplankton (microscopic plants)
Reproduction	Separate sexes release sperm and eggs into the water

Keyhole limpet

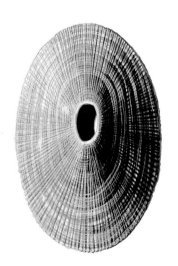

Like many marine gastropods, keyhole limpets lead a grazing lifestyle, moving slowly around on the sea floor in shallow waters, in search of algae and organic detritus. Their muscular foot also allows them to anchor to rocks and withstand heavy waves. The distinctive hole in the keyhole limpet's shell is a vent where air passed through the gills, as well as the animal's waste, can be ejected cleanly. When alive, the limpet's internal body casing, called the mantle, spreads out from under the shell's edges to cover most of the shell. The hole in the top of the shell is just a hint that the keyhole limpet is a quite different animal from the true limpet.

Scientific name	*Megathura crenulata*
Classification	Phylum Mollusca; Class Gastropoda;
	Order Archaeogastropoda; Family Fissurellidae
Size	Up to 125mm (5in) long
Distribution	Californian coast
Habitat	Sea floor from intertidal zone to 33m (100ft)
Diet	Algae and tunicates
Reproduction	Separate sexes release sperm and eggs into the water

Painted topshell

Topshells are small, pretty, and widespread gastropods, with very attractive shells. The shells are conical with very straight sides and usually pink or purple patterning, while the exposed lip of the interior is covered in iridescent nacre (mother-of-pearl). A horny valve called the opercula can block the shell's opening if it is left stranded on the shore when the tide goes out. Topshells in general reproduce sexually by releasing eggs and sperm into the water, where they fertilize to form a larval stage. The painted topshell, however, has no larval stage – instead miniature versions of the adults are born directly onto the shore.

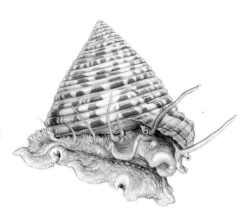

Scientific name	*Calliostoma zizyphinum*
Classification	Phylum Mollusca; Class Gastropoda;
	Order Archaeogastropoda; Family Trochidae
Size	Up to 3cm (1in) long
Distribution	Eastern North Atlantic; Mediterranean; North Sea
Habitat	Rocky shorelines from low water to 300m (1000ft) depth
Diet	Algae; organic detritus; corals
Reproduction	Sexual, through free-swimming gametes

Common periwinkle

Periwinkles are grazing gastropods with snail-like shells, often with bright colours. They are widespread on shores around the North Atlantic, and some have gills partly converted to lungs so they can venture far up the beach. Although periwinkles will venture onto sand, they spend most of their time attached to hard rock surfaces, where they can glue themselves down and survive harsh conditions such as the absence of water or even being frozen. In order to reproduce, the male creeps up alongside the female to insert a packet of sperm under her shell, and the female then produces a sac of larvae that develop in the sea.

Scientific name	*Littorina littorea*
Classification	Phylum Mollusca; Class Gastropoda;
	Order Mesogastropoda; Family Littorinidae
Size	Up to 50mm (2in) tall
Distribution	North Atlantic
Habitat	Rocky coasts and shorelines; estuaries
Diet	Algae and organic detritus
Reproduction	Sexual

Architectonica

The genus architectonica contains some of the most attractive of all shelled molluscs, with black, white, yellow, and brown concentric markings. Often these animals are called sundials, for obvious reasons. The conical shell is unusually flattened, and formed from a series of expanding whorls that leaves a large chamber, the umbilicus, open in the centre. The animal can close up this aperture with a button-shaped operculum. Sundial larvae can spend a very long time in their free-swimming 'veliger' stage, allowing them to cross large distances floating among the plankton, and giving them a wide distribution.

Scientific name	*Architectonica perspectiva*
Classification	Phylum Mollusca; Class Gastropoda;
	Order Heterogastropoda; Family Architectonicidae
Size	Up to 50mm (2in) across
Distribution	Indo-Pacific
Habitat	Shallow sandy sea beds
Diet	Small invertebrates
Reproduction	Sexual – larvae have a prolonged free-swimming stage

Turret shell

Although they look very different from the more snail-like gastropods, turret shells in fact follow the same basic pattern, simply stretched out into a long conical form. The animal can retreat far inside the shell when threatened, closing the door behind it with a horny flexible operculum that itself withdraws well inside the shell. Turritellidae burrow into the sand in search of organic particles, and they can also consume microscopic food from the water by passing it through their gills. The Turritellidae, also called deep sea augers, look superficially very similar to another group called 'auger shells,' the Terebridae.

Scientific name	*Turritella communis*
Classification	Phylum Mollusca; Class Gastropoda;
	Order Mesogastropoda; Family Turritellidae
Size	Up to 50mm (2in) long
Distribution	Mediterranean
Habitat	Sandy sea beds
Diet	Algae
Reproduction	Sexual – free-swimming sperm released by male are taken in by female

Worm shell

Perhaps the most distorted forms of gastropod shell are found in the worm shells or Vermetidae. These animals begin life with a tightly wound but elongated shell resembling a turret shell, but their conical structure gradually 'unwinds' as the mollusc grows older. In extreme cases, the shell completely loses any resemblance to a normal gastropod, as the animal inside changes the direction in which it builds new chambers. The worm shells also have an unusual method of feeding, using secretions from the muscular foot to produce a web-like mucous filter that is held up by tentacles on the feet to trap small planktonic animals.

Scientific name	*Vermetus* species
Classification	Phylum Mollusca; Class Gastropoda;
	Order Mesogastropoda; Family Vermetidae
Size	Up to 40cm (16in) long
Distribution	Tropical waters
Habitat	Below shoreline, frequently inside sponges
Diet	Plankton
Reproduction	Sexual – male produces free-floating sperm packet

Violet sea snail

One of the most unusual marine gastropods, the violet sea snail floats near the surface of tropical waters. In order to do this, it secretes a 'raft' of mucus from its foot, which traps air bubbles and allows the snail to stay buoyant. In order to reduce its weight, the snail only has a papery shell, and it is also blind. With few natural defences, these sea snails rely on camouflage to protect them from predators – their shells are shaded so they are difficult to see from above or below. However, because it has evolved to be permanently in the water, the snail lacks an operculum to close its shell, and dries out rapidly if washed up on land.

Scientific name	*Janthina janthina*
Classification	Phylum Mollusca; Class Gastropoda;
	Order Mesogastropoda; Family Janthinidae
Size	Up to 2cm (⅘in)
Distribution	Cosmopolitan
Habitat	Ocean surface in tropical waters
Diet	Floating hydrozoan colonies
Reproduction	Sexual

Precious wentletrap

The precious wentletrap gets its unusual name from the Dutch word for a spiral staircase, on account of its complex geometrical shape. Its porcelain-like shell has made it a favourite with shell collectors ever since its discovery in the seventeenth century – at first it was such a rarity that shells frequently sold for small fortunes. Some Chinese craftsmen even constructed fakes from a paste made with rice. Today the habits of the wentletrap are better known – it feeds on anemones, often inserting a proboscis into their bodies and feeding on them for hours at a time – and so it is easily found by fishermen.

Scientific name	*Epitonium scalare*
Classification	Phylum Mollusca; Class Gastropoda;
	Order Heterogastropoda; Family Epitoniidae
Size	5–6cm (2–2½in)
Distribution	Japan; South Pacific
Habitat	Sea floors at 20–50m (65–165ft)
Diet	Anemones
Reproduction	Sexual, by copulation

Pelican's foot shell

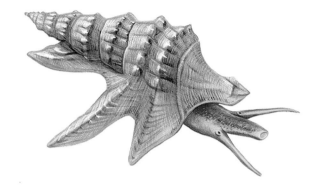

The pelican's foot shell is so-called after the shape formed by the distinctive outer lip of its shell. Only seven species are known from around the world, living at medium depths on the sea floor. The lip only develops in mature adults, and acts as a shield for the animal's head as it crawls around on sandy and muddy sea floors, digging out its prey using a specially strengthened and lengthened operculum (shell door). The muscular foot is also particularly powerful and active, allowing the animal an unusual method of movement – the body rears up off the ground and the heavy shell falls forward, so the mollusc takes 'steps' along the seabed.

Scientific name	*Aporrhais pespelecani*
Classification	Phylum Mollusca; Class Gastropoda;
	Order Caenogastropoda; Family Aporrhaidae
Size	Up to 50mm (2in)
Distribution	North Atlantic and Mediterranean
Habitat	Sandy sea floors down to around 200m (660ft)
Diet	Small invertebrates
Reproduction	Sexual; eggs develop to larvae inside female's shell, then float free

Spider shell

The spider shells are representatives of the strombid family of gastropods, which include conches and tibias. These molluscs all have highly developed eyes on the end of their front tentacles, which emerge through a special notch in the front of the shell. They are active animals, with a unique way of moving, hopping across the seabed by rapidly flipping their operculum (the horny 'doorway' to the shell) open and closed. Spider shells, also sometimes called scorpion shells, have developed their characteristic spines to help anchor them on sandy and muddy sea floors, and as a defence against predators.

Scientific name	*Lambis chiragra*
Classification	Phylum Mollusca; Class Gastropoda;
	Order Mesogastropoda; Family Strombidae
Size	Around 15cm (6in) long
Distribution	Indo-Pacific
Habitat	Sandy sea floors at moderate depths
Diet	Algae and organic debris
Reproduction	Sexual; internal fertilization; eggs laid in long strands

Rooster-tail conch

Conches have some of the most attractive shells found in nature, and have been prized for many centuries across their home territory of the Caribbean. They have been used in art and decoration, as musical instruments, and as food. Unfortunately this means they are overfished and are already a protected species in US waters. The name conch (pronounced 'konk') comes from the Greek for shell, and they have lent it to the entire hobby of shell-collecting: conchology. Conches are sea floor grazers, feeding on algae and organic debris. They live for several years, and during the breeding season lay huge strings of eggs up to 20m (66ft) long.

Scientific name	*Strombus gallus*
Classification	Phylum Mollusca; Class Gastropoda;
	Order Mesogastropoda; Family Strombidae
Size	10–15cm (4–6in)
Distribution	Caribbean
Habitat	Sea floor at moderate depths; coral reefs
Diet	Algae; organic debris
Reproduction	Sexual; female fertilized internally by male

Tibia

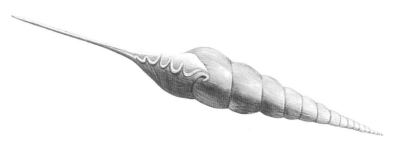

Although they look very different from other strombidae shells, tibias (also known as spindle shells) are members of the same group, and share the unique trait of hopping along with powerful flicks of their operculum. Like all strombids, they also have highly developed eyes – the best of any gastropod. *Tibia fusus*, from the Indo-Pacific, has a particularly long and fragile 'spike' – in fact a channel along which the animal can extend its siphon, the chemical-detecting equivalent of a 'nose'. It has a lustrous brown coloration, and five small 'fingers' along the outer lip of the shell opening.

Scientific name	*Tibia fusus*
Classification	Phylum Mollusca; Class Gastropoda;
	Order Mesogastropoda; Family Strombidae
Size	Up to 30cm (12in) long
Distribution	Indo-Pacific
Habitat	Sandy and muddy sea floors, down to 60m (200ft) depth
Diet	Algae and small invertebrates
Reproduction	Sexual, by copulation

Carrier shell

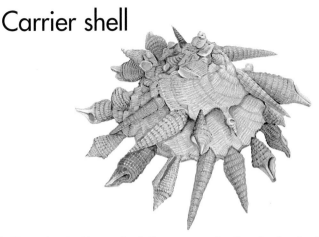

The strange-looking carrier shells are a type of mollusc that has developed unique behaviour – though no one is quite sure why! The xenophora (the name means 'bearers of foreigners') collect the shells of other dead molluscs and carefully cement them to their own shell. This laborious process may take several hours for each shell. The most likely explanation is that they do this as a form of camouflage, but several carrier shells live in the deep sea where no light penetrates. Alternatively they might be strengthening their own shells. *Xenophora pallidula* is even more of a puzzle, since it usually has a sponge attached to the apex of its shell.

Scientific name	*Xenophora pallidula*
Classification	Phylum Mollusca; Class Gastropoda; Order Caenogastropoda; Family Xenophoridae
Size	Around 8cm (3⅛in)
Distribution	Indo-Pacific; South Africa
Habitat	Muddy sea floor
Diet	Possibly algae and organic debris
Reproduction	Sexual

Carinaria

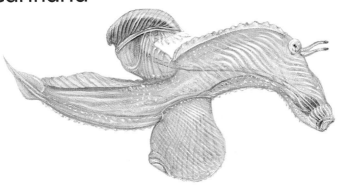

Two different groups of gastropods have developed independently towards a free-floating lifestyle away from the ocean floor. One is the pteropods or 'wing-feet,' while the other is the heteropods or 'different feet'. The common name 'sea butterflies' is often applied to both. Heteropods such as *Carinaria mediterranea* are related to the violet sea snail. They have reduced their shells to a minimum and in some cases lost them completely, while the foot has become a modified fin, held upright in the water. *Carinaria* holds its body rigid with a dorsal crest, and swims with an undulating movement through the water.

Scientific name	*Carinaria mediterranea*
Classification	Phylum Mollusca; Class Gastropoda;
	Order Caenogastropoda; Family Carinariidae
Size	20–40mm (⅘–1⅗in)
Distribution	Atlantic; Mediterranean
Habitat	Free-swimming at a range of depths
Diet	Larvae; plankton
Reproduction	Sexual; young are all born male, but some turn female

Cowrie

Cowries have beautiful glossy shells, covered with pebble-like patterns that make them highly prized among collectors. In life, though, the patterns on the shell are usually covered by two flaps of the mantle (the mollusc's body case) that wrap around it. These flaps are sometimes dull in colour, but in the case of the sieve cowry (*Cypraea cribraria*), the mantle is bright red, while the shell has distinctive white spots and a white rim around its edge. Cowrie shells have a long, narrow opening in their shells, folded in on itself along the shell's underside and with serrated edges. Thus the animal inside can retreat a long way if threatened.

Scientific name	*Cypraea cribraria*
Classification	Phylum Mollusca; Class Gastropoda;
	Order Mesogastropoda; Family Cyrpraeidae
Size	Around 25mm (1in)
Distribution	Indo-Pacific
Habitat	Coral reefs
Diet	Algae; organic matter; sponges
Reproduction	Sexual; internal fertilization; female guards eggs after laying them

Warted egg cowry

The egg cowries or ovulidae are closely related to the true cowries, and come in an equally wide variety of shapes and colorations, some of which are highly specialized. The warted egg cowry inhabits coral reefs in the Indian Ocean and West Pacific, where it wanders over the surfaces of mushroom leather corals, feeding on polyps. Its spotted mantle blends perfectly with the coral skeleton when the polyps are retracted, and as an additional defence, this cowry tastes extremely unpleasant to predators. Beneath the mantle, the shell is smooth and white, kept shiny by the constant back-and-forth motion of the mantle.

Scientific name	*Calpurnus verrucosa*
Classification	Phylum Mollusca; Class Gastropoda;
	Order Mesogastropoda; Family Ovulidae
Size	Up to 4cm (1⅗in)
Distribution	Indo-Pacific
Habitat	Shallow coral reefs
Diet	Coral polyps
Reproduction	Sexual; internal fertilization

Shuttlecock volva

Some egg cowries have unusually extended ends to their shells, giving them a spindle-like appearance. However, the main distinction between egg cowries and true cowries is that the inward curling edge of the shell is not serrated. The shuttlecock volva is one example. Like all egg cowries it feeds on corals, and its carnivorous habit gives a pink or pale purple colouring to its shell. In life, the shuttlecock volva's body is covered by a mottled brown and white mantle that confuses potential predators. Volvas live buried in the sea floor or hidden among the soft corals on which they feed.

Scientific name	*Volva volva*
Classification	Phylum Mollusca; Class Gastropoda;
	Order Mesogastropoda; Family Ovulidae
Size	Up to 135mm (5⅜in) long
Distribution	Indo-Pacific
Habitat	Sandy seabeds; coral reefs
Diet	Coral polyps
Reproduction	Sexual; internal fertilization

Naticaria

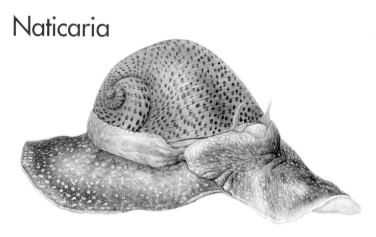

The Naticidae or moon shells get their name from their half-moon-shaped shell aperture. They are highly evolved burrowing predators that move along the sea floor ploughing into the surface layers with their powerful muscular foot. When they find a suitable victim (often a bivalve mollusc), the foot clasps the shell while the mollusc's toothed radula saws a neat hole through it and shreds the creature inside. Some species have pores in the foot which allow water to pump in and out, expanding and contracting the foot to increase its efficiency as a plough. *Naticarius millepunctatus* is a Mediterranean example.

Scientific name	*Naticarius millepunctatus*
Classification	Phylum Mollusca; Class Gastropoda;
	Order Mesogastropoda; Family Naticaridae
Size	Around 40mm (1⅝in) tall
Distribution	Mediterranean and Eastern North Atlantic
Habitat	Sandy seabeds
Diet	Bivalve molluscs
Reproduction	Sexual, by copulation

Fig shell

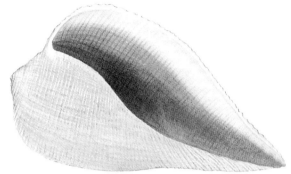

The delicate shells of the family ficidae have a bulging body and a flattened top that gives them their resemblance to figs. They were once known as pear shells or pyrulae, and live on sandy seabeds in tropical regions. The gastropod that lives inside has a large and distinctive head which it sticks out through the end of the shell, sniffing around with its siphon in search of organic detritus and small prey. Its most remarkable feature, though, is its large foot, which is truncated across the front, giving it a triangular shape. Ficus shells are thin and delicate – they rarely survive being washed up on the shore.

Scientific name	*Ficus filosa*
Classification	Phylum Mollusca; Class Gastropoda;
	Order Caenogastropoda; Family Ficidae
Size	Up to 100mm (4in)
Distribution	Indo-Pacific
Habitat	Sandy seabeds, 100–200m (330–660ft)
Diet	Detritus and small invertebrates
Reproduction	Sexual

Helmet shell

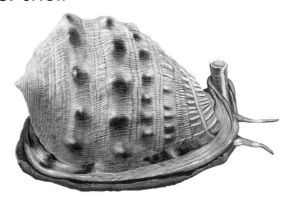

The cassidae or helmet shell family include some of the largest of all molluscs, and are named from their supposed resemblance to the helmets of Roman gladiators. They inhabit tropical warm waters, where they feed on other molluscs and poisonous sea urchins, pouncing on them by rearing up and then abruptly dropping and engulfing the prey, whose shell is then dissolved by secretions of strong sulphuric acid. *Cypraecassis rufa*, the bullmouth helmet, was much prized for its decorative value in Roman times, and was used by craftsmen for carving cameos. Julius Caesar is said to have given one as a gift to Queen Cleopatra of Egypt.

Scientific name	*Cypraecassis rufa*
Classification	Phylum Mollusca; Class Gastropoda; Order Caenogastropoda; Family Cassidae
Size	Around 16cm (6⅓in)
Distribution	Indo-Pacific
Habitat	Sandy seabeds; coral reefs
Diet	Molluscs; sea urchins
Reproduction	Sexual, through copulation

Tun shell

Tun shells are named from an old word for a cask or barrel. These molluscs have large but relatively thin shells, often surrounded by spiral ribs that resemble the metal bands around an old wooden barrel. They have huge ranges because their larval stages can float among the plankton for months at a time before settling to the seabed. *Tonna galea*, the giant tun, spends most of the day in the sands off coral reefs and islands, emerging at night to hunt crustaceans, small fish, and sea urchins, whose poison it counters with a paralysing chemical of its own. The females lay thousands of eggs in a ribbon-like mat on the seabed.

Scientific name	*Tonna galea*
Classification	Phylum Mollusca; Class Gastropoda;
	Order Caenogastropoda; Family Tonnidae
Size	Up to 25cm (10in)
Distribution	Atlantic; Mediterranean; Indo-Pacific
Habitat	Sandy seabeds, down to 40m (130ft)
Diet	Crustaceans; sea urchins; small fish
Reproduction	Sexual, through copulation

Triton

Tritons are found throughout the world in tropical and warm temperate waters – there are some 150 different species, and the shells of many have been used for centuries as trumpets. The shells frequently have strong ridges and bumps, but when alive, the triton extends its mantle over the outside, and many species have extensions called papillae which give them a 'hairy' appearance. Most species live on shallow sandy sea floors, hunting invertebrates with a similar technique to the helmet shells – but some live at great depths. Tritons produce a paralysing anaesthetic which stops their prey struggling as it is swallowed whole.

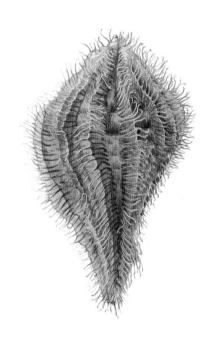

Scientific name	*Cymatium parthenopaeum*
Classification	Phylum Mollusca; Class Gastropoda;
	Caenogastropoda; Family Cymatiidae
Size	Around 80mm (3⅛in)
Distribution	Worldwide in tropical waters
Habitat	Sandy sea floors
Diet	Molluscs, worms, sea cucumbers and other invertebrates
Reproduction	Sexual, through copulation

Spiny murex

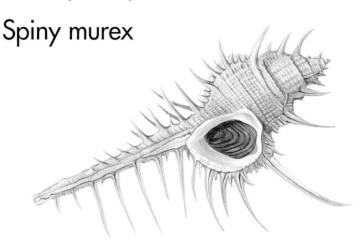

The murexes are neogastropods – the most highly evolved type of marine snails. They are voracious predators with a unique method of feeding, drilling holes into the shells of bivalves and other creatures by a variety of chemical and mechanical means. Once it has penetrated the shell, the murex consumes the flesh of the animal inside with its toothed tongue, the radula. Murexes are widely distributed around the world – this spiny variety is found around Micronesia. Other murexes from around the Mediterranean were used in ancient times to make a dye called Tyrian purple.

Scientific name	*Murex troscheli*
Classification	Phylum Mollusca; Class Gastropoda;
	Order Neogastropoda; Family Muricidae
Size	Up to 150mm (6in) long
Distribution	Indo-Pacific
Habitat	Sandy seabeds
Diet	Other molluscs; barnacles
Reproduction	Sexual

Baler shell

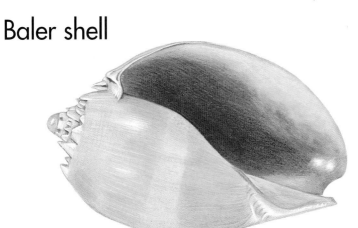

Balers have some of the largest shells of any gastropods, growing up to 45cm (18in) long. Found around Australian coasts, they were used by fishermen as useful buckets to bale out their canoes. The baler lays very few eggs for a gastropod – typically just a few dozen – because it makes a large investment in them. Each egg is wrapped into a gelatinous sac before birth, and the whole mass is glued down to a rock or other hard surface. The eggs are glued together in a spiral stack, which can take several weeks to build, and the young develop inside the egg mass until they are strong enough to break free.

Scientific name	*Melo amphora*
Classification	Phylum Mollusca; Class Gastropoda;
	Order Neogastropoda; Family Volutidae
Size	Around 45cm (18in) long
Distribution	Australian coastal waters
Habitat	Sandy sea floors
Diet	Other molluscs and invertebrates
Reproduction	Sexual; small number of eggs laid in egg mass

Olive shell

The olive shells are widespread in tropical seas. They exhibit some of the most complex patterning in nature – overlapping zigzags and v-shapes that confuse predators and prey alike – and no two shells are identical. For this reason, they have been treasured as ornaments and jewellery for centuries. Inside the shell lives a large mollusc, and the shell is usually covered by the animal's mantle. They spend most of the day buried in the sand, emerging at night to prey on smaller molluscs, crustaceans and worms, which they hold down with their large foot. They are among the fastest-moving gastropods.

Scientific name	*Olivia porphyria*
Classification	Phylum Mollusca; Class Gastropoda;
	Order Neogastropoda; Family Olividae
Size	Around 35mm (1⅜in)
Distribution	Tropical Atlantic
Habitat	Sandy seabeds at moderate depths
Diet	Small invertebrates
Reproduction	Sexual, mating throughout the year

Harp shell

The harp shells or harpidae are close relatives of the olividae, but their shells have strong ribs that resemble a harp. These molluscs are shoreline predators, spending the time while the tide is out buried in the sand to preserve moisture, since they have no operculum to close their shells. The animal inside is quite oversized for its shell, and has an enormous foot that extends when it is on the hunt, trapping unwary invertebrates in the same way as the olive shells. If a predator attempts to grab the rear portion of the foot, the harp shell can simply shed it and retreat into its shell, rather as some lizards drop their tails when cornered.

Scientific name	*Harpa amouretta*
Classification	Phylum Mollusca; Class Gastropoda;
	Order Neogastropoda; Family Harpidae
Size	Up to 60mm (2⅜in)
Distribution	Indo-Pacific
Habitat	Sandy shores, burrowing up to 3m (10ft) down
Diet	Small invertebrates
Reproduction	Sexual, mating throughout the year

Mitre shell

Although their shell looks more like a bullet than anything, mitre shells are named from their supposed resemblance to a bishop's ceremonial hat. There are over 500 species known, and most are found around the world buried in offshore sand and mud. Mitres stay on the sea floor in the day, sheltering beneath rocks and coral. At night they burrow into the sand to hunt other invertebrates, such as clams and worms, but unusually they kill their prey before eating it. They can do this because they have specially evolved glands in the radula that inject poison.

Scientific name	*Mitra zonata*
Classification	Phylum Mollusca; Class Gastropoda;
	Order Neogastropoda; Family Mitridae
Size	Around 50mm (2in)
Distribution	Mediterranean; Eastern North Atlantic
Habitat	Sandy offshore seabeds at depths 30–150m (100–500ft)
Diet	Other invertebrates
Reproduction	Sexual, through copulation

Cone shell

The cone shells are an extremely widespread family, found around the world in tropical waters with a wide variety of variations on the basic design. They are prized by collectors, but trying to catch live ones can be a risky business because they have a highly venomous bite. The cone shell 'sniffs out' prey with an organ called the siphon, then spears it with harpoon-like teeth and injects a paralysing poison. It then swallows the unfortunate animal whole and digests it within its distended stomach. Cone shell venom has been known to kill humans, but it is now attracting the attention of medical researchers investigating painkillers.

Scientific name	*Conus marmoreus*
Classification	Phylum Mollusca; Class Gastropoda;
	Order Neogastropoda; Family Conidae
Size	Around 100mm (4in)
Distribution	Indo-Pacific
Habitat	Sandy shorelines down to depths of 30m (100ft)
Diet	Small fish; molluscs
Reproduction	Sexual; several hundred eggs laid in a capsule

Actaeon

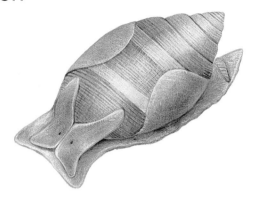

Most gastropods have a body that has been modified in order to fit it into the shell – the organs have been twisted and rearranged and it is said to be 'in torsion'. However, one group, called the opisthobranchiae, show evolution towards escaping their shells – they have less twisted body plans and some have reduced or even absent shells. The actaeonidae show the first step on this path – the shell is still large, usually with a pinkish coloration and three white bands around it, but the animal has a spade-like head specially adapted for burrowing, and the internal body plan is less distorted than usual.

Scientific name	*Actaeon tornatilis*
Classification	Phylum Mollusca; Class Gastropoda;
	Order Cephalaspidea; Family Actaeonidae
Size	Up to 25mm (1in)
Distribution	Eastern North Atlantic; Mediterranean
Habitat	Sandy shorelines below low-tide mark
Diet	Algae and organic debris
Reproduction	Sexual

Sea hare

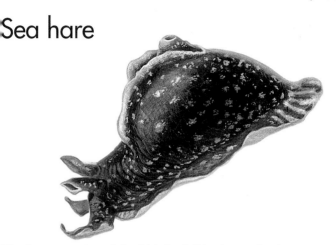

Sea hares are gastropods in which the shell has become less important – evolutionarily they are midway between marine snails and sea slugs. Their name comes from the two ear-like erect tentacles on the front of the head, and they appear to have lost their shells. But although the shell is not obvious from the outside, it is still present inside the animal's body. Sea hares have also developed limb-like folds in their bodies, called parapodia, which enable them to swim around in the seaweed beds where they feed on algae. When threatened, they can squirt a purple ink to confuse predators and cover their escape.

Scientific name	*Aplysia punctata*
Classification	Phylum Mollusca; Class Gastropoda;
	Order Anaspidea; Family Aplysiidae
Size	Up to 30cm (12in)
Distribution	Eastern North Atlantic and Mediterranean
Habitat	Offshore seabeds with rich vegetation
Diet	Algae
Reproduction	Individuals are hermaphrodites, but form chains to mate

Sea butterfly

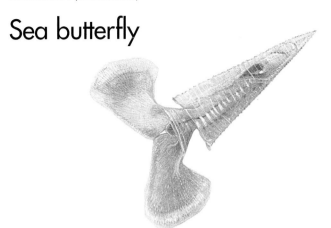

The pteropods are a group of opisthobranchs that have developed a free-floating lifestyle, but have taken a different evolutionary pathway from the undulating heteropods. These molluscs have a reduced shell and an extended foot, spreading out and forming two wing-like lobes, which they flap back and forth in order to 'fly' through the water. The scientific name for these animals, pteropods, means wing-feet, appropriately enough. *Hyalocylis striata* belongs to a group called the Thecosomata, which are plant-eaters, floating in the water and feeding on minute algae and organic particles they find there.

Scientific name	*Hyalocylis striata*
Classification	Phylum Mollusca; Class Gastropoda;
	Order Thecosomata; Family Cavolinidae
Size	Up to around 8mm (⅜in)
Distribution	Atlantic and Caribbean
Habitat	Free-swimming at 100–300m (330–1000ft)
Diet	Algae and organic particles
Reproduction	Sexual

Naked sea butterfly

Some pteropods have lost their shells completely, transforming into plankton-like predators. *Clione limacina* has a small shell as a larva, but loses this by the time it reaches adulthood. It mostly hunts shelled, plant-eating pelagic snails. When it finds a victim it grips the shell with a set of specially adapted tentacles, turning it until the shell's opening faces its mouth. It then extends large hooks that dig into the shell and evict the occupant, which the clione swallows whole. Most species of clione flourish in cold waters, and can happily survive among pack-ice – hence one nickname for them is 'the angel of the ice floes.'

Scientific name	*Clione limacina*
Classification	Phylum Mollusca; Class Gastropoda;
	Order Gymnosomata; Family Clionidae
Size	Up to 40mm (1⅜in)
Distribution	Worldwide in polar and cold temperate seas
Habitat	Free-swimming at a range of depths
Diet	Plankton, other marine snails
Reproduction	Sexual

Hypselodoris

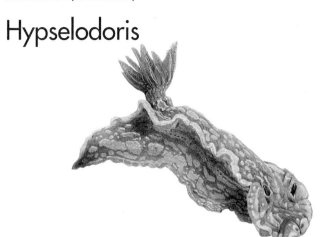

Nudibranchs are marine gastropods that have completely shed their shells and roam the seabed looking like larger versions of their terrestrial cousins, the slugs. They carry a flower-like secondary set of gills on their backs, and two sense organs called rhinophores on their heads, which can retract into sheaths when threatened. These rhinophores act as the nudibranch's nose, detecting chemical changes in the water. *Hypselodoris picta*, from the Mediterranean and Atlantic, comes in a variety of colours, but all have bright yellow marking to indicate that they are poisonous to predators.

Scientific name	*Hypselodoris picta*
Classification	Phylum Mollusca; Class Gastropoda;
	Order Nudibranchia; Family Chromodorididae
Size	Up to 20cm (8in)
Distribution	Mediterranean; Eastern North Atlantic
Habitat	Rocky seabeds in warm, shallow seas
Diet	Sponges
Reproduction	Sexual, through copulation

Chromodoris

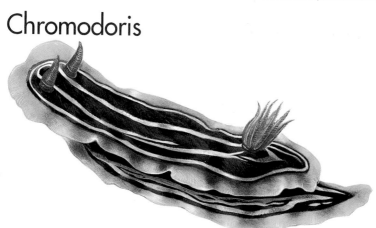

One of the most brilliantly coloured of all ocean animals is *Chromodoris quadricolor* (its very name means four-coloured). It is just one of more than 150 brightly coloured species that make up the genus Chromodoris, and always has the same pattern – orange rhinophores, gills, and mantle edge, blue and black stripes along the back, and a thin white band around the inner edge of the mantle. Without protective shells, many nudibranchs have evolved bright or complex colours, either to warn off predators or to camouflage themselves. Their eyesight is so poor that these colours cannot be for communication between nudibranchs.

Scientific name	*Chromodoris quadricolor*
Classification	Phylum Mollusca; Class Gastropoda;
	Order Nudibranchia; Family Chromodorididae
Size	Up to 30mm (1¼in)
Distribution	Red Sea, Western Indian Ocean
Habitat	Rocky sea floors
Diet	Sponges
Reproduction	Sexual, through copulation

Dendronotus

Several different animals took different evolutionary pathways to end up as sea slugs, so there is a lot of variety in their appearance. As well as the flat Chromodorididae, there are more slender forms with multiple gill attachments called cerata on their backs. Dendronotids are one of these forms – they have evolved branched cerata along their backs which give them a strong resemblance to floating seaweed. In *Dendronotus arborescens*, the rhinophores are also placed on long stalks and surrounded by further bushy growths called papillae, adding to the camouflage. The animal may be found in a number of colour variations.

Scientific name	*Dendronotus frondosus*
Classification	Phylum Mollusca; Class Gastropoda;
	Order Nudibranchia; Family Dendronotinae
Size	Up to 100mm (4in)
Distribution	North Atlantic; North Pacific; Arctic
Habitat	Coastal waters with plentiful plant life
Diet	Sea anemones; corals
Reproduction	Sexual

Flabellina

Flabellina is a member of the Aeolidiidae, a group of sea slugs with a more slender plan from the flat Chromodorididae. They are distinguished by longer rhinophores on their heads, tentacles beneath their mouths, and multiple gills (cerata) on their backs. The cerata do not attach directly to the animal's back, but instead emerge from 'trunks' that sprout from the back. *Flabellina affinis* (the name means 'fan-like') is a small but distinctive purple sea slug that is common throughout the Mediterranean. It specializes in feeding on colonies of a bushy polyp called eudendrium.

Scientific name	*Flabellina affinis*
Classification	Phylum Mollusca; Class Gastropoda;
	Order Nudibranchia; Family Flabellinidae
Size	Up to 50mm (2in)
Distribution	Eastern North Atlantic; Mediterranean
Habitat	Coastal waters with plentiful plant life
Diet	Sea anemones; corals
Reproduction	Sexual

Coryphella

Yet another animal known as a 'sea butterfly,' *Coryphella verrucosa* is a distinctive and attractive nudibranch, related to Flabellina and found throughout the North Atlantic and its neighbouring seas. Its body has a translucent white colour offset by large numbers of tube-shaped red, brown, or black cerata with white tips. It shows a great deal of variation in colour, partly because the pigmentation of its cerata depends on diet (its favourite food is a hydroid called *Tubularia*). Coryphella females lay their eggs in a very distinctive way, producing a spiral thread on the seabed.

Scientific name	*Coryphella verrucosa*
Classification	Phylum Mollusca; Class Gastropoda;
	Order Nudibranchia; Family Flabellinidae
Size	Around 25mm (1in)
Distribution	Eastern North Atlantic; Baltic; Mediterranean
Habitat	Shallow offshore waters
Diet	Hydroids
Reproduction	Sexual

Grey sea slug

The grey sea slug is a large example of the aeolid group, with a flattened body covered with thick, dense cerata that give it a furry appearance. Although grey is the most common colour, these nudibranchs can vary from purple to pale orange, usually with an inverted white V-shaped marking on the head. Grey sea slugs roam the coastal sea floor hunting for a wide range of sea anemones. They release a variety of secretions that cause the anemone's stinging cells to fire prematurely, neutralizing it so that the oral tentacles beneath the slug's mouth can grab hold of the prey safely.

Scientific name	*Aeolidia papillosa*
Classification	Phylum Mollusca; Class Gastropoda; Order Nudibranchia; Family Aeolidiidae
Size	Up to 12cm (4⅔in)
Distribution	Worldwide in temperate waters
Habitat	Offshore seabeds down to 800m (2600ft)
Diet	Sea anemones
Reproduction	Sexual

Nut shell

The nut shells or Nuculanidae are simple bivalves – that is, they live inside a pair of shells, hinged together on one side by a powerful muscle. In more advanced bivalves, the gills are used to filter food from the water and pass it to the gut, but in nut shells they are simply for extracting oxygen from the water. This mollusc lives buried in the sandy or muddy sea floor, using its muscular foot for burrowing. Tentacles collect food particles such as algae that have collected on the sand around it, and pass them to a pair of sense organs called the labial palps which sort food from waste and pass the former to the mouth.

Scientific name	*Nuculana acinacea*
Classification	Phylum Mollusca; Class Bivalvia; Order Nuculoida; Family Nuculanidae
Size	Around 10mm (⅜in)
Distribution	Indo-Pacific
Habitat	Sandy shallow seabeds
Diet	Algae and organic debris
Reproduction	Sexual, with external fertilization

Zebra ark shell

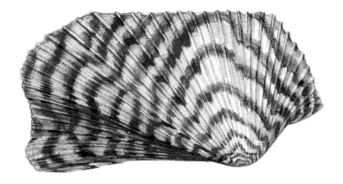

The ark shells are roughly square or rectangular in shape, with radial ribs running across their surface. When the animal is alive, the exterior of the shell is covered by velvety tissue called a periostracum. Pigments in this tissue help the shell blend with its surroundings, making it look like a stone. The zebra ark gets its name from its striped shell, and is also known as the turkey wing, which it resembles when splayed out. It secures itself to rocks and other hard substances by means of a byssus – a cluster of strong filaments produced from secretions in a special gland of the body.

Scientific name	*Arca zebra*
Classification	Phylum Mollusca; Class Bivalvia;
	Order Arcoida; Family Arcidae
Size	Up to 10cm (4in)
Distribution	Western Atlantic
Habitat	Shallow seabeds
Diet	Algae and organic debris
Reproduction	Sexual, with external fertilization

Horse mussel

Mussels are pear-shaped bivalves that form large colonies. Some attach themselves to rocks while others bury themselves in sand by means of their byssus filaments. Both types rely on strong flows of water to bring them food, which they filter out with their gills, so they are usually found at or just below the surface on wave-beaten shores. Horse mussels are not edible, but other types are a major food source and edible mussels are farmed widely. However, their feeding method means that toxins can become highly concentrated in them – algal blooms can sometimes render whole populations poisonous.

Scientific name	*Modiolus difficilis*
Classification	Phylum Mollusca; Class Bivalvia;
	Order Mytiloida; Family Mytilidae
Size	Up to 20cm (8in)
Distribution	Indo-Pacific
Habitat	Sandy seabeds in shallow waters
Diet	Algae and organic debris
Reproduction	Sexual, with external fertilization

Pearl oyster

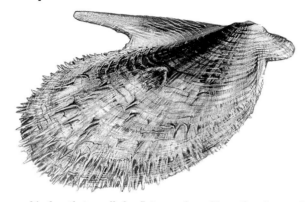

Oysters are bivalves that usually lay flat on rocks and have therefore evolved one flat shell and one convex one. However, they still retain their byssus gland, which allows them to attach to corals, rocks, or anything handy. The family Pteriidae are famous as the oysters that form pearls. They are easily distinguished by the long straight extension of the hinge, giving the shell a wing-like shape. This is particularly pronounced in *Pteria hirundo*, the Mediterranean pearl oyster. Pearls form when a piece of grit lodges in the oyster – the oyster secretes layers of nacre (the substance with which its inner shell is coated) onto the grit to neutralize it.

Scientific name	*Pteria hirundo*
Classification	Phylum Mollusca; Class Bivalvia;
	Order Pterioida; Family Pteriidae
Size	Around 11cm (4⅜in)
Distribution	Mediterranean; Eastern North Atlantic
Habitat	Sandy seabed just offshore
Diet	Floating detritus and algae
Reproduction	Sexual, with external fertilization

Hammer oyster

Since oysters have evolved to attach themselves to rocks by one valve, the foot of these molluscs has fallen out of use and become vestigial, while the shape of the shell has become elongated, especially in more advanced oysters. Perhaps the ultimate example are the hammer oysters, which have evolved into a long T-shape, with the crossbar of the T forming the hinge, and the downstroke enclosing the main body of the animal. *Malleus malleus,* from the Indo-Pacific, lives around coral reefs and has a white shell with a pearly blue lining near the hinge, where the mollusc itself lives.

Scientific name	*Malleus malleus*
Classification	Phylum Mollusca; Class Bivalvia;
	Order Pterioida; Family Malleidae
Size	Up to 17cm (6⅔in)
Distribution	Indo-Pacific
Habitat	Shallow waters around coral reefs
Diet	Algae and organic debris
Reproduction	Sexual, with external fertilization

Pen or ear shell

One of the largest bivalves in the world, *Pinna nobilis* is found across the Mediterranean. Typically it grows to around 30–50cm (12–20in), but it continues to grow throughout its life and specimens have been found as large as 1.2m (48in) long. The shell grows with up to one third of its body buried beneath the sand, in coastal waters overgrown with plants. It attaches itself to the sea floor by a byssus of long, fine threads – the Romans used threads from this creature to weave light and transparent textiles. The shell frequently becomes overgrown with smaller organisms, and two small crab species have evolved to live in symbiosis with it.

Scientific name	*Pinna nobilis*
Classification	Phylum Mollusca; Class Bivalvia;
	Order Pterioida; Family Pinnidae
Size	Typically around 30–50cm (12–20in)
Distribution	Mediterranean
Habitat	Coastal waters with plant life, down to 30m (100ft)
Diet	Algae and organic debris
Reproduction	Hermaphroditic, alternating sexes; external fertilization

Regal thorny oyster

Despite their common name, thorny oysters are not oysters at all – they are a type of scallop. Important differences include the nature of their hinge – while oysters (and most bivalves) have a relatively simple toothed hinge held together by muscle, the thorny oysters have a more complex ball-and-socket arrangement. Like scallops, they also have a more complex nervous system, with eyes on the end of tentacles peering out from the shell in all directions, and sending information back to a simple brain. Thorny oysters seem to have developed their thorns for camouflage – they encourage the growth of other marine creatures on them.

Scientific name	*Spondylus regius*
Classification	Phylum Mollusca; Class Bivalvia;
	Order Pterioida; Family Spondylidae
Size	Around 150mm (6in)
Distribution	Indo-Pacific
Habitat	Coral reefs
Diet	Algae and organic debris
Reproduction	Sexual, with external fertilization

Queen scallop

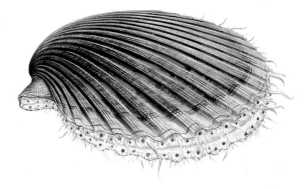

With their attractive fan-shaped shells, scallops have been a motif of art around the world for centuries, in everything from Boticelli's 'Birth of Venus' to oil company logos. But these molluscs fully deserve the attention – they are highly evolved with several unique features. Eyes on tentacles around the edge of the creature's mantle allow it to sense light and dark, and probably more complex patterns. They are also fast movers – by rapid opening and closing of their shells, they can pump water in and out of siphons near the hinge, allowing them to make a jet-propelled escape from predators.

Scientific name	*Chlamys opercularis*
Classification	Phylum Mollusca, Class Bivalvia, Order Pterioida, Family Pectinidae
Size	Up to around 9cm (3⅗in)
Distribution	North Atlantic
Habitat	Rocky offshore waters
Diet	Algae and organic debris
Reproduction	Sexual, with external fertilization

Oyster

The edible oysters of the family Ostreidae are found around the world and widely used as a source of food. They are designed for an adult life permanently fixed to a single rock, typically in estuaries and shallow coastal waters, and can be readily farmed if suitable conditions can be produced. *Ostrea edulis* is the common European species, and has been introduced to oyster farms around the world, but is threatened in some of its native habitats by the faster-growing Pacific oyster. *O. cristagalli* from the Pacific is a more unusual species, with several large folds in its valves and special anchoring spines to secure itself to rocks.

Scientific name	*Ostrea cristagalli*
Classification	Phylum Mollusca; Class Bivalvia;
	Order Ostreoida; Family Ostreidae
Size	Up to around 9cm (3⅔in)
Distribution	Indo-Pacific
Habitat	Coral reefs
Diet	Algae and organic debris
Reproduction	Sexual, with external fertilization

Oxheart cockle

Cockles are bivalves whose two shells are identical and symmetrical. They live buried half-in and half-out of sandy seabeds, and their shells are rigid and hard enough to resist most predators. The valves frequently have strong well-defined ridges across them. *Glossus humanus*, the oxheart cockle, covers its shell with a red periostracum when alive. It is distinctive because of the two curved 'umbones' at the back of the shell by the hinge. Interestingly, many shallow-water species of bivalve are thought to have triangular shells because it aids them in re-embedding if they are torn loose from the seabed by storms.

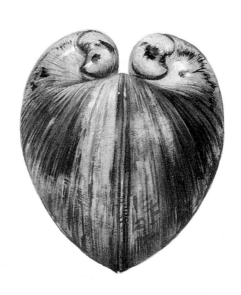

Scientific name	*Glossus humanus*
Classification	Phylum Mollusca; Class Bivalvia;
	Order Glossoidea; Family Glossidae
Size	Typically around 7cm (2⅔in), but up to 12cm (4¾in)
Distribution	Eastern North Atlantic and Mediterranean
Habitat	Sandy off shore seabeds
Diet	Algae and organic debris
Reproduction	Sexual, with external fertilization

Spiny cockle

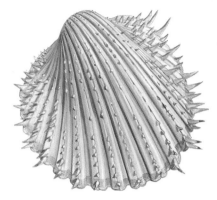

The European spiny cockle, *Acanthocardia aculeata*, is common in waters off Europe and West Africa. Its strong radial ridges are marked with rows of thorn-like spines, and separated by wide channels. The shell itself is relatively thin, and because the thorns are relatively fragile, they rarely survive intact when the shell is washed ashore. However, the live animal, which roots itself to sandy seabeds at depths down to around 100m (330ft), typically has many ranks of evenly arranged spines. Spiny cockles are popular edible shellfish around the Mediterranean, and are now being harvested in Scandinavia too.

Scientific name	*Acanthocardia aculeata*
Classification	Phylum Mollusca; Class Bivalvia;
	Order Heterodonta; Family Cardiidae
Size	Typically 6–7cm (2⅜–2⅞in), up to 10cm (4in)
Distribution	Eastern North Atlantic; Mediterranean
Habitat	Sandy seabeds down to 100m (330ft)
Diet	Algae and organic debris
Reproduction	Sexual, with external fertilization

Giant clam

Giant clams are the largest and most impressive bivalves, growing to more than a metre (40in) across, and weighting up to 270 kilos (600lb). But although they may look fearsome, and have a reputation from numerous horror stories, giant clams are far too slow-moving to attack divers, or even fish. They dwell on coral reefs, and feed off organic nutrients produced by the algae colonies that grow on their mantle. During the day, the mollusc extends its mantle out over the lips of its shell to absorb sunlight that allows the algae to flourish. The algae in turn flourish on the clam's waste products.

Scientific name	*Tridacna gigas*
Classification	Phylum Mollusca; Class Bivalvia;
	Order Heterodonta; Family Tridacnae
Size	Up to 1m (40in) across
Distribution	Indo-Pacific
Habitat	Coral reefs
Diet	Nutrients from algae, some filter-feeding in harsh conditions
Reproduction	Sexual, with external fertilization

Venus clam

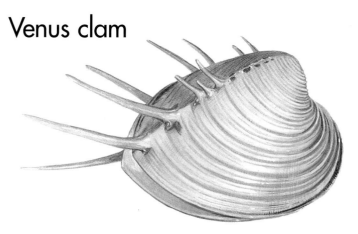

The Venus clams are the largest family of bivalves, and generally live buried in sandy seabeds at a wide variety of depths. They have a hard porcelain-like shell, and show a wide variety of variations in both the sculpture and the coloration of their shells. However, as rules of thumb they usually have a polished surface and often have rounded, concentric ribs. They also have an inward curve near the rear of the shell, called a pallial sinus. This curve, present on both valves, creates an opening for an organ called the siphon, which pumps water down through the sand. Spiny forms such as *Pitar lupanaria* are relatively rare.

Scientific name	*Pitar lupanaria*
Classification	Phylum Mollusca; Class Bivalvia;
	Order Veneroida; Family Veneridae
Size	Up to 6cm (2⅜in)
Distribution	Eastern Pacific
Habitat	Sandy offshore seabeds
Diet	Algae and organic debris
Reproduction	Sexual, with external fertilization

Wedding-cake Venus shell

Venus shells often exhibit elaborate patterns, as shown by the many-tiered appearance of the wedding-cake Venus shell, *Bassina disjecta*. The individual layers of the mollusc's shell, called lamellae, form a typical pattern of concentric but irregularly shaped rings, and extend outward in fragile, thin frills that may provide some form of defence. The wedding-cake Venus can grow up to 65mm (2½in) long, and is found in waters around southern Australia and Tasmania. It is typically a pale pinkish-white colour, and lives buried in shallow muddy sands just offshore, filtering water and food through its gills via its siphon.

Scientific name	*Bassina disjecta*
Classification	Phylum Mollusca; Class Bivalvia;
	Order Veneroida; Family Veneridae
Size	Around 40mm (1⅗in)
Distribution	Australia
Habitat	Sandy coastal seabeds
Diet	Algae and organic debris
Reproduction	Sexual, with external fertilization

Common razor shell

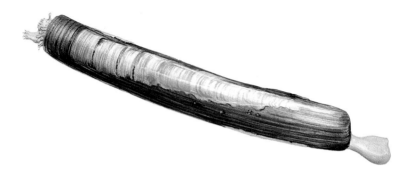

Razor shells are unusual bivalves that have evolved a tube-like shape. The two valves fit together with a gap at both ends, and the animals bury themselves upright in the sand, with either the tip of one end or just the breathing and feeding siphons sticking up through the seabed. Razor shells leave a distinctive keyhole-shaped hole in the sand when they withdraw their siphons, which can reveal their presence. If threatened, the razor shell can dig down through the sand at high speed, expanding and contracting its foot rapidly to push sand out of the way. When it settles, the foot extends from the bottom of the shell and forms an anchor.

Scientific name	*Ensis ensis*
Classification	Phylum Mollusca; Class Bivalvia;
	Order Veneroida; Family Pharidae
Size	Up to 20cm (8in)
Distribution	Eastern North Atlantic; Mediterranean
Habitat	Sandy shores just below low-water mark
Diet	Algae and organic debris
Reproduction	Sexual, with external fertilization

Otter shell

Otter shells are large bivalves found around most European and North African coasts. They can grow up to 12cm (4¾in) across, and their shells are stretched into a long elliptical shape with only shallow sculpting on the surface. The shell itself is a creamy white colour, and in life is covered with a brownish periostracum. The otter shells' stretched shape is an adaptation to a deep burrowing lifestyle – they bury themselves up to 25cm (10in) deep in the sand and feed and breathe through a long siphon. Other members of their family, the Mactridae, are even capable of burrowing into rock.

Scientific name	*Lutraria lutraria*
Classification	Phylum Mollusca, Class Bivalvia,
	Order Veneroida, Family Mactridae
Size	Typically around 10cm (4in), up to 12.5cm (5in)
Distribution	Eastern North Atlantic, Mediterranean
Habitat	Sandy seabeds just offshore
Diet	Algae and organic debris
Reproduction	Sexual, with external fertilization

Ship worm

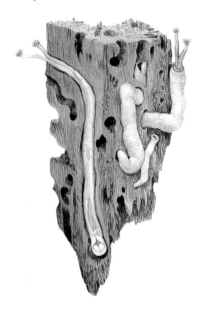

Shipworms have evolved for a highly specialized lifestyle, boring into driftwood and, in more modern times, ships' hulls and shipwrecks. They are sometimes referred to as marine woodworm, but are in fact unusual bivalves. They penetrate wood when they are still larvae, and as they grow they bore holes through the wood simply by rocking their abrasive shell back and forth. As they move through the wood, they line their tunnel with a white chalky deposit. Since the signs of entry are microscopic, timbers can frequently be riddled with shipworm before the first signs of damage are noticed.

Scientific name	*Teredo navalis*
Classification	Phylum Mollusca; Class Bivalvia;
	Order Myoida; Family Teredinidae
Size	Up to 20cm (8in) long
Distribution	Eastern North Atlantic
Habitat	Waterlogged wood
Diet	Wood, algae and organic debris through filter-feeding
Reproduction	Hermaphroditic, alternating sexes; fertilization takes place in water

Giant watering-pot shell

The Clavagellidae are bizarre bivalves that have long-outgrown their original shells, and all but abandoned them in favour of life in a chalky tube somewhat similar to those formed by shipworms. *Penicillus giganteus*, known as the giant watering-pot shell, has tiny vestigial valves at the bottom end of an upright tube that it builds from its own calcium-rich secretions. The tube sticks up through shallow, sandy seabeds in the Indian and Pacific oceans, and allows the animal to filter-feed on algae and organic particles. Clavagellidae are just one family within the Pholadomyoida, which frequently show unusual adaptations.

Scientific name	*Penicillus giganteus*
Classification	Phylum Mollusca; Class Bivalvia;
	Order Pholadomyoida; Family Clavagellidae
Size	Up to 20cm (8in) long
Distribution	Indo-Pacific
Habitat	Sandy shallow seabeds
Diet	Algae and organic debris
Reproduction	Sexual, with external fertilization

Angelwing clam

Angelwing clams have some of the most attractive shells of all bivalves. They are members of the family Pholadidae, the piddocks, and are specialized burrowers. Piddock shells do not actually join up to enclose the animal itself – rather they are attached to either side and are slowly rotated to grind through sand, wood, or even rock. The shells themselves are sturdy and have blunt forward-facing spikes to increase their grinding ability. Young angelwings bury themselves as much as 90cm (3ft) into sand or mud. They grow rapidly, feeding through a long siphon to the seabed, and are unable to rebury themselves if disturbed as adults.

Scientific name	*Cyrtopleura costata*
Classification	Phylum Mollusca; Class Bivalvia;
	Order Myoida; Family Pholadidae
Size	Up to 20cm (8in)
Distribution	Western Atlantic
Habitat	Sandy shores in intertidal zone, or just below low-tide mark
Diet	Algae and organic debris
Reproduction	Sexual, with external fertilization

Pearly nautilus

The bizarre, prehistoric-looking nautilus is a cephalopod (from the Greek for 'head-foot'), like squids and octopuses. It is the most primitive member of the class still surviving, with a strong resemblance to the famous fossil ammonites. Nautiluses have ninety tentacles on their heads, rather than the eight or ten in other living cephalopods, and they are also the only cephalopods with shells. However, their shells are very different from those of other molluscs – they consist of a number of chambers, and the animal itself lives in the outermost one. By pumping gas in or out of the inner chambers, the nautilus can rise or sink like a submarine.

Scientific name	*Nautilus pompilius*
Classification	Phylum Mollusca; Class Cephalopoda;
	Order; Family Nautilidae
Size	Up to 20cm (8in) diameter
Distribution	Western Pacific
Habitat	Free-swimming at depths to 500m (1600ft)
Diet	Crustaceans; small fish
Reproduction	Sexual, through copulation

Spirula

The curious cephalopod spirula seems to be a half way stage between nautiloids and cuttlefish and squid proper. It has a curling, horn-shaped spiral shell within its body, with similar 'buoyancy tanks' to those found in the nautilus, allowing it to rise and sink in the water column. However, it has eight short arms and two longer tentacles, plus a pair of tiny fins to help it swim. The animal lives in the outermost chamber of its shell, and can withdraw completely when threatened, closing the shell up with its mantle flaps. It has also abandoned any shell structures that would help it to remain 'horizontal', and is quite content to hang head-down in the water.

Scientific name	*Spirula spirula*
Classification	Phylum Mollusca; Class Cephalopoda;
	Order Sepioidea; Family Spirulidae
Size	1.5–5cm (⅔in–2in)
Distribution	Worldwide
Habitat	Free-swimming at depths from 100–1000m (330–3300ft)
Diet	Plankton
Reproduction	Sexual, through transfer of sperm

Common cuttlefish

Cuttlefish are remarkably complex cephalopods, commonly found in shallow offshore waters. Their bodies are built around a curved internal 'bone' (in fact the remnant of their ancestors' shells). They have well-developed eyes and brains, and move around by a form of 'jet propulsion' – pumping water through their siphons at high speed. One of their most impressive features is their ability to rapidly alter their skin pigmentation – for courtship, camouflage, help in hunting, or even, it seems, purely depending on their mood. They catch their prey by ambush, suddenly unleashing their two long tentacles to grab unwary fish and crustaceans.

Scientific name	*Sepia officinalis*
Classification	Phylum Mollusca; Class Cephalopoda;
	Order Sepiida; Family Sepiidae
Size	Body up to 40cm (16in) 6 FT. +
Distribution	Eastern North Atlantic, Mediterranean
Habitat	Shallow offshore waters
Diet	Small fish; molluscs; crustaceans; other cuttlefish
Reproduction	Sexual, through transfer of sperm

Lycoteuthis

The small deep-sea squid *Lycoteuthis diadema* has all the typical features of a squid – eight arms and two longer tentacles, large eyes, muscular mouthparts with a horny beak for tearing prey, a bullet-shaped body with fins and a shield-like blade called the pen (the remnant of the shell) to which muscles attach. What makes *L. diadema* unusual, though, is the bioluminescent patches or photophores that dot the bodies of both males and females. The two sexes have distinctly different patterns and colours, and scientists think they must indulge in elaborate mating rituals, although these have never been observed.

Scientific name	*Lycoteuthis diadema*
Classification	Phylum Mollusca; Class Cephalopoda;
	Order Teuthida; Family Lycoteuthidae
Size	Body up to 15cm (6in)
Distribution	Pacific
Habitat	Free-swimming in deep seas, migrating upwards at night
Diet	Crustaceans; small fish
Reproduction	Sexual, through transfer of sperm

Common squid

The common squids of the genus *Loligo* are widespread around the world, and are fished for food in many countries. They are long and slender, with broad fins at the rear creating a diamond-shape. Like all squids, they hunt with their tentacles, which have clubs on the end and are covered with grasping suckers. Once captured, prey is pulled towards the mouth, where the shorter arms help manoeuvre it into the mouth, and the beak and radula (a tooth-lined tongue), help shred it into digestible chunks. Another common feature shared by all advanced cephalopods is the ability to squirt ink when threatened to cover their escape.

Scientific name	*Loligo vulgaris*
Classification	Phylum Mollusca; Class Cephalopoda;
	Order Teuthida; Family Loliginidae
Size	Body up to 55cm (22in) long
Distribution	Eastern North Atlantic; Mediterranean
Habitat	Shallow coastal waters
Diet	Small fish; crustaceans
Reproduction	Sexual, through transfer of sperm

Histioteuthis

Squids of the Histioteuthidae family are distinguished in various ways – their arms are long and thick compared to those of most squids, and can be folded back over the head for protection, while their bodies are short, with only small fins. However, the strangest feature of this family by far is their unequal eyes. The left eye is much larger than the right, and angled upward to look for the silhouettes of animals against the weak light from the surface. It even incorporates a filter that is thought to help it spot creatures using bioluminescence to disguise themselves. The right eye, by comparison, is normal-sized and often ringed by bioluminescent spots.

Scientific name	*Histioteuthis reversa*
Classification	Phylum Mollusca; Class Cephalopoda;
	Order Teuthida; Family Histioteuthidae
Size	Body up to 33cm (13in)
Distribution	North Atlantic; Mediterranean; Eastern South Atlantic
Habitat	Free-swimming at moderate depths
Diet	Small fish; crustaceans
Reproduction	Sexual, through transfer of sperm

Sandalops

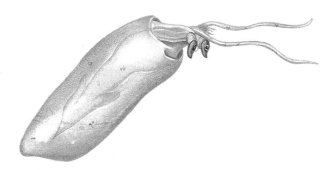

The Cranchiidae family of squids have an unusually enlarged buoyancy chamber, which tends to give them a bloated appearance. They also have shorter arms than some other squids, which they sometimes fold back over their heads in a little-understood 'cockatoo posture'. Many cranchiids are transparent – probably because they spend the early part of their lives near the sunlit top of the water column. As they grow older and larger, they also sink downwards, and some species undergo distinct biological changes – for example, the eyes of *Sandalops melancholicus* change radically to adapt to the lower light levels.

Scientific name	*Sandalops melancholicus*
Classification	Phylum Mollusca, Class Cephalopoda, Order Teuthida, Family Cranchiidae
Size	Up to 12cm (4⅔in)
Distribution	Worldwide in tropical waters
Habitat	Free-swimming in mid-ocean depths
Diet	Plankton, small crustaceans
Reproduction	Sexual, through transfer of sperm

Chiroteuthis

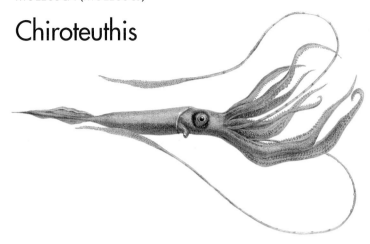

The chiroteuthids are medium-sized deep-sea squids with a very long and slender body pattern. They have long arms, and very long tentacles that can retract completely into a sheath. All chiroteuthids pass through a larval form called the doratopsis stage, which has an extremely long and highly complex tail. The adult squids are frequently semi-transparent, with complex light-producing photophores lining the arms and parts of the body. In order to function well in deep seas, they have 'flotation chambers' filled with ammonium chloride. This fluid is lighter than water, and balances the weight of the squid's other tissues, giving it neutral buoyancy.

Scientific name	*Chiroteuthis picteti*
Classification	Phylum Mollusca; Class Cephalopoda;
	Order Teuthida; Family Chiroteuthidae
Size	Body up to 40cm (16in) long
Distribution	Indo-Pacific
Habitat	Free-swimming in deep seas
Diet	Small fish; crustaceans
Reproduction	Sexual, through transfer of sperm

Giant squid

The giant squid is a living legend, a beast from sailor's stories that turned out to be real – but it is only known from dead specimens washed ashore, dredged up, or found in the stomachs of stranded sperm whales . It has never been seen alive. However, the animal is truly impressive – the largest invertebrate that has ever lived, with a body length of up to 5 metres (16ft), and total length up to 18m (60ft) – and the largest specimens may still not have been found! Despite their size, Architeuthids have relatively small fins. However, they have the largest eyes in the animal kingdom, some 25cm (10in) across, adapted to work in the inky deep seas.

Scientific name	*Architeuthis dux*
Classification	Phylum Mollusca; Class Cephalopoda;
	Order Teuthida; Family Architeuthidae
Size	Up to 18m (60ft) in known specimens
Distribution	Worldwide
Habitat	Free-swimming at depths of 200–1000m (660–3300ft)
Diet	Fish, other squid
Reproduction	Sexual – females lay up to a million eggs

Vampire squid

Looking like something from a nightmare, *Vampyroteuthis infernalis*, 'the vampire squid from hell' is not quite so fearsome as its name suggests, but it is fascinating nonetheless. A small and usually black gelatinous creature with the consistency of a jellyfish and a large collection of photophores, it actually lies in an order of its own, between squids and octopuses. *Vampyroteuthis* has eight arms almost completely joined by webbing, but also two long sensory filaments that it extends into the water. When a prey animal disturbs the filaments, *Vampyroteuthis* pounces, using its large fins to 'fly' through the water.

Scientific name	*Vampyroteuthis infernalis*
Classification	Phylum Mollusca; Class Cephalopoda;
	Order Vampyromorphida; Family Vampiroteuthidae
Size	Body up to 18cm (6in) long
Distribution	Worldwide in tropical and temperate seas
Habitat	Free-swimming at around 400–1000m (1300–3300 ft)
Diet	Small fish; crustaceans
Reproduction	Sexual, details unknown

Opisthoteuthis

The comical-looking octopus *Opisthoteuthis extensa* and its relatives are sometimes called 'flapjack devilfish' on account of their extremely flattened appearance. Like all octopuses, they have eight arms, but in this case the arms are engulfed and disguised by a broad web. Opisthoteuthids are also unusual in that their internal vestigial shell is quite large and well developed. They spend most of their time scuttling along the sea floor or swimming just above it by contractions of the web. The radula (toothed tongue) is nearly useless, which means they have to swallow prey whole.

Scientific name	*Opisthoteuthis extensa*
Classification	Phylum Mollusca; Class Cephalopoda;
	Order Octopoda; Family Opisthoteuthidae
Size	Up to 1.5m (5ft) total length
Distribution	Eastern Pacific
Habitat	Sea floors at depths of 800–1500m (2700–5000ft)
Diet	Small fish; crustaceans
Reproduction	Sexual, through transfer of sperm

Bolitaena

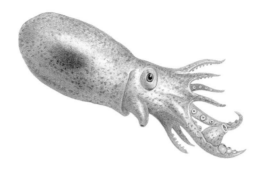

Bolitaenids are small octopuses found around the world in deep warm seas. They are semitransparent, with relatively short arms, each lined with a single row of suckers. Like most octopods, they are incirrate – they have lost their fins – and uniquely for this group they are bioluminescent. The females develop a large glowing ring around the mouth when they reach sexual maturity, which gives off a yellow-green light of a very specific wavelength that is difficult for predators to see. Presumably they use this to attract males, while the males have an enlarged saliva gland that may produce pheromones to attract the females.

Scientific name	*Bolitaena pygmaea*
Classification	Phylum Mollusca; Class Cephalopoda;
	Order Octopoda; Family Bolitaenidae
Size	Body around 25mm long
Distribution	Worldwide in tropical and subtropical seas
Habitat	Free-swimming at 800–1400m (2700–4600ft)
Diet	Plankton
Reproduction	Sexual, through transfer of sperm

Amphitretus

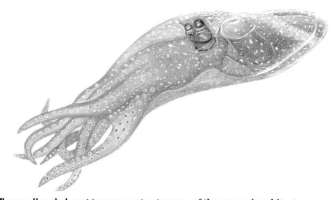

The small and almost transparent octopuses of the genus *Amphitretus* are thought to inhabit the mid-depths of all the world's tropical oceans. Floating in dark waters, they look more like plankton than advanced cephalopods. They have a single row of suckers along each tentacle, doubling near the tip. As with most cephalopods, one arm in the males is specially adapted to form a hectocotylus – a special organ used in conveying a packet of sperm from the male into the female's body. Another unusual feature of this *Amphitretus* are its eyes – simple cylindrical organs far removed from those of other cephalopods.

Scientific name	*Amphitretus pelagicus*
Classification	Phylum Mollusca; Class Cephalopoda; Order Octopoda; Family Opisthoteuthidae
Size	Body around 9cm (3⅗in)
Distribution	Worldwide in tropical and subtropical waters
Habitat	Free-swimming in moderate to deep seas
Diet	Plankton
Reproduction	Sexual, through transfer of sperm

Blue-ringed octopus

Small but fearsome predators, the blue-ringed octopuses live in Indo-Pacific waters and especially around Australia. Normally they are an undistinguished brown colour, but when threatened, they flush cream with blue rings or stripes (depending on species). This is a warning to predators that the octopus has a fearsome defence, usually used in hunting crustaceans and molluscs. After piercing skin or shell with its beak, it can spit a paralysing poison from modified saliva glands into its prey's bloodstream. This poison has been known to kill unwary people on Australian beaches by inhibiting respiration and suffocating the victim.

Scientific name	*Hapalochlaena maculosa*
Classification	Phylum Mollusca; Class Cephalopoda;
	Order Octopoda; Family Octopodidae
Size	Around 10cm (4in)
Distribution	Australia
Habitat	Seashores and offshore reefs to 40m (130ft)
Diet	Crustaceans; molluscs
Reproduction	Sexual, through transfer of sperm

Common octopus

The common octopus is a remarkable and highly successful creature. It is found all around the world, favouring sandy seabeds where it can dig itself a burrow and disguise itself when not feeding or breeding. Octopuses feed on a variety of creatures, including crustaceans and molluscs, and stockpile food supplies. When threatened, they can rapidly change their body shape or colour, or squirt ink to cover a rapid, jet-propelled escape. Fascinatingly, octopuses are highly intelligent – laboratory experiments have shown that they are skilled problem-solvers and good at remembering the solutions to previous problems when presented with new ones.

Scientific name	*Octopus vulgaris*
Classification	Phylum Mollusca; Class Cephalopoda; Order Octopoda; Family Octopodidae
Size	Up to around 90cm (36in) long
Distribution	Worldwide in tropical and temperate seas
Habitat	Shallow coastal waters, preferring sandy seabeds
Diet	Molluscs; crustaceans; fish
Reproduction	Sexual – females lay up to a quarter of a million eggs

Giant Pacific octopus

One of the largest and without doubt the heaviest of all cephalopods, the giant Pacific octopus is a creature of myth and legend. However, true giants are rare – these creatures reach sexual maturity at a weight of only 15 kilos (33lb), and have short lifespans, usually around five years. Females brood their huge clutches of eggs and die as the eggs begin to hatch. In many respects, giant octopus behaviour is similar to that of the common octopus – they hunt similar prey and make use of burrows for protection and brooding. Nevertheless, some truly monstrous examples have been found, with weights up to 270kg (600lb) and arms over 9m (30ft) across.

Scientific name	*Enteroctopus dofleini*
Classification	Phylum Mollusca; Class Cephalopoda;
	Order Octopoda; Family Octopodidae
Size	Typically around 2m (80in) across, up to 9m (30ft)
Distribution	Pacific Ocean
Habitat	Coastal seabeds from low tide to 750m (2500ft)
Diet	Molluscs; crustaceans; fish
Reproduction	Sexual, through transfer of sperm

Paper argonaut

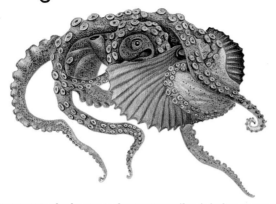

The paper argonaut, also known as the paper nautilus, is in fact an unusual form of octopus. Females are typically 20cm (8in) across, and wrap themselves in a thin shell secreted by a special web between two of their arms. This shell acts as a brooding pouch for the female's eggs, and larvae only emerge once hatched. The males are equally unusual – they are typically just a fraction of the size of the female, and have a sperm-handling third arm (hectocotylus) that breaks off during breeding and wriggles into the female's mantle to fertilize the eggs. Thus, unusually for cephalopods, females can reproduce many times while males only breed once.

Scientific name	*Argonauta argo*
Classification	Phylum Mollusca; Class Cephalopoda;
	Order Octopoda; Family Octopodidae
Size	Females around 20cm (8in), males around 2cm (⅘in)
Distribution	Global in tropical and subtropical waters
Habitat	Free-swimming at moderate depths
Diet	Plankton – sometimes parasitize jellyfish
Reproduction	Sexual, through transfer of sperm

Errant polychaete, or scaleworm

Polychaetes are primitive segmented worms – creatures whose bodies are divided into a number of segments called metameres. The front segment forms the head, while the rearmost contains the anus, and the ones in between are usually identical. Polychaetes are divided into Errantia and Sedentaria – names suggested by their different lifestyles. Errantia are mobile, using appendages called parapodia, formed by folds in the body wall, for propulsion. They are also covered in irritating and sometimes poisonous spines. The heads of Errantia also have well-developed eyes and sensory tentacles for finding their way along the seabed.

Scientific name	*Hermione hystrix*
Classification	Phylum Annelida; Class Polychaeta (Polychaete worms); Subclass Errantia
Size	Up to around 25cm (10in) long
Distribution	Eastern North Atlantic; Mediterranean
Habitat	Rocky seabeds and shores
Diet	Predatory on small invertebrates
Reproduction	Sexual, by external fertilization

Sea mouse

The sea mouse, so called because of its furry appearance, is a bottom-dwelling polychaete worm that normally lies buried head-first in the sand. Yet this unassuming creature has one unique feature – the iridescent threads or setae that emerge from its scaled back. Normally, these have a red sheen, warning off predators, but when light shines on them perpendicularly, they flush green and blue. The setae are made of millions of submicroscopic crystals that reflect and filter the faint light of the ocean depths – physicists hope that something similar might one day be used in light-based optical computers.

Scientific name	*Aphrodite aculeata*
Classification	Phylum Annelida, Class Polychaeta (Polychaete worms),
	Subclass Errantia, Family Aphroditae
Size	Up to 15–20cm (6–8in)
Distribution	Eastern North Atlantic, Mediterranean
Habitat	Muddy sea floors down to around 2000m (6600ft)
Diet	Predatory on small invertebrates
Reproduction	Sexual

Palolo worm

Palolo worms have one of the most unique reproduction methods in all of nature. Normally they live burrowed into shallow-water coral reefs, and are considered a delicacy by the natives of the Polynesian islands. However, on a couple of nights each year, tied precisely to the phase of the moon, the worms detach their rear segments, called gametophores. These swim to the surface and form a huge swarm, releasing eggs and sperm which fertilize to form larvae. After a few days floating on the surface, the surviving larvae swim down to the reefs below. The adults, meanwhile, continue feeding as normal, and regenerate a new gametophore so they can repeat the cycle next year.

Scientific name	*Eunice viridis*
Classification	Phylum Annelida; Class Polychaeta (Polychaete worms);
	Subclass Errantia; Family Eunicidae
Size	Around 40cm (16in) long
Distribution	South Pacific
Habitat	Coral reefs
Diet	Corals
Reproduction	Sexual with external fertilization by separate gametophores

Fireworm

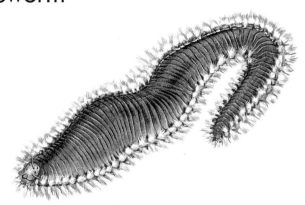

Coloured bright red or brown as a warning to other animals, fireworms are not to be trifled with – their bristles or setae are hollow and contain powerful venom. In fact, the name fireworm is a general term for a variety of venomous polychaetes. The most dangerous and voracious of these are the Amphinomidae, which live on reefs and feed on the coral by engulfing its tips and digesting the living tissue off the skeleton. These fireworms have setae designed to break off and embed themselves in a predator. The fibres are extremely brittle and difficult to remove, and humans who have touched them have sometimes lost fingers as a result.

Scientific name	*Hermodice Carunculata*
Classification	Phylum Annelida; Class Polychaeta (Polychaete worms);
	Subclass Errantia; Family Amphinomidae
Size	Up to 30cm (12in)
Distribution	Mediterranean; Caribbean
Habitat	Coral reefs
Diet	Corals
Reproduction	Sexual, with external fertilization

Myrianida

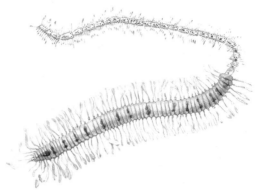

Myrianida is another errant polychaete with a bizarre reproductive system. Like several others, it practises 'schizogamy' for reproduction – an organism that separates from the main body of the worm is responsible for breeding. In Myrianida's case, each worm produces a series of sexual 'buds' from the penultimate (preanal) segment, which later release sperm and eggs. Myrianida normally dwells on the sea floor, rising to the surface to reproduce. Some species have developed into ectoparasites – for example *Myrianida pinnigera* fastens onto sea squirts and feeds off their bodily fluids.

Scientific name	*Myrianida sp.*
Classification	Phylum Annelida; Class Polychaeta (Polychaete worms);
	Subclass Errantia; Family Syllidae
Size	Variable
Distribution	Eastern North Atlantic; Mediterranean
Habitat	Sea floor at varying depths
Diet	Corals and other small invertebrates; some species parasitic
Reproduction	Schizogamy and reproduction by sexual buds

Fan worm

Sedentary polychaetes, as their name suggests, remain in one location throughout their lives. Typically they live either buried in the sand or hidden in tubes of their own construction. Fan worms occupy papery tubes made of their own solidified secretions, anchored either in the sand or preferably on rocks or shells. Their parapodia (rudimentary appendages) are underdeveloped or absent, but they have a spiral-shaped fan of brightly coloured tentacles around their head that are used to capture plankton, and retract instantly on contact or when the light is blocked by a shadow. *Spirographis* is a very successful colonist, having invaded Australian coastal waters attached to the hulls of ships.

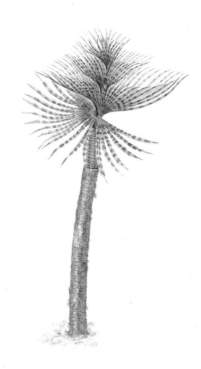

Scientific name	*Spirographis spallanzani*
Classification	Phylum Annelida; Class Polychaeta (Polychaete worms);
	Subclass Sedentaria; Family Sabellidae
Size	Tube up to 50cm (20in) long; fan up to 20cm (8in) radius
Distribution	Mediterranean; Australia
Habitat	Shallow subtidal waters and deep sheltered waters
Diet	Floating organic matter
Reproduction	Sexual, with external fertilization

Peacock worm

The peacock worm is one of the most beautiful fan worms, with striped feathery tentacles that give it its name. The worm itself lives buried in the sand, protected by a hardened tube of sand particles cemented with its own mucus. Only the top few centimetres protrude from the sand, along with the tentacles which spread out and act like gills to absorb oxygen. They also capture particles of organic and inorganic matter. Inorganic grains are used to repair and extend the protective tube, while organic ones are transported to the worm's mouth along grooves in the tentacles, each of which is lined with tiny hairlike cilia that wave back and forth to move the food downward.

Scientific name	*Sabella pavonina*
Classification	Phylum Annelida; Class Polychaeta (Polychaete worms);
	Subclass Sedentaria; Family Sabellidae
Size	Up to 40cm (16in) long
Distribution	Eastern North Atlantic; Mediterranean
Habitat	Shallow and tidal waters, including some freshwaters
Diet	Plankton, suspended organic particles
Reproduction	Sexual, with external fertilization

Encrusting polychaete

Some sedentary worms secrete harder materials to form pinkish-white calcareous (limestone) tubes. Some species even build a trapdoor (operculum) in the mouth of the tube for protection. Their trumpet-shaped tubes form reefs on solid rock outcrops or shells embedded in the seabed, which can act as habitats for other creatures. The worm that lives within the tube has red tentacles and a funnel-shaped head. Unlike other worms which are entirely passive feeders, these ones gently stir their tentacles to generate a current and bring food to them, allowing them to live in becalmed waters where other species could not survive.

Scientific name	*Serpula vermicularis*
Classification	Phylum Annelida, Class Polychaeta (Polychaete worms)
	Subclass Sedentaria, Family Sabellidae
Size	Up to 7cm (2⅗in) long
Distribution	Eastern North Atlantic, Mediterranean
Habitat	Extreme low tide down to 250m (800ft)
Diet	Floating organic matter
Reproduction	Sexual, with external fertilization

Lugworm

The lugworm is best known as the favourite bait of many fishermen, and as the creature responsible for leaving casts on the sand as the tide goes out. Lugworms are a form of sedentary polychaete that burrow up to 60cm (24in) into the sand for protection, only emerging to feed on decayed organic matter that has fallen to the seabed. Because they gulp down food and sand together, they must excrete large amounts of waste in the form of casts. Lugworms are hermaphrodites, with the reproductive parts of both sexes. However, they do not reproduce on their own – their eggs must still be fertilized by another worm's sperm, and vice versa.

Scientific name	*Arenicola Marina*
Classification	Phylum Annelida; Class Polychaeta (Polychaete worms);
	Subclass Sedentaria; Family Arenicolidae
Size	Up to about 25cm (10in)
Distribution	North Atlantic and surrounding waters
Habitat	Tidal sands
Diet	Organic debris
Reproduction	Sexual; hermaphroditic but still require external fertilization

Chaetopterus

Also known as parchment worms, polychaetes of the genus Chaetopterus live in tough, U-shaped tubes of papery material they secrete themselves. Both ends of the tube taper, so the stout worm is permanently trapped inside. It feeds by using specially adapted parapodia (body folds) that wave back and forth and pump water and food particles through the tube. Among the most successful species is *Chaetopterus variopedatus*, which has two unique features – it can glow in the dark by bioluminescence, and more remarkably it can regenerate itself completely from even a single segment.

Scientific name	*Chaetopterus variopedatus*
Classification	Phylum Annelida; Class Polychaeta (Polychaete worms); Subclass Sedentaria; Family Chaetopteridae
Size	Up to 25cm (10in) long
Distribution	North Atlantic; North Pacific
Habitat	Sandy and rocky seabeds and rock outcrops to middle depths
Diet	Plankton
Reproduction	Sexual, with external fertilization

Spoon worm

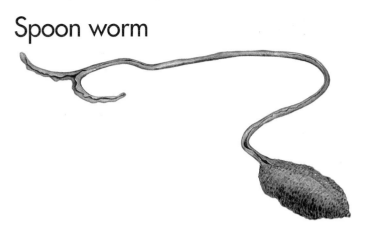

The spoon worms or Echiuridae are a group closely related to annelids, but which are only segmented as larvae – their segments disappear when they reach the adult stage. For this reason they are considered a separate phylum, but grouped with annelids and other close relatives in a larger group, the Trochozoa. *Bonellia viridis* has a bizarre reproduction method – the female is large and round, with a long proboscis. The male is tiny and lives permanently inside the female. Another unusual spoon worm is the innkeeper worm *Urechis caupo*, which lives in a U-shaped burrow and 'trawls' the water with a net of mucus.

Scientific name	*Bonellia viridis*
Classification	Group Trochozoa, Phylum Echiura
	Family Bonellidae
Size	Female up to 1m (40in) including proboscis; male about 2mm (⅙in)
Distribution	North Atlantic; Mediterranean; Red Sea
Habitat	Rocky seabeds
Diet	Plankton
Reproduction	Male lives permanently inside female

Giant rift worm

Growing taller than a man, and unknown to science until the 1970s, giant rift worms are among the strangest creatures of the deep ocean floor. They live in huge colonies around undersea thermal vents, where molten lava wells up from inside the earth and heats the waters. Rift worms look like giant tube worms, but grow from both ends, and have a segmented front and an unsegmented rear from which the tentacles burrow into the sand. Strangest of all, the worms have no digestive system – their food is provided by a colony of bacteria living within them, and the worms have developed to deliver oxygen and sulphur-rich compounds to these bacteria, allowing them to do their work.

Scientific name	*Vestimentefera sp.*
Classification	Group Trochozoa; Phylum Pogonophora;
	Class Vestimentifera or Phylum Vestimentifera (still disputed)
Size	Up to 3m (10ft) long
Distribution	Global mid-ocean trenches
Habitat	Deep sea thermal vents
Diet	Carbon compounds produced by internal bacteria
Reproduction	Larvae drift between thermal vents on deep-ocean currents

Ostracods

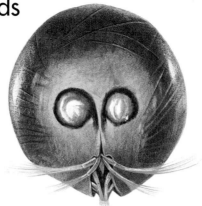

The seed shrimps or ostracods are a group of simple crustaceans that live in hinged spherical shells. They can retreat into these shells for protection with only their feathery antennae sticking out to find food. The antennae also propel the shrimp about by 'rowing' through the water. Over 12,000 species of these minute creatures are known – most dwelling on the sea floor but many floating freely in the oceans. Some species can cope with extremes of temperature and salinity, while others signal to each other in flashes of light. The largest of all, *Gigantocypris agassizi*, is a deep-ocean dweller with the most sensitive eyes in all of nature.

Scientific name	*Gigantocypris agassizi*
Classification	Phylum Crustacea; Class Ostracoda;
	Order Myodocopina; Family Cypridinidae
Size	Up to 3cm (1¼in)
Distribution	Cosmopolitan
Habitat	Deep oceans at 1000–3000m (3300–10,000ft)
Diet	Copepods and small fish
Reproduction	Sexual, through copulation

Copepods

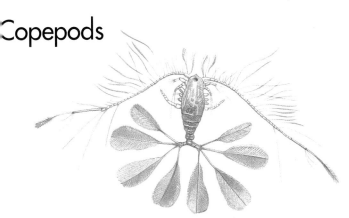

Copepods are often called the insects of the sea, because their 12,000 species are incredibly widespread and constitute the majority of the ocean's living mass. They form the basis of the marine food chain in many parts of the world, and occupy every available niche from the deep-ocean floor to freshwater rivers and lakes. About a third of known species are parasitic. Free-swimming copepods are called calanoids – they have long and elaborate feathery antennae ideal for swimming, despite their tiny size (most are just a couple of millimetres long). Many migrate to the surface waters at night, and sink back to the depths in daylight.

Scientific name	*Calocalanus pavo*
Classification	Phylum Crustacea; Class Copepoda;
	Order Calanoida; Family Paracalanidae
Size	A few millimetres across
Distribution	Cosmopolitan
Habitat	Free-swimming in the water column
Diet	Free-floating algae
Reproduction	Male mates by attaching a sperm package to the female

Goose barnacle

The goose barnacle is a member of the cirripedia – a specialized group of crustaceans that live all their adult lives without moving. Goose barnacles attach themselves to rocks and other hard surfaces by a stalk called a peduncle, that can grow up to 90cm (36in) long. Their main body consists of two plates that open to allow their six pairs of legs to emerge, and these legs act as a net, filtering the waters for food particles. Barnacle young swim freely until they attach themselves to a suitable surface, which may include floating debris, larger animals, and ships – a buildup of barnacles on a ship's hull can reduce its speed by up to 30%.

Scientific name	*Lepas anatifera*
Classification	Phylum Crustacea; Class Cirripedia; Order Thoracica; Family Lepadidae
Size	Up to 10cm (4in) long
Distribution	Worldwide in warm and temperate waters
Habitat	Intertidal zone
Diet	Plankton
Reproduction	Sexual, through copulation

Common barnacle

There are at least 800 known species of barnacle around the world, occupying a wide range of habitats from tidal shores down to the deep seas. Common barnacles, also called acorn barnacles, attach themselves directly to a hard surface and grow a cone of fixed limestone plates from which a pair of flexible opening plates emerges. They filter the water for food with their legs, in the same way as goose barnacles. Some barnacles are parasitic, burrowing into corals, other crustaceans, or the shells of molluscs. Others, while not parasitic, have evolved to live on larger animals such as whales.

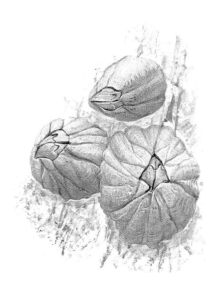

Scientific name	*Balanus crenatus*
Classification	Phylum Crustacea; Class Cirripedia;
	Order Thoracica; Family Balanidae
Size	Up to 25mm (1in) across
Distribution	Cosmopolitan
Habitat	Intertidal zone to deep sea
Diet	Plankton
Reproduction	Sexual, through copulation

Nebalia

The Malacostraca are the crustacean group that contains most of the familiar crustaceans – shrimps, prawns, lobsters, and crabs. The most primitive members of this group are the Leptostraca, of which Nebalia are the most successful genus. Malacostrans are usually distinguished by a body with 20 segments (six fused ones forming the head, eight the thorax, and six the abdomen). The Nebaliidae are unusual because they have an extra segment in their abdomen. They also have a large transparent carapace in two parts – the rear section encloses and protects most of their body, while the smaller front section forms a visor to protect the head.

Scientific name	*Nebalia bipes*
Classification	Phylum Crustacea; Class Malacostraca;
	Order Leptostraca; Family Nebaliidae
Size	A few millimetres
Distribution	Northeast US coast; Europe
Habitat	Coastal waters, particularly with vegetation
Diet	Organic waste and vegetation
Reproduction	Sexual, through copulation

Mantis shrimp

The mantis shrimp's name is doubly misleading – it is unrelated to preying mantises, and is also not a shrimp. Mantis shrimps are fearsome hunters, capable of attacking prey much larger than themselves. They bear either vicious spines or heavily calcified clubs on their forelimbs, and these limbs are capable of lightning quick movements, lashing out faster than 10m/s (30ft/s) to spear or club passing prey. The largest 'smasher' mantis shrimps pack a punch equivalent to that of a .22 calibre bullet. The mantises also have the natural world's most complex eyes, with no less than 16 different types of photoreceptors for detecting light.

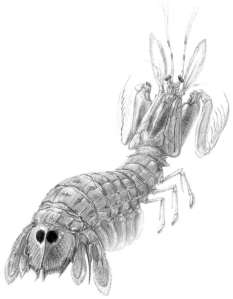

Scientific name	*Squilla empusa*
Classification	Phylum Crustacea; Class Malacostraca;
	Order Stomatopoda; Family Squillidae
Size	Up to 30cm (12in)
Distribution	Gulf of Mexico and surrounding waters
Habitat	Offshore seabeds down to 150m (500ft)
Diet	Small fish, crustaceans, and other invertebrates
Reproduction	Sexual, through copulation, with a complex mating ritual

Opossum shrimp

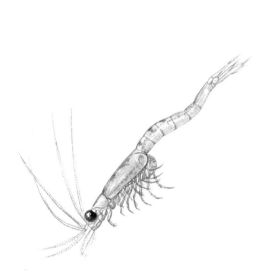

The opposum shrimps, or Mysidacea, are so-called because they carry their eggs and young in a pouch, like marsupial mammals. They are largely filter-feeders, swimming in huge swarms just above the sea floor and filtering plankton from the water. Sometimes they will also scavenge on dead animals and even hunt animals smaller than they are. Opposum shrimps are almost entirely transparent, except for their black compound eyes, making them difficult for predators to see. They propel themselves using a fan on the end of their tail. The tail also has a pair of statocysts – cavities containing a particle floating in fluid, which act as balance sensors.

Scientific name	*Mysis relicta*
Classification	Phylum Crustacea; Class Malacostraca;
	Order Mysidacea; Family Mysidae
Size	Up to 1.6cm (⅝in)
Distribution	Europe, Asia, North America
Habitat	Shallow seafloor and cold freshwater lakes
Diet	Plankton; carrion
Reproduction	Young brooded in a pouch

Diastylis

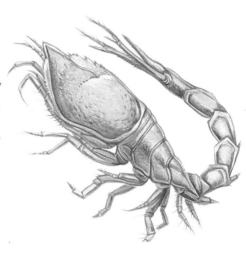

The cumaceans are a group of crustaceans that live on the sea floor, burrowing into mud (frequently backwards), in search of particles of organic matter. Areas of the sea floor with rich pickings are often densely colonized by cumaceans. They have five pairs of legs and three pairs of feeding appendages around their mouths, emerging from a large protective carapace that is often highly ornamented. Because the carapace is hard, juvenile cumaceans must shed it several times as they grow. In the case of *Diastylis rathkei*, all the juveniles moult at the same time, resulting in huge swarms coming to the surface at once.

Scientific name	*Diastylis rathkei*
Classification	Phylum Crustacea; Class Malacostraca; Order Cumacea; Family Diastylidae
Size	Up to 2cm (¾in)
Distribution	Arctic waters
Habitat	Ocean floor at depths less than 200m (660ft)
Diet	Organic matter on sand grains; plankton
Reproduction	Probably mate while swimming freely at night

Sea slater

Sea slaters, also called sea cockroaches, are in fact shoreline animals – malacostracans who have shed their carapace, and whose gills have adapted to a terrestrial lifestyle. They are fast-moving scavengers with an obvious resemblance to woodlice, to which they are related. Other adaptations include sessile eyes (in other words, eyes on the exterior of the head, and not on stalks), and simple thoracic legs without pincers. Isopods form swarms of large numbers and only venture below the tidemark when the waters are out. They brood about 80 young in a pouch on the thorax, and release them as miniature versions of the adult.

Scientific name	*Ligia oceanica*
Classification	Phylum Crustacea; Class Malacostraca;
	Order Isopoda; Family Ligiidae
Size	Up to 3cm (1¼in)
Distribution	Coasts worldwide
Habitat	Rocky seashores
Diet	Organic detritus
Reproduction	Male fertilizes female using specialized spike on abdominal legs

Amphipod

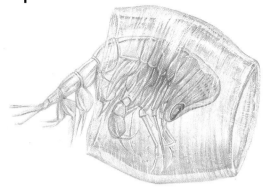

Like the sea slaters, amphipods are specially adapted malacostracans which have shed their carapaces. Their bodies have become flattened from side to side, and they are curved into a C-shape, which their long antennae continues. Their 4,000 species include the widespread sand hoppers and the skeleton shrimps, and are found in a variety of habitats around the world, ranging from the shorelines to the deep seas. *Phronima sedentaria* is a member of the family Hyperiidae, and like many of its relatives, it is parasitic, living, feeding, and bearing live young inside jellyfish and tunicates such as sea squirts and salps.

Scientific name	*Phronima sedentaria*
Classification	Phylum Crustacea; Class Malacostraca;
	Order Amphipoda; Family Hyperiidae
Size	A few millimetres
Distribution	Mid-ocean waters
Habitat	Living within gelatinous zooplankton
Diet	Internal tissues of hosts
Reproduction	Bears live young which can swim away to a new host

Skeleton shrimp

Skeleton shrimps are amphipods with slender, elongated bodies, a long thorax and a very short abdomen. They resemble a cross between a shrimp and a praying mantis and are equipped with fearsome-looking pincers on their front limbs. The shrimps do not swim, but instead climb around on sponges and seaweeds, scraping algae and polyps from the surface. They can also anchor themselves with their hind legs, reaching out to grab passing plankton with their foreclaws. Like opposum shrimps, they brood their young in a pouch. Several hundred species of skeleton shrimp are known, and close relatives live a similar lifestyle on the skins of whales and dolphins.

Scientific name	*Caprella linearis*
Classification	Phylum Crustacea; Class Malacostraca;
	Order Amphipoda; Family Caprellidae
Size	Up to 2cm (¾in)
Distribution	Eastern North Atlantic and surrounding waters
Habitat	Seaweeds and sponges at depths of up to 4m (13ft)
Diet	Algae; polyps; plankton
Reproduction	Females brood young in a pouch

Krill

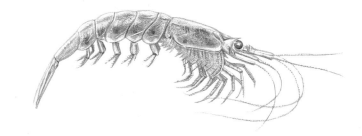

One of the most important species on the planet, the shrimp-like krill form the base of many food chains around the world. These small transparent, bioluminescent creatures form huge swarms in the cold waters around both poles. In the Southern Ocean particularly, Antarctic krill replace small fish as a major food supply for larger animals. Nordic krill (as illustrated) feed on copepods, giving them a distinctive red gut, while their larger southern cousins feed on phytoplankton floating at the water's surface. Krill migrate daily, spending the daylight hours in deep waters and coming to the surface to feed and lay eggs at night.

Scientific name	*Meganyctiphanes norvegica*
Classification	Phylum Crustacea; Class Malacostraca;
	Order Eucarida; Family Euphausiacea
Size	Up to 3cm (1¼in)
Distribution	Cold northern waters
Habitat	Free-swimming; sheltering among ice floes
Diet	Copepods
Reproduction	Eggs are laid at surface, but sink to deep waters

Penaeus prawn

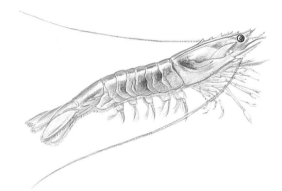

Prawns or shrimps of the genus *Penaeus* are among the most widely harvested – they include the large tiger prawns that are eaten throughout Asia, and the pink 'shrimp' from the western Atlantic. Pink shrimps spend the day burrowed into the seabed, emerging on the darkest nights to hunt and scavenge. They mate throughout the year, with the eggs hatching in the water and the young migrating to shallow inshore 'nursery grounds' rich in vegetation. As shrimps grow, they must cast off their old shells, leaving their soft bodies exposed until a new larger shell hardens around them. This is the time at which females lay their eggs.

Scientific name	*Penaeus duorarum*
Classification	Phylum Crustacea; Class Malacostraca;
	Order Decapoda; Family Penaeidae
Size	Around 17.5cm (7in) long – females larger than males
Distribution	Western Pacific, Carolina to Uruguay
Habitat	Sandy shallow seabeds down to 100m (330ft)
Diet	Algae; organic particles; plankton
Reproduction	Sexual, through copulation; breeds throughout the year

Common prawn

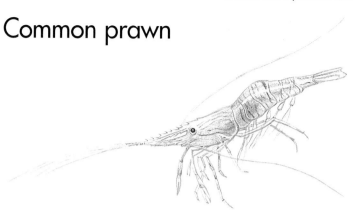

The common prawn, *Palaemon serratus*, is found throughout the eastern North Atlantic, Mediterranean, and surrounding seas, where it is harvested in large amounts for food. It has a semitransparent carapace, extending forward above the eyes to form a head-shield or rostrum. The head turns upwards and has several indentations, and the rostrum splits in two at the end. The common prawn also has extremely long antennae, up to one and a half times its body length, which it uses to sense danger. Like most prawns and shrimps, its preferred habitat is a shallow seabed where it can hide among rocks and seaweed.

Scientific name	*Palaemon serratus*
Classification	Phylum Crustacea; Class Malacostraca;
	Order Decapoda; Family Palaemonidae
Size	Up to around 10cm (4in)
Distribution	Eastern North Atlantic; Mediterranean
Habitat	Shallow rocky seabeds with plentiful seaweed
Diet	Plankton; organic debris
Reproduction	Sexual, through copulation; breeds throughout the year

Common shrimp

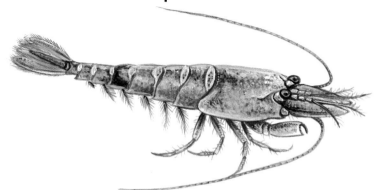

Although the terms prawn and shrimp are often used interchangeably, one way of separating them is by their body plan – shrimps are wide and flat while prawns are tall and narrow. The common shrimp *Crangon crangon* is widespread throughout Europe. It is an active predator, hunting planktonic animals, molluscs, and other shrimps in the hours of darkness on shallow seabeds, and eating up to ten per cent of its own body weight per day. In daylight the shrimp buries itself in sand with only its eyes and antennae sticking out. Shrimps also make an annual migration down the shore into deeper, warmer waters with the onset of winter.

Scientific name	*Crangon crangon*
Classification	Phylum Crustacea; Class Malacostraca; Order Decapoda; Crangonidae
Size	Up to 10cm (4in)
Distribution	Eastern North Atlantic; Mediterranean; Baltic
Habitat	Shallow shorelines, just below low-tide mark
Diet	Plankton; small invertebrates
Reproduction	Sexual, 4–5 times a year

Pistol or snapping shrimp

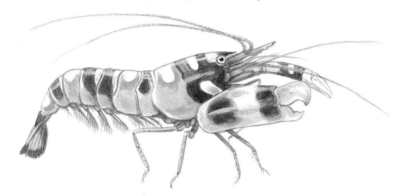

The pistol shrimp is a remarkable little shrimp with an unusual weapon. By snapping its claw closed at high speed, the shrimp can create a powerful sound wave that can stun or even kill nearby prey or predators. It is not the snapping of the claw itself which makes the sound, but the expansion and collapse of nearby microscopic air bubbles disturbed by the high-speed water movements. Pistol shrimps also form unusual symbiotic relationships with goby fish, often allowing the fish to share their burrows and food. In exchange the fish uses its better eyesight to watch for danger and alert the shrimp.

Scientific name	*Alpheus randalli*
Classification	Phylum Crustacea; Class Malacostraca; Order Decapoda; Family Alpheidae
Size	About 2.5cm (1in) long
Distribution	Indo-Pacific
Habitat	Shallow sandy seabeds, mostly around reefs
Diet	Small fish; crustaceans
Reproduction	Sexual, through copulation

Common lobster

L obsters are large, ten-footed crustaceans with huge claws on their first pair of legs. They are normally blue-black in colour (cooked lobster is red because its pigments are destroyed by boiling), and can grow to around 1m (40in) long, from tail to claw. In order to keep growing, the lobster must slough its shell every year, after which it expands slightly before its new shell hardens (shrimps, prawns, and crabs moult in a similar way). Lobsters usually live among rocks and crevices, ideally with sandy seabeds that they can burrow in. They hunt a wide variety of food, using some ingenious hunting strategies, and will also scavenge.

Scientific name	*Homarus gammarus*
Classification	Phylum Crustacea; Class Malacostraca; Order Decapoda; Family Astacidae
Size	Up to 75cm (30in) long
Distribution	Eastern North Atlantic; Mediterranean
Habitat	Rocky seabeds
Diet	Crustaceans; fish; molluscs
Reproduction	Sexual; spawns once a year after moulting

piny lobster or crayfish

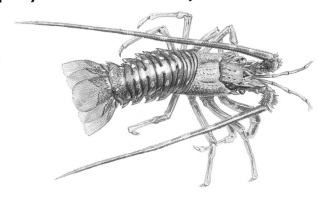

The spiny lobster, crayfish, or crawfish is a close relative of the common lobster, but with several differences – most noticeably the front three pairs of legs all ear claws (though much smaller than the lobster's), and most of the body is overed in sharp defensive spines. Crayfish are generally smaller than lobsters and here are freshwater as well as marine species. They are usually orange on the back nd white on their underside. Crayfish hide in the sand during the day, and emerge hunt a variety of prey at night. If threatened, they can escape at high speed with icks of their tail.

Scientific name	*Palinurus elephas*
Classification	Phylum Crustacea; Class Malacostraca; Order Decapoda; Family Astacidae
Size	Up to 60cm (24in)
Distribution	Eastern North Atlantic and Mediterranean
Habitat	Rock-strewn, sandy seabeds
Diet	Crustaceans; fish; molluscs
Reproduction	Sexual; spawns once a year after moulting

Hermit crab

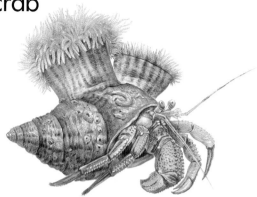

Hermit crabs are unusual malacostracans because they have an unarmoured and vulnerable abdomen. To protect themselves, they hide these body parts in empty gastropod shells, which they hold onto with their reduced hind legs. When threatened, the crab can withdraw into the shell completely, with just one enlarged pincer left outside. When the crab outgrows the shell, it simply finds another one, evicting the mollusc inside it if necessary. Hermit crabs often allow anemones and polyps to grow on their shells – the crab benefits from camouflage and protection, while the squatters feed on the debris of the crab's meals.

Scientific name	*Eupagurus bernhardus*
Classification	Phylum Crustacea; Class Malacostraca;
	Order Decapoda; Family Paguridea
Size	Up to 10cm (4in) total
Distribution	Eastern North Atlantic; Mediterranean
Habitat	Shallow sea floors down to 30m (100ft)
Diet	Small crustaceans; molluscs; carrion
Reproduction	Sexual; males carry females during mating

Striated hermit crab

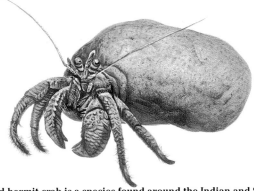

The striated hermit crab is a species found around the Indian and South Pacific oceans, marked with distinctive red and white stripes and a highly developed left claw. It is frequently completely disguised by a red-coloured sponge, *Suberites domuncula*, or covered by several anemones whose stings may help to fend off predators. Hermit crabs are not true crabs – they are members of a sub-order called Anomura, in which the two hindmost legs are reduced, leaving only six main legs for walking and two for pincers. In hermit crabs, the two reduced legs help hold on to the shell.

Scientific name	*Dardanus arosor*
Classification	Phylum Crustacea; Class Malacostraca;
	Order Decapoda; Family Diogenidae
Size	Up to 14cm (5⅜in) long
Distribution	Indo-Pacific
Habitat	Seafloors down to 200m (660ft)
Diet	Small crustaceans; molluscs; carrion
Reproduction	Sexual; males carry females during mating

Porcelain crab

Porcelain crabs, like hermit crabs, are anomurans, with a reduced fifth pair of legs that are often completely hidden. These tiny crabs are typically less than 2.5cm (1in) across, and cling tightly to rocks with spikes that prevent them being washed away. *Porcellana platycheles*, the broad-clawed porcelain crab of Europe, has a 'hairy' shell and two very well-developed front pincers. It collects mud on the hairs, which help to disguise its body, and it can filter organic matter from the sediment on its pincers. It can also use the claws to snip chunks of flesh from any dead animals that it finds on the beach.

Scientific name	*Porcellana platycheles*
Classification	Phylum Crustacea; Class Malacostraca;
	Order Decapoda; Family Diogenidae
Size	Up to 15mm (⅗in)
Distribution	Eastern North Atlantic; Mediterranean
Habitat	Among stones on muddy shores
Diet	Organic debris; carrion
Reproduction	Sexual, through copulation

Squat lobster

Squat lobsters are another group of crab-like decapods with reduced back legs. They have flattened bodies that allow them to lurk among rocks in the day, emerging at night to feed. They are mainly scavengers, though they are also well equipped to catch their own prey. *Galathea strigosa*, the spiny squat lobster from western Europe, has distinctive colouring with a rusty-orange carapace and blue stripes. The large front claws are covered in defensive spines, while much of the abdomen is folded back under the body, from where it can uncurl rapidly to flip the lobster through the water if it is threatened.

Scientific name	*Galathea strigosa*
Classification	Phylum Crustacea; Class Malacostraca;
	Order Decapoda; Family Galatheidae
Size	Up to around 15cm (6in) long
Distribution	Eastern North Atlantic; Mediterranean
Habitat	Rocky seabeds
Diet	Carrion; small crustaceans
Reproduction	Sexual, breeding once a year

Common or shore crab

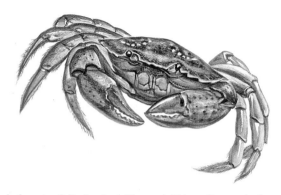

True crabs have ten fully developed legs – eight to walk on and a front pair of 'chelipeds' equipped with pincers. Not all can swim, but those that can, such as the shore crab *Carcinus maenas*, have paddle-shaped hind limbs that help propel them through the water. Shore crabs are scavengers and predators, generally scuttling sideways on the seashore or in shallow water, searching for anything they can find. When the crab moults, it leaves an empty shell, complete with claw cases, on the beach. Shore crabs are extremely successful and adaptable animals, and have become an invasive species in some areas of North America.

Scientific name	*Carcinus maenas*
Classification	Phylum Crustacea; Class Malacostraca;
	Order Decapoda; Family Portunidae
Size	About 10cm (4in) across
Distribution	Eastern North Atlantic; Mediterranean
Habitat	Seashores and shallow waters
Diet	Crustaceans; molluscs; carrion
Reproduction	Sexual, through copulation

Blue swimmer crab

While many crabs can paddle a little, swimming crabs have specially adapted back legs equipped with broad spoon-shaped scoops. The blue swimmer from Pacific and Indian Ocean waters has attractive colouring (orange and brown in females, blues and purples in males), distinctive long straight pincers, and a carapace with spikes to either side. *Portunus pelagicus* is a highly skilled swimmer that cannot survive for long out of water – it normally hunts in shallow offshore seaweed beds, but comes inshore at high tides to catch fish in coastal mangrove swamps and estuaries.

Scientific name	*Portunus pelagicus*
Classification	Phylum Crustacea; Class Malacostraca;
	Order Decapoda; Family Portunidae
Size	Up to about 21cm (8¼in) across the carapace
Distribution	Indo-Pacific
Habitat	Offshore seaweed beds; mangrove swamps
Diet	Crustaceans; fish
Reproduction	Sexual, mating once a year

Scorpion spider crab

Spider crabs are so-called because they often have extremely long legs, and their pincered front legs (known as chelipeds) are usually not much longer than the other pairs. They are also distinguished by the triangular shape of their carapace, which narrows towards the front and often extends to form a pointed rostrum. The scorpion spider crab, *Inachus dorsemensis*, has relatively stout pincers compared to most. Spider crabs frequently use their pincers to pick up sponges, seaweed, and other materials, which they attach to hooks on their carapace and claws to decorate and disguise themselves.

Scientific name	*Inachus dorsemensis*
Classification	Phylum Crustacea; Class Malacostraca;
	Order Decapoda; Family Majidae
Size	Carapace up to 30mm (1¼in) long
Distribution	Eastern North Atlantic
Habitat	Plant-strewn seabeds at depths up to 200m (660ft)
Diet	Algae; molluscs; carrion
Reproduction	Sexual, through copulation

Homolid spider crab

A few spider crabs, in the family Homolidae, have roughly square carapaces with forward-facing spikes along the upper front edge. Homolid crabs also have pincer-bearing chelipeds that are shorter than their other legs. Another distinctive feature is that the back pair of legs are thin, short and doubled back on themselves, ending in a hook-shaped 'subchela' claw. Homolids live on the seabed in moderate to deep waters and, like other spider crabs, they frequently attach shells and living polyps to their shells for camouflage. They have even been know to change their decorations when forced to change habitat.

Scientific name	*Homola barbata*
Classification	Phylum Crustacea; Class Malacostraca; Order Decapoda; Family Homolidae
Size	Carapace up to 5cm (2in) long
Distribution	Mediterranean and into North Atlantic
Habitat	Shelly, sandy, and muddy seabeds from 40–400m (130–1300ft)
Diet	Algae; molluscs; carrion
Reproduction	Sexual, through copulation

Pea crab

The pea crabs are tiny crustaceans with bodies about the size of a garden pea. They are found living within a wide variety of bivalves, including mussels, oysters, and pen shells (pinnidae), and frequently have bizarre life cycles – the females are much larger than the males and are sometimes mistaken for a different species entirely. Apart from a brief period spent swarming in open water during the mating season, the crabs spend all their lives inside the shell of the host, thriving on nutrients produced by the mollusc. They are usually commensal (doing no harm to the host), but some are damaging parasites.

Scientific name	*Pinnotheres piscum*
Classification	Phylum Crustacea; Class Malacostraca;
	Order Decapoda; Family Pinnotheridae
Size	Carapace up to 1cm (⅜in)
Distribution	Eastern North Pacific; Mediterranean
Habitat	Commensal with bivalves
Diet	Algae and organic particles filtered by the bivalve; chemicals from host
Reproduction	Sexual, through copulation

Fiddler crab

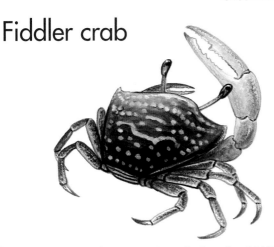

There are at least 75 known species and subspecies of fiddler crab, united by one distinctive feature – the single huge claw on the male that can account for up to half of its body weight. The crabs live largely on land, burrowing in mud and feeding on organic matter buried in sediments. They feed only with their smaller claw, and its rapid movements back and forth to the mouth while the larger claw is held still give fiddler crabs their name. The huge claw, which can be either on the left or the right side, is used in mating rituals for attracting females and challenging rival males, though grappling matches between males rarely end in injury.

Scientific name	*Uca vocans*
Classification	Phylum Crustacea; Class Malacostraca; Order Decapoda; Family Ocypodidae
Size	Around 3–4 cm (1⅕–1⅗in) across
Distribution	Indian Ocean
Habitat	Shallow sandy coastlines and mangrove swamps
Diet	Algae and organic debris
Reproduction	Sexual; female broods eggs until they hatch

King or horseshoe crab

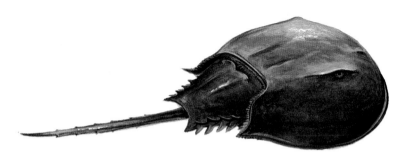

The prehistoric-looking horseshoe crab is a living fossil, and a close relative of the trilobites that roamed the sea floors and beaches 300 million years ago. Despite its name, it is closer to an arachnid than a true crab, though it does not bear much resemblance to any other living animal. The head and thorax form a single unit, with six pairs of appendages attached – two feeding palps, two claws, and eight walking legs. The head and abdomen are covered by a horseshoe-shaped shell and a long tail extends from the back. Horseshoe crabs are burrowing predators that feed on molluscs, worms, and other invertebrates.

Scientific name	*Limulus polyphemus*
Classification	Phylum Arthropoda; Class Merostomata;
	Order Xiphosura; Family Limulidae
Size	Up to 60cm (24in) long, including tail
Distribution	Western North Atlantic
Habitat	Shallow sandy offshore seabeds
Diet	Molluscs, worms, and other small invertebrates
Reproduction	· Spawns on land annually; eggs laid on sand are then fertilized by males

Sea spider

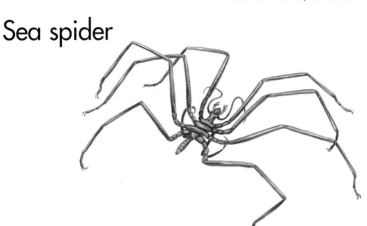

The sea spiders or Pycnogonida are marine arthropods, cousins of terrestrial arachnids. Most have eight walking legs, though some have more. Although a few types can swim, most sea spiders live on the ocean floor, feeding on corals, hydroids, and anemones, by puncturing the outer membrane with a long proboscis and sucking at the internal tissues. Sea spiders have no gut and no gills – they absorb and release gases, nutrients, and waste by simple diffusion through their thin body walls. Most are tiny, but some, such as *Colossondeis colossea*, reach monstrous proportions.

Scientific name	*Colossondeis colossea*
Classification	Phylum Arthropoda; Class Pycnogonida;
	Order Pantopoda; Family Colossendeidae
Size	Up to 75cm (30in) across
Distribution	Cosmopolitan
Habitat	Deep-ocean seabeds
Diet	Corals; hydroids; sea anemones
Reproduction	External fertilization; male gathers fertilized eggs and broods them

Sea lily

Sea lilies are members of the echinoderm phylum – a group of unique animals with no brain but a remarkably sophisticated system for moving and feeding. All show a five-fold symmetry and have a body made largely of calcium carbonate plates. The sea lilies are actually crinoids, the simplest echinoderms. They live permanently attached to the sea floor by a root called a peduncle, and complex branching arms collect floating organic debris with water-filled tentacles called tube feet. Food is then passed down the arms to the upturned mouth, where it is digested. In between the tube feet on each arm are numerous defensive spines of calcium carbonate.

Scientific name	*Rhizocrinus lofotensis*
Classification	Phylum Echinodermata; Class Crinoidea;
	Order Bourgueticrinida; Family Bourgueticrinidae
Size	Around 15cm (6in) long
Distribution	Seabeds down to 2000m (6600ft)
Habitat	Eastern North Atlantic; Baltic
Diet	Algae and organic debris
Reproduction	Sexual with external fertilization

Sea lily

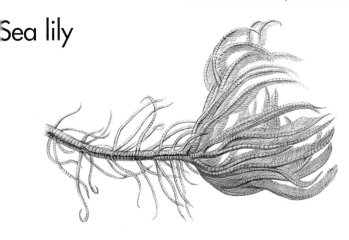

Sea lilies display a unique echinoderm anatomical feature called a water vascular system. It consists of water-carrying vessels spreading out from a ring around the mouth, along the arms, and into the tube feet, controlled by a loose network of nerves that runs throughout the creature's body. A sea lily is rather like an upside-down starfish attached by a stalk to the ocean floor, and like some starfish, it is capable of self-repair, regrowing an arm if it is lost. *Cenocrinus asterius* is a large form, found at moderate depths in the Caribbean and Atlantic, growing on a long peduncle with its arms spreading out in a fan shape.

Scientific name	*Cenocrinus asteria*
Classification	Phylum Echinodermata; Class Crinoidea;
	Order Isocrinida
Size	Up to 25cm (10in) long
Distribution	Caribbean; Central Atlantic
Habitat	Seabed and outcrops, 300–900m (1000–3000ft)
Diet	Algae and organic debris
Reproduction	Sexual with external fertilization

Mediterranean feather star

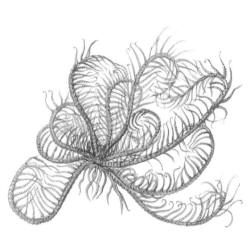

Feather stars are close relatives of sea lilies, but are free-swimming. They start life as larvae rooted to the ground by a peduncle, but detach from this on reaching maturity and float free in the water, or crawl along the seabed by pumping water in and out of their tube feet. They usually display ten arms and resemble lightweight starfish. A set of small appendages on the opposite side to the mouth (normally the underside) can temporarily attach the feather star to the seabed. The Mediterranean feather star is found in several colours, including red, yellow and white. It often hides among coral reefs, extending only its arms to feed.

Scientific name	*Antedon mediterranea*
Classification	Phylum Echinodermata; Class Crinoidea; Order Comatulida; Family Antedonidae
Size	Up to 20cm (8in) across
Distribution	Mediterranean
Habitat	Shallow seabeds and surface waters
Diet	Algae and organic debris
Reproduction	Sexual with external fertilization

Sea cucumber

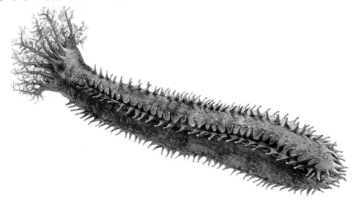

Although sea cucumbers look very different from other echinoderms, they are in fact members of the group, with the same five-fold symmetry in their bodies, calcareous plates, and flexible tube feet covering their bodies. They are mostly filter feeders, and make their way along the sea floor with movements of their lower tube feet. A group of adapted tentacles around the mouth catch food from the water. Most sea cucumbers have separate sexes, but *Cucumaria planci* is hermaphroditic, and capable of asexual reproduction by cell division. All sea cucumbers have the ability to regenerate missing limbs.

Scientific name	*Cucumaria planci*
Classification	Phylum Echinodermata; Class Holothuroidea;
	Order Dendrochirotida; Family Cucumariidae
Size	Up to around 15cm (6in) long
Distribution	Mediterranean
Habitat	Shallow seabeds among plantlife
Diet	Algae and organic particles
Reproduction	Sexual and asexual

Swimming sea cucumber

Not all sea cucumbers are restricted to crawling along the sea floor. A few species have become free-swimming, by evolving extended tentacles around the mouth joined together by a thin web-like membrane. This allows them to swim around rather like jellyfish, and their bodies are similarly soft, light and gelatinous. Swimming sea cucumbers feed on plankton and organic debris suspended in the water. They are a stark contrast to the seabed-dwelling species, which mostly feed on organic matter in the sediment, and frequently end up with bodies full of mud, from which their digestive system extracts any available nutrients.

Scientific name	*Pelagothuria ludwigi*
Classification	Phylum Echinodermata; Class Holothuroidea;
	Order Elasipodida; Family Pelagothuridae
Size	Around 15cm (6in) across
Distribution	Cosmopolitan in deep seas
Habitat	Free-swimming a few metres above seabed
Diet	Plankton and organic particles
Reproduction	Sexual, with external fertilization

Burrowing sea cucumber

Some sea cucumbers have abandoned their mobile lifestyles in favour of a safer static or sessile one. One example is *Rhopalodina lageniformis*, which buries itself in the seabed with just a proboscis protruding through the sand or mud. The animal's entire body plan has changed, so both its mouth and its anus are at the end of the proboscis, and it feeds by filtering microorganisms from the water. Other sea cucumber defences include an ability in some species to produce fine threads of toxic mucus, or in others a reflex that violently expels the viscera (internal soft parts), allowing the animal to escape and regrow them later.

Scientific name	*Rhopalodina lageniformis*
Classification	Phylum Echinodermata; Class Holothuroidea;
	Order Dactylochirotida; Family Rhopalodinidae
Size	Around 10cm (4in) across
Distribution	South-east Atlantic
Habitat	Buried in the sand of deep seabeds
Diet	Plankton; organic debris
Reproduction	Sexual, with external fertilization

Melon urchin

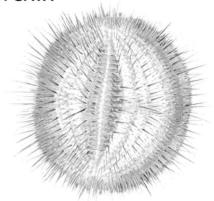

Sea urchins are armless echinoderms in which the body plates form a sphere or near-sphere called a 'test'. They graze over seabeds with their mouth downward and dispose of waste through an anus in their upper surface. Sea urchins are covered in long calcareous spines, with suckered tube feet in between them and tentacles around the mouth for crawling slowly over the sea floor. *Echinus melo*, the melon urchin, is yellow-white in colour and has fewer spines than some species, though they are of two different sizes and colours – short pale green ones and longer yellow ones.

Scientific name	*Echinus melo*
Classification	Phylum Echinodermata; Class Echinoidea;
	Order Echinoida; Family Echnidae
Size	Up to around 14cm (5½in)
Distribution	Eastern North Atlantic, Mediterranean
Habitat	Rocky seabeds and coral reefs, 20–100m (65–330ft)
Diet	Algae; seaweed; polyps; hydroids
Reproduction	Sexual, with external fertilization

Purple sea urchin

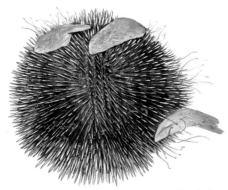

ot all sea urchins are perfectly spherical – most are slightly flattened, such as the purple sea urchin *Sphaerechinus granularis,* whose sex organs are a delicacy in parts of the Mediterranean. All urchins have a unique feeding apparatus known as 'Aristotle's lantern', first noted by the Greek philosopher more than two millennia ago. The organ does indeed look like a five-sided glass lantern, with calcareous struts supporting it on the inside of the body, and only the sharp teeth sticking out through the creature's mouth. It is worked by a total of 60 internal muscles and is very efficient at mashing up the urchin's food.

Scientific name	*Sphaerechinus granularis*
Classification	Phylum Echinodermata; Class Echinoidea;
	Order Temnopleuroida; Family Toxopneustidae
Size	Up to around 15cm (6in)
Distribution	Northern Mediterranean; Eastern North Atlantic
Habitat	Shallow rocky and weed-strewn sea floors
Diet	Algae and organic debris
Reproduction	Sexual, with external fertilization

Pencil urchin

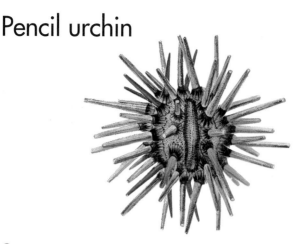

Some sea urchins do not use their spines defensively – they have thicker, sturdier and relatively sparse spines that look like pencil leads, with the shell clearly visible between them. *Eucidaris tribuloides*, the pencil urchin of the Caribbean, is a small example. It uses its spines to wedge itself into cavities in rocks and coral reefs during the day, only emerging at night to feed. Other sea urchins employ a variety of defensive tactics – many dig burrows or bore holes into rock for shelter, while some have poison sacs on the end of their spines. Some even seem to use deliberate camouflage, collecting a variety of other organisms on their spines.

Scientific name	*Eucidaris tribuloides*
Classification	Phylum Echinodermata; Class Echinoidea;
	Order Cidaroida; Family Cidaridae
Size	Up to 7.5cm (3in) across
Distribution	Caribbean; Western Atlantic
Habitat	Reefs and turtle grass beds
Diet	Algae; hydroids; molluscs; carrion
Reproduction	Sexual, with external fertilisation

Sand dollar

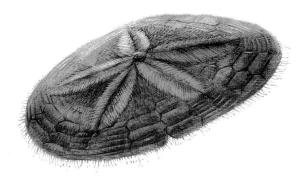

The flattened sand dollars or Scutellidae are close relatives of sea urchins that have evolved specifically for burrowing in sandy and muddy seabeds. Their spines are shortened into a bristly coat of 'fur' and collect food which hair-like cilia transfer on the underside of the body to the mouth. When the creature dies, it loses its outer skin of spines and only the internal discoid skeleton washes up on the beach, looking rather like it is moulded from compressed sand. The undersides of these skeletons clearly show five symmetrical apertures where water enters the animal to be used in the water vascular system.

Scientific name	*Echinarachnius parma*
Classification	Phylum Echinodermata; Class Echinoidea;
	Order Clypeasteroida; Family Scutellidae
Size	Around 8cm (3⅛in) across
Distribution	Atlantic and Pacific coasts of North America
Habitat	Sandy seabeds to 1500m (5000ft) deep
Diet	Algae; organic debris
Reproduction	Sexual, with external fertilization

Sea-potato

Sea potatoes are heart urchins – unevenly shaped burrowing urchins that are covered in short, spiny fur that is generally swept towards the back of the animal. As their name suggests, they have a distinctive heart shape, with a dip at the front. They are burrowers, digging some 10–15cm (4–6in) beneath sandy seabeds, where their position can be revealed by circular depressions on the surface. The sea potato feeds on organic detritus, passing food back to the mouth using its front-facing sets of tube feet. Burrowing urchins perform an important function in seabed communities by churning over and mixing the upper layers of the sand.

Scientific name	*Echinocardium cordata*
Classification	Phylum Echinodermata; Class Echinoidea; Order Spatangoida; Family Loveniidae
Size	Around 7cm (2⅜in) across
Distribution	Cosmopolitan in warm and temperate seas
Habitat	Sandy and muddy seabeds
Diet	Organic detritus
Reproduction	Sexual, with external fertilization

Brittle-star

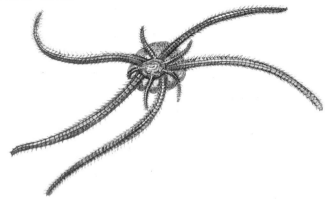

Brittle-stars are echinoderms related to starfish, but with a less robust structure, their five arms emerging from a central disc-shaped body. They are highly mobile, and are either scavengers or suspension feeders – they are too fragile to hunt larger prey on their own. They are capable of regenerating lost body parts, and in some cases an entire new brittle-star can also grow from a severed limb. They reproduce sexually – some species are hermaphrodites but most have two genders. In a few cases, such as *Ophiocomina nigra*, a smaller male attaches to the female during breeding.

Scientific name	*Ophiocomina nigra*
Classification	Phylum Echinodermata; Class Ophiuroidea;
	Order Laemophiurina; Family Ophiacanthidae
Size	Body disc up to 25mm (1in) across; arms about 12.5cm (5in) long
Distribution	Eastern North Atlantic
Habitat	Sheltered seabeds to 400m (1300ft)
Diet	Plankton; organic detritus; carrion
Reproduction	Sexual, with external fertilization

Brittle-star

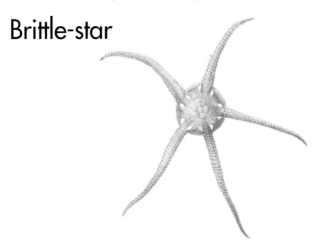

Brittle-stars such as *Ophiura albida* are members of the class Ophiuroidea, along with their close relatives the basket stars. These have multiple branching arms which they hold aloft to filter the waters. *O. albida* is widespread around western Europe, and is distinctive because of the heart-shaped plates where its arms join the central disc. The arms are lined with spines, but they lie almost flat against the arm itself – brittle-stars often catch their food on a sticky mucus strung between the spines, rather than relying on spiking food directly. As with many of its relatives, this small brittle-star grows at a remarkably slow rate, taking up to six years to reach maturity.

Scientific name	*Ophiura albida*
Classification	Phylum Echinodermata; Class Ophiuroidea;
	Order Chilophiurina; Family Ophiuridae
Size	Body disc up to 1.5cm (⅗in) across; arms about 6cm (2⅗in) long
Distribution	Eastern North Atlantic; Mediterranean
Habitat	Sandy seabeds down to 1,000m (3,300ft)
Diet	Plankton; organic detritus; carrion
Reproduction	Sexual, with external fertilization

Common brittle-star

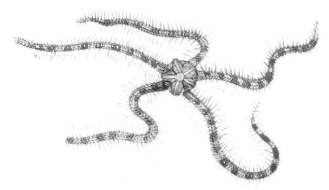

The common brittle-star *Ophiothrix fragilis* is relatively large (its central disc reaches around 20mm [⅘in] across, and its arms grow to 10cm [4in] long) and is very widespread around the entire eastern Atlantic. It comes in a variety of colours, is covered in untidy spines, and, as the name suggests, is very fragile – though it can regenerate lost limbs and spawn new individuals from the severed limbs themselves. As well as being found scattered on coastlines, it accumulates in huge offshore beds where tidal flows can bring it a plentiful supply of food. Within these beds, up to 2000 individuals can occupy a square metre (11 sq ft) of sea floor.

Scientific name	*Ophiothrix fragilis*
Classification	Phylum Echinodermata; Class Ophiuroidea;
	Order Ophiurida; Family Ophiotrichidae
Size	Body disc up to 2cm (⅘in) across; arms about 10cm (4in) long
Distribution	Eastern Atlantic Ocean
Habitat	Coastlines and offshore sea floor
Diet	Plankton; organic detritus; carrion
Reproduction	Sexual, with external fertilization

Striped brittle-star

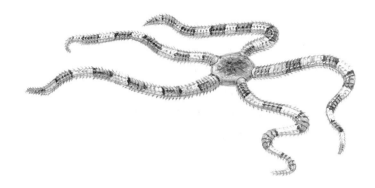

The striped brittle-star *Ophioderma longicauda* is a large and active western European species, distinguished by the broad stripes along its arms. This brittlestar is relatively fast-moving and will hide from bright sources of light. Scientists think most echinoderms sense light through reactive pigments in their skin, but a recently discovered species turns out to have a much more sophisticated visual system – its entire body is covered in crystalline lenses that are thought to function as an all-directional compound eye. Scientists are now looking again at whether other echinoderms have similar surprises in store.

Scientific name	*Ophioderma longicauda*
Classification	Phylum Echinodermata; Class Ophiuroidea;
	Order Ophiurida; Family Ophidermatidae
Size	Body disc up to 40mm (1⅝in) across; arms about 12cm (4¾in) long
Distribution	Eastern North Atlantic; Mediterranean
Habitat	Shallow sandy and rocky seabeds with plant life
Diet	Plankton; organic detritus; carrion
Reproduction	Sexual, with external fertilization

Crown-of-thorns starfish

The crown-of-thorns starfish is an unusual and very efficient predator on corals. It has up to twenty arms, all protected with extremely sharp thorn-like spines, coated with a venomous nerve toxin. It feeds on corals in a remarkable way – it crawls on top of them, then uses its tube feet to pull its large yellow stomach out of its mouth and spread it over the coral. The digestive juices dissolve the living tissue from the coral, leaving just the skeleton behind. Since the 1960s this starfish has had a devastating effect on coral reefs – it is thought to have thrived due to overfishing of one of its main predators, the triton gastropods.

Scientific name	*Acanthaster planci*
Classification	Phylum Echinodermata; Class Asteroidea;
	Order Valvatida; Family Acanthasteridae
Size	Around 40cm (16in) across
Distribution	Indo-Pacific
Habitat	Coral reefs
Diet	Coral polyps
Reproduction	Sexual, with external fertilization

Common sunstar

Sunstars, also called rose stars, are some of the most attractive echinoderms, with a brightly coloured centre and striped arms in a variety of colours. Starfish bodies are more robust than those of brittlestars, with strong muscles in their arms, and spikes and powerful suckers on their undersides. They typically have 10–12 arms, but can have fewer or still more. Although the body retains its pentagonal symmetry, the varying number of arms is a result of their willingness to grow (and grow back). Sunstars swallow their prey – such as shellfish, sea anemones and other echinoderms – whole, ejecting the inedible parts later.

Scientific name	*Crossaster papposus*
Classification	Phylum Echinodermata; Class Asteroidea;
	Order Velatida; Family Solasteridae
Size	Up to around 25cm (10in) across
Distribution	North Atlantic and Pacific
Habitat	Hard sea floors down to 50m (165ft)
Diet	Shellfish; sea anemones; sea cucumbers
Reproduction	Sexual, with external fertilization

Combtooth starfish

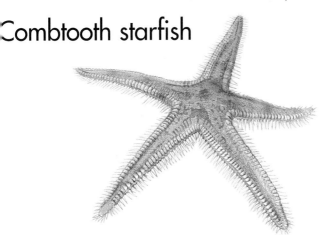

The classic starfish have five arms and are vividly coloured, such as the combtooth starfish *Astropecten aranciacus*. This large starfish from the Mediterranean grows up to 30cm (12in) across, and has a bright orange upper surface and a yellow underside. The arms are lined with two rows of long spines that stick out to either side and give the creature its common name. Between them runs a long furrow from the centre which contains two lines of unsuckered tube feet for locomotion. Each arm contains a set of sexual organs, and the tips also have a chemical-sensing olfactory organ and a light-sensitive red spot.

Scientific name	*Astropecten aranciacus*
Classification	Phylum Echinodermata; Class Asteroidea;
	Order Paxillosida; Family Astropectinidae
Size	Up to 30cm (12in) across
Distribution	Mediterranean and surrounding waters
Habitat	Sandy and rocky seabeds down to around 25m (85ft)
Diet	Molluscs; corals; anemones
Reproduction	Sexual, with external fertilization

Blue starfish

The brightly coloured blue starfish *Linckia laevigata* is found on coral reefs throughout the western Pacific and Indian oceans, frequently hiding in crevices in the rock or coral. It usually has five arms, though occasionally it may have between three and seven, and can grow up to 30cm (12in) across. Blue starfish are omnivores, feeding largely on algae and organic detritus by extruding their stomach and rolling it over the reef surface. Blue starfish are particularly good at regenerating themselves, with new animals growing from the very tip of a severed arm (sometimes called a 'comet').

Scientific name	*Linckia laevigata*
Classification	Phylum Echinodermata; Class Asteroidea;
	Order Valvatida; Family Ophidiasteridae
Size	Up to 30cm (12in) across
Distribution	Indo-Pacific
Habitat	Coral reefs
Diet	Algae; plankton; organic detritus
Reproduction	Sexual, with external fertilization

Spotted starfish

The genus *Fromia* consists of medium-sized, slow-moving and generally placid starfish, most of which dwell on reefs across the Indian and Pacific oceans. Most are red or orange in colour, usually with blue or white symmetrical patterns of spots on their upper surfaces. In some species these spots form scales that end up covering most of the arms. *Fromia ghardaqana*, from the Red Sea, generally has sparse white (but sometimes blue) spots. It is an omnivore, generally feeding on algae and detritus that collects on the living reef, but also on any other creature too slow to escape.

Scientific name	*Fromia ghardaqana*
Classification	Phylum Echinodermata; Class Asteroidea;
	Order Valvatida; Family Ophidiasteridae
Size	Up to 7.5cm (3in) across
Distribution	Red Sea and surrounding waters
Habitat	Coral reefs
Diet	Algae; organic detritus; carrion
Reproduction	Sexual, with external fertilization

Common European sea squirt

They do not look it, but sea squirts represent a key evolutionary stage between invertebrates and vertebrates such as fishes and mammals. They are the most primitive chordates – the wider group that takes in vertebrates. Adult sea squirts consist simply of a translucent bag, attached to a rock, with two syphons, or spouts. They are filter-feeders: internal hairs waft water into the side syphon and out of the top, and small plankton – floating animals and plants – are filtered from it. The clue to their evolutionary position comes from their larvae, which look like transparent tadpoles; they have a stiff rudimentary backbone called a notochord.

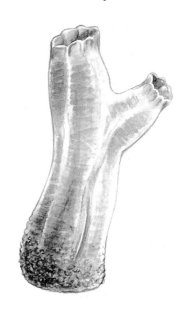

Scientific name	*Ciona intestinalis*
Classification	Class Ascidiacea (sessile tunicates or sea squirts)
	Order Enterogona (solitary sea squirts)
Size	Adult up to about 15cm (6in) long
Distribution	North-western European coastal waters; Mediterranean
Habitat	Shallow water up to low-water mark, attached to rock or seaweed
Diet	Plankton
Reproduction	Hermaphrodite. Eggs fertilized within syphon; larvae free-swimming

Hagfish

Direct descendants of the armoured ostracoderms of 500 million years ago, the jawless fishes are the most primitive living fish species. They have no scales, no skull, no paired fins, no backbone – in fact, no true bones – although (like sea-squirt larvae; *see opposite*) they do have a stiffening notochord made of cartilage. They have a simple mouth opening in place of jaws. Hagfish are slimy, eel-like blind fish that find their food prey by smell. They are scavengers, attacking dead and dying fish on the seabed and rasping at the flesh with two horny 'teeth' attached to a tongue-like piston. They often clean out the prey's insides, leaving the skin intact.

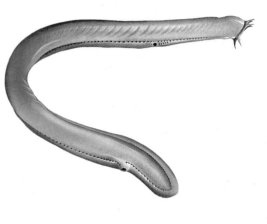

Scientific name	*Myxine glutinosa*
Classification	Superclass (formerly class) Agnatha; class Myxini
	Order Myxiniformes (formerly Cyclostomata); family Myxinidae
Size	Up to about 60cm (24in) long
Distribution	Eastern and western North Atlantic; western Mediterranean; Arctic
Habitat	Cool coastal waters; burrows in bottom mud, to 600m (2000ft) deep
Diet	Invertebrates; internal tissues and organs of dead or sickly fish
Reproduction	Spawns in summer; few (20–30) very large eggs up to 25mm (1in) long

Sea lamprey

With their sucker-like mouth, lampreys are classed with the hagfish (*see p.173*) as jawless fishes, but scientists now believe that lampreys evolved separately later than hagfish. Adult lampreys are parasites. They attach themselves to a fish, rasp at its skin with their short, sharp teeth, then suck its blood (aided by a chemical that prevents the victim's blood clotting). Sea lampreys live mainly in coastal water but swim up a river to spawn and then die. Eggs hatch into small worm-like larvae, which take several years to develop into adults and return to the sea. Sea lampreys kill many commercially caught fish – especially in the North American Great Lakes.

Scientific name	*Petromyzon marinus*
Classification	Superclass (formerly class) Agnatha; class Cephalaspidomorphi
	Order Petromyzoniformes (ex-Cyclostomata); family Petromyzonidae
Size	Up to about 90cm (3ft) long
Distribution	Eastern, western North Atlantic; western Mediterranean; Great Lakes
Habitat	Coastal waters; some fresh water lakes; breeds in rivers
Diet	External parasite: sucks host fish's blood and body fluids
Reproduction	Spawns in sandy or gravelly river bottom; larvae mature after 3–6 years

Port Jackson shark

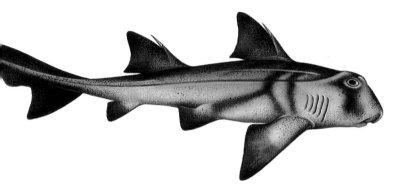

Sharks, rays and other members of the Chondrichthyes are called cartilaginous fishes because their skeleton is made of gristly cartilage instead of bone; but they do have a proper jaw, unlike the hagfish and lampreys (*see pp.173–174*). The Port Jackson shark is one of the most primitive types, closely related to shark species that lived in the Devonian period, 400 million years ago. It has a bulbous head and a venomous spine on each dorsal fin. Its upper jaw is fixed to the cranium, so it does not have the big bite of many other sharks. It is a rather slow swimmer, but migrates each year to the same shallow, rocky waters to deposit its cased eggs in crevices.

Scientific name	*Heterodontus portusjacksoni*
Classification	Order Heterodontiformes (horn or bullhead sharks)
	Family Heterodontidae
Size	Up to 1.5m (5ft) long
Distribution	Southern Pacific and Indian oceans; Southern Ocean
Habitat	Temperate offshore waters to 200m (650ft) deep; breeds in shallows
Diet	Molluscs; crustaceans; sea urchins
Reproduction	Spawns winter; 10–16 eggs in helix-shaped cases; hatch in 10–12 months

Spotted wobbegong

Wobbegong is the Australian aboriginal name for this relatively small, bottom-living shark. Its irregularly spotted back pattern gives it good camouflage as it rests or creeps across the sea-floor; the closely related ornate wobbegong (*Orectolobus ornatus*) is very similar but with a different pattern. Wobbegongs are usually inactive by day and forage at night, using their beard-like fringe of nasal barbels, or feelers, to help them to locate food. They are not aggressive, but often rest in tide pools and reef shallows, and may well attack anyone who accidentally disturbs them. They have long, dagger-like teeth and can inflict serious wounds.

Scientific name	*Orectolobus maculatus*
Classification	Order Orectolobiformes (carpet sharks and allies)
	Family Orectolobidae
Size	Up to 3m (10ft) long, usually less
Distribution	Coasts of south-western, southern and south-eastern Australia
Habitat	Cool waters, up to 110m (360ft) deep, with sandy or rocky bottom
Diet	Bottom-living fish, crustaceans, octopuses and other invertebrates
Reproduction	Ovoviviparous; up to about 20 young; gestation period uncertain

Whale shark

The whale shark is the biggest known fish, but it poses little threat to people. It is a slow-swimming filter-feeder that sucks in seawater and filters from it masses of small crustaceans, squid and other floating organisms. Sometimes it bobs up and down vertically, holding its mouth open at the surface to allow food-bearing water to rush in. The food is separated out by stacks of spongy tissue in the gaps between the fish's gill bars; the filtered water passes out through the gill slits. The whale shark's mouth is so big that it often takes in fish – mainly small sardines, anchovies and mackerel, but sometimes species as big as tuna – as well as smaller creatures.

Scientific name	*Rhincodon typus*
Classification	Order Orectolobiformes (carpet sharks and allies)
	Family Rhincodontidae
Size	Up to about 15–18m (50–60ft) long; up to 20 tonnes
Distribution	Tropical parts of Atlantic, Indian and Pacific oceans
Habitat	Warm surface waters
Diet	Plankton; fish
Reproduction	Ovoviviparous; details little known, but one female had 301 embryos

Basking shark

Second only in size (among sharks and all fish) to the whale shark (*see p.177*), the basking shark is also a normally peaceful plankton-feeder. It has a huge mouth which it holds wide-open while cruising at all depths, from the surface to 550m (1800ft), to catch shrimps and other small crustaceans. It has more than 150 rows of tiny hooked teeth, but filters its food with bristle-like attachments to its gills. Its gill slits are extremely long, almost meeting behind the head. Basking sharks have long been hunted in both the Atlantic and Pacific, mainly for their huge liver, which may yield over 2000 litres (440 Imp gal; 530 US gal) of oil, rich in vitamin A, per fish.

Scientific name	*Cetorhinus maximus*
Classification	Order Lamniformes (mackerel sharks and allies)
	Family Cetorhinidae
Size	Up to about 9–11m (30–36ft) long; up to 4 tonnes
Distribution	Worldwide except tropics
Habitat	Warm temperate to cold waters, from inshore to open ocean
Diet	Plankton – mainly small crustaceans
Reproduction	Ovoviviparous; number of young uncertain; gestation probably 3½ years

Sand-tiger shark

The sand-tiger – also known as the sand shark, in Australia as the grey nurse, and in South Africa as the spotted ragged-tooth – is one of the best-known aquarium and marine-park sharks. It looks streamlined and aggressive as it cruises with mouth open, exposing its long, sharp teeth. Yet it is rather slow-moving and docile, and most experts agree that it probably attacks only when provoked. It is certainly less dangerous than the similarly named tiger shark (*Galeocerdo cuvier*). In one respect, sand-tigers are aggressive: the largest or strongest embryo in each part of a female's uterus (womb) eats all the other embryos, ensuring that only two are born.

Scientific name	*Carcharias* (or *Odontaspis* or *Eugomphodus*) *taurus*
Classification	Order Lamniformes (mackerel sharks and allies)
	Family Carchariidae; sometimes included in Odontaspididae
Size	Up to about 3.2m (10½ft) long
Distribution	Widely distributed in temperate and tropical seas
Habitat	Shallow waters, to 200m (650ft): reefs, sandy bays, estuaries
Diet	Many kinds of fish; squid; crabs and lobsters
Reproduction	Ovoviviparous; two young (*see above*) born after 9–12 month gestation

Thresher shark

Thresher sharks are instantly recognizable by their long, whip-like tail, which accounts for half of the fish's total length. When feeding, they thrash the water with their tail to herd their prey, which consists mainly of fish such as herring, mackerel, sardines, bluefish and other small species that naturally swim in schools or groups. Threshers have even been observed using their tail to stun a fish. Where prey is abundant, they often hunt in pairs or small groups. The tail also helps thresher sharks to swim powerfully, and they sometimes leap completely clear of the water. They are caught by both sports fishermen and commercial fishing fleets.

Scientific name	*Alopias vulpinus*
Classification	Order Lamniformes (mackerel sharks and allies)
	Family Alopiidae
Size	Up to 5.5m (18ft) or even 6m (20ft) long; up to about 350kg (770lb)
Distribution	Almost worldwide in cool and warm seas
Habitat	Coastal zone to far offshore in tropical and temperate waters
Diet	Mainly schooling fish; also squid, octopuses and deep-sea crustaceans
Reproduction	Ovoviviparous; usually two to four young; gestation period unknown

Great white shark, or white pointer

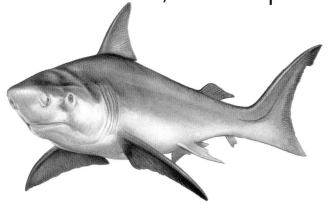

No shark is more feared than the great white, 'star' of the film *Jaws*, although the tiger shark (*Galeocerdo cuvier*) comes a close second as a threat to swimmers. In some ways, the great white's reputation is unjustified; shark experts believe that it has no more taste for human flesh than any other shark – in fact, it rarely consumes its human victims – but its size, speed and aggression make it the most dangerous species. It accounts for one-third of all deaths from shark attacks, despite the fact that it prefers relatively cold water where few people swim. The truth is that it will attack almost anything smaller than itself – seals, whales and people included.

Scientific name	*Carcharodon carcharias*
Classification	Order Lamniformes (mackerel sharks and allies)
	Family Lamnidae
Size	Up to about 7.5m (25ft) long and about 2.5 tonnes, sometimes more
Distribution	Almost worldwide between 60°S and 60°N
Habitat	Mainly cool temperate inshore and offshore waters; also open ocean
Diet	'Top' predator, eating vast range of marine organisms, including sharks
Reproduction	Ovoviviparous; usually up to ten young born after 12-month gestation

Lesser spotted dogfish

The common dogfish of British waters, this species is caught commercially and sold as rock fish or 'rock salmon'. It is not a true dogfish but one of the catshark family – whose members are distinguished by the number and position of their fins. Unlike true dogfishes (*see p.186*), they have an anal fin under the body, just in front of the tail, and both dorsal (back) fins are set well back, behind the pelvic fins. The lesser spotted dogfish is a well-camouflaged bottom-dweller. Its egg cases have fine tendrils that anchor them to seaweed until the young fish, about 10cm (4in) long, hatch. The cases, sometimes washed up on shore, are known as mermaids' purses.

Scientific name	*Scyliorhinus canicula*
Classification	Order Carcharhiniformes (ground or whaler sharks)
	Family Scyliorhinidae
Size	Up to about 1m (3¼ft) long
Distribution	Eastern North Atlantic; Mediterranean
Habitat	Shallow waters (to 400m [1300ft] in Mediterranean) over sand or mud
Diet	Bottom-living invertebrates; small fish
Reproduction	Eggs encased in horny, tendrilled cases; hatch after 5–11 months

Smooth hammerhead shark

The extraordinary T-shaped 'hammer' forming the head of hammerhead sharks makes them unmistakable, but has long puzzled shark biologists. The sharks' eyes and nostrils are located at the ends of the hammer, and many experts believe that they increase the sensitivity and accuracy of the fishes' senses – in particular allowing them to smell in 'stereo', as it were. This idea is backed up by the fact that hammerheads are often the first on the scene when a fishery ship dumps offal. Its second purpose is probably to increase the sharks' manoeuvrability, acting like a hydroplane, or underwater wing. Hammerhead sharks sometimes attack people.

Scientific name	*Sphyrna zygaena*
Classification	Order Carcharhiniformes (ground or whaler sharks)
	Family Sphyrnidae
Size	Up to about 4.25m (14ft) long
Distribution	Atlantic, Indian and Pacific oceans; Mediterranean
Habitat	Coastal to open seas, to 400m (1300ft), in tropical and temperate waters
Diet	Mostly fish (especially rays); offal and other dead matter
Reproduction	Ovoviviparous; up to 37 young, born with 'hammer' folded

Blue shark

With its streamlined body and long, wing-like pectoral fins, the blue shark is a long-range cruiser and voracious feeder that sometimes attacks swimmers. But it mostly travels the open oceans, far from land, commonly migrating 3000km (1900 miles) or more with the seasons. The longest journey on record is nearly 6000km (about 3700 miles), and its maximum speed 69km/h (43mph). It usually travels near the surface, its dorsal and tail fins often projecting. It has big eyes that help it to spot prey – which is scarce in mid-ocean – and finger-like gill-rakers to filter any small fish and other creatures from water passing from its mouth out through its gills.

Scientific name	*Prionace glauca*
Classification	Order Carchariniformes (ground or whaler sharks)
	Family Carcharhinidae
Size	Up to about 4m (13ft) long; up to 206kg (455lb)
Distribution	Worldwide, in temperate and tropical zones
Habitat	Mostly open ocean, near surface, but sometimes inshore
Diet	Fish (especially schooling species); squid and other invertebrates
Reproduction	Viviparous; often 50 or more young, born after 9–12 month gestation

Frilled shark

With its slim body and fins, the frilled shark looks more like a giant eel than a typical shark. It is a deep-water species, rarely seen except in the nets of deep-sea trawlers, and may be responsible for some stories of 'sea-serpents'. Like a few other species of the related order Hexanchiformes (the cow sharks; within which some biologists also place the frilled shark), it has six gill-slits – resembling a frilled collar – on each side; most sharks have five. The mouth opening is at the very front of its head, rather than being underslung as in most sharks; inside it has about 300 teeth, each three-pointed like a trident. It probably strikes snake-like at its prey.

Scientific name	*Chlamydoselachus anguineus*
Classification	Order Chlamydoselachiformes (frilled sharks) or Hexanchiformes
	Family Chlamydoselachidae
Size	Male up to 2m (6½ft) long; female up to 1.5m (5ft)
Distribution	Worldwide, in temperate and tropical zones
Habitat	Offshore waters, to 1200m (4000ft) deep
Diet	Mainly deep-sea squid and fish
Reproduction	Ovoviviparous; 4–12 young born after up to 3½-year gestation

Spiny dogfish

This true dogfish is distinguished from the lesser spotted dogfish (*see p.182*) by its lack of an anal fin, and by the sharp spines at the front of each dorsal fin, which also lie farther forward on its body. Also known as the spurdog or piked dogfish, it has been fished commercially for centuries (and sold as rock fish, 'rock salmon' or flake), but is probably still the most abundant shark. However, fishermen have to beware the spines; they are not venomous, but are coated with slime that contains bacteria, and these can cause illness if the skin is punctured. It is very long-lived (to at least 70 years), and females cannot breed until they are 20 or more years old.

Scientific name	*Squalus acanthias*
Classification	Order Squaliformes (dogfish sharks)
	Family Squalidae
Size	Male up to about 1m (3¼ft) long; female up to 1.2m (4ft)
Distribution	Worldwide in cool temperate seas; also deep tropical waters
Habitat	Mainly bottom-living, in coastal to oceanic waters, to 900m (3000ft)
Diet	Schooling and bottom-living fish; marine invertebrates
Reproduction	Ovoviviparous; up to 20 young born after 18–24 month gestation

Angular rough shark

This strange-looking small shark is sometimes caught by fishermen, but is rarely seen otherwise, because it is a relatively deep-sea species living on the sea-floor mainly near the edge of the continental shelf. It has a stout, high-backed body that is almost triangular in cross-section, and a short blunt snout; in several European languages it is known as the pigfish or sea-pig. It has two large but rather floppy dorsal fins, each with a large spine embedded within the fin. Large, prickly scales give its skin a rough surface, and at one time this was sometimes used as sandpaper. A similar fish living off South Africa may belong to the same or a different species.

Scientific name	*Oxynotus centrina*
Classification	Order Squaliformes (dogfish sharks)
	Family Oxynotidae or Dalatiidae
Size	Up to about 1.5m (5ft) long, usually much less; male smaller than female
Distribution	Eastern Atlantic; Mediterranean
Habitat	Warm temperate and tropical offshore waters, at 60–660m (200–2200ft)
Diet	Mainly invertebrates such as molluscs and worms
Reproduction	Ovoviviparous; usually seven or eight young; gestation period uncertain

European angel shark

Angel sharks look rather like a cross between a shark and a ray (*see pp.190–196*) with their flattened body and wide pectoral and pelvic fins. But the position of the mouth (near the tip of the snout) and gill slits (on the back) shows that they are true sharks; in rays, both are on the ventral (lower) surface. Angel sharks spend long periods – often days or even weeks – lying half-buried in the sand or mud of the seabed waiting for prey to come within range of their snapping jaws. Although sometimes called the monkfish, this Mediterranean and eastern Atlantic species is quite different from the angler fish (*see p.230*) sold by fishmongers as monkfish.

Scientific name	*Squatina squatina*
Classification	Order Squatiniformes (angel sharks)
	Family Squatinidae
Size	Usually up to 1.8m (6ft) long; female sometimes 2m (6½ft) or more
Distribution	Eastern North Atlantic; Mediterranean
Habitat	Bottom-living in temperate coastal waters, to 100m (330ft) or more
Diet	Bottom-living fish; crustaceans; octopuses; shellfish
Reproduction	Ovoviviparous; up to 25 young, born after about 10-month gestation

Smalltooth or greater sawfish

Sawfishes are instantly recognizable by their long, toothed rostrum, or snout – the 'saw' – which they use to stun prey and stir up the seabed when feeding. In other ways they look more like sharks than rays, but the position of the gills and mouth (*see opposite*) shows that they belong among the rays. The smalltooth sawfish is one of the biggest species, with 24 to 32 pairs of teeth on its saw. It used to be common in many areas, including the western Atlantic and Gulf coast, but like several other sawfishes is now seriously endangered. It was often caught in fishing nets, and its sword was (and still is) collected as a souvenir and for its supposed magical powers.

Scientific name	*Pristis pectinata*
Classification	Order Pristiformes (sawfishes)
	Family Pristidae
Size	Up to 7.5m (25ft) long including saw, but 5.5m (18ft) more common
Distribution	Worldwide (but rare) in tropical and temperate waters
Habitat	Mainly shallow coastal waters and estuaries; sometimes enters rivers
Diet	Schooling fish; bottom-living invertebrates
Reproduction	Ovoviviparous; 15–20 young, born with soft saw in protective sheath

CHONDRICHTHYES (CARTILAGINOUS FISHES)

Common guitarfish

The guitarfishes are another group that are intermediate in shape between rays and sharks. The head and front part of the body are flattened and arrow-shaped with a pointed snout and wing-like pectoral fins, while the hind part looks shark-like. However, the mouth and gill slits are on the underside of the body, showing that they are rays. The common guitarfish of European and west African waters is a rather slow-moving, bottom-living fish that often half-buries itself in the sand or mud of the seabed, up to 100m (330ft) down. It is caught commercially in some part of the Mediterranean. There are more than 40 other species worldwide.

Scientific name	*Rhinobatos rhinobatos*
Classification	Order Rhinobatiformes or Rhynchobatiformes (shovelnose guitarfishes); sometimes included in Rajiformes. Family Rhinobatidae
Size	Up to 1m (3¼ft) long
Distribution	Eastern Atlantic; Mediterranean
Habitat	Shallow tropical and warm temperate waters, near coasts
Diet	Bottom-living fish and invertebrates
Reproduction	Ovoviviparous; one or two litters per year of 4–10 young

Eyed electric ray, or common torpedo

Electric rays have a small kidney-shaped muscular organ on each side of their body that, when contracted, can generate an electrical discharge. Some species generate only 20 or 30 volts, but the common torpedo, or eyed electric ray, can inflict a shock of up to 200 volts. (A related species, the much larger *Torpedo nobiliana*, generates as much as 220 volts.) The electricity is used to defend the ray against attack, and also to stun or kill its prey, and is quite enough to give a person a shock if touched. All electric rays have a rounded shape, with a blunt snout and thick body. The common torpedo has five bright blue, dark-edged spots on its back.

Scientific name	*Torpedo torpedo*
Classification	Order Torpediniformes (electric rays and allies)
	Family Torpedinidae
Size	About 60cm (24in) long; female smaller
Distribution	Eastern Atlantic; Mediterranean
Habitat	Tropical and warm temperate waters, usually near coasts
Diet	Mainly small fish; also bottom-living invertebrates
Reproduction	Ovoviviparous; up to 21 young born after about 5-month gestation

Brown ray

In true rays and skates (order Rajiformes) and stingrays (order Myliobatiformes; *see pp.194–196*), the body and the wing-like pectoral and pelvic fins together form a single distinct unit called the pectoral disc; it is usually diamond-shaped. There is no real biological distinction between rays and skates, but skates generally have a much more pointed snout than rays; both have a slender tail. The brown ray, one of many species found in European and African waters, has a distinctive bright blue and yellow 'eye' spot on each side of the brown, speckled back. Males have three rows of spines along the centre of the tail, females five rows. The underside is white.

Scientific name	*Raja miraletus*
Classification	Order Rajiformes (skates and true rays)
	Family Rajidae
Size	Up to about 63cm (25in) long; female slightly smaller
Distribution	Eastern Atlantic; Mediterraean; south-western Indian Ocean
Habitat	Warm temperate and tropical waters, to depth of 300m (1000ft)
Diet	Bottom-living fish and invertebrates; remains of dead creatures
Reproduction	Up to 70 eggs, laid in 4.5cm (1¾in) capsules on sandy or muddy flats

Common Atlantic skate

Known also as the grey or blue skate, this has long been an important catch for commerical fishermen off European shores as far north as Iceland and the Arctic coast of Norway. Most of those traditionally caught were relatively small and young fish living on the floor of the continental shelf at depths of up to 200m (650ft), but catches there have dropped sharply and the common skate is now seriously endangered in these areas. Fishermen have switched to deeper waters – where some much larger specimens live – in an effort to maintain catches, but experts believe that the species will be in serious danger unless catches are curtailed.

Scientific name	*Dipturus* (or *Raja*) *batis*
Classification	Order Rajiformes (skates and true rays)
	Family Rajidae
Size	Female up to 2.4m (8ft) long; male up to 2m (6½ft); up to 98kg (216lb)
Distribution	Eastern and northern North Atlantic; western Mediterranean
Habitat	Temperate and cold waters, from shallows to depth of 600m (2000ft)
Diet	Mainly bottom-living fish and crustaceans; also other fish and octopuses
Reproduction	About 40 eggs per year, laid in capsules up to 24.5cm (9½in) long

Common stingray

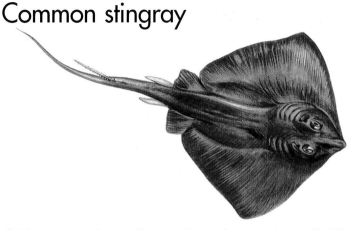

Slightly more rounded in outline than skates and true rays (*see pp.192–193*), the common eastern Atlantic and Mediterranean stingray lives mostly in shallow water. One or sometimes two saw-toothed poisonous barbs, up to 35cm (14in) long, project from the top of its tail. If disturbed – perhaps by a diver or fisherman, or even by a bather stepping on the half-buried fish – it will lash with its tail and may stab or cut its victim seriously. Poison entering the wound from a gland at the base of the barb causes intense pain, but deaths are rare. However, a few people die worldwide each year from stingray stings, usually to the upper part of their body.

Scientific name	*Dasyatis* (or *Trygon*) *pastinaca*
Classification	Order Myliobatiformes (great rays); sometimes included in Rajiformes
	Family Dasyatidae
Size	Up to about 60cm (24in) wide; to about 1.5m (5ft) long
Distribution	Eastern Atlantic; Mediterranean
Habitat	Temperate coastal waters and estuaries, to 200m (650ft) deep
Diet	Bottom-living fish; crustaceans; molluscs
Reproduction	Ovoviviparous; four to seven young, born after 4-month gestation

Common eagle ray

Eagle rays are aptly named, because their pectoral disc (*see p.192*) is much wider than it is long, so the fish appears to have wings. They are also (along with the manta rays; *see p.196*) much more active swimmers than most other rays, 'flying' through the water by flapping their wings, and even leaping into the air. They use their flat, plate-like teeth to crush crustacean and mollusc shells. The common eagle ray has a venomous tail spine but is not regarded as dangerous. The huge spotted eagle ray (*Aetobatus narinari*), up to 3m (10ft) wide, is common worldwide in the tropics and subtropics, including waters off the southern and south-eastern USA.

Scientific name	*Myliobatis aquila*
Classification	Order Myliobatiformes (great rays); sometimes included in Rajiformes
	Family Myliobatidae
Size	Up to about 1.8m (6ft) wide
Distribution	Eastern Atlantic from British Isles to South Africa; Mediterranean
Habitat	Temperate and tropical coastal waters; offshore to 300m (1000ft) deep
Diet	Mainly bottom-living crustaceans and molluscs; also fish
Reproduction	Ovoviviparous; three to seven young born after 6–8-month gestation

Giant manta ray

This is truly the giant among rays, the biggest living species and one of the so-called devil rays. Apart from its size, it is distinguished by strange paddle-like fins or lobes projecting from the front of its body, with the eyes on either side. These lobes look threatening, but are merely scoops that the ray unfurls when feeding, to direct food into its large, rectangular mouth. Giant mantas feed mainly on plankton – floating organisms – and some schooling fish. In spite of their size, they sometimes leap out of the water; this may be part of their courtship ritual, which is known to involve one or more males chasing a female for up to 30 minutes before mating.

Scientific name	*Manta birostris*
Classification	Order Myliobatiformes (great rays); sometimes included in Rajiformes
	Family Myliobatidae; sometimes placed in separate family Mobulidae
Size	Up to 8m (26ft) wide and possibly up to 3 tonnes
Distribution	Worldwide in tropical and subtropical seas
Habitat	Surface waters, mainly near shores and reefs, but also in open ocean
Diet	Mainly plankton; some small and medium-sized fish
Reproduction	Ovoviviparous; one or two young; gestation period uncertain

Rat or rabbit fish

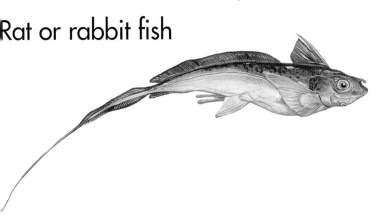

Most fishes with a skeleton of cartilage rather than bone belong to the rays and sharks (suborder Elsamobranchii; *see pp. 175–196*). A small second group, the chimaeras of the suborder Holocephali, are strange-looking deep-sea fishes. The rat fish or rabbit fish (not to be confused with the warm-water rabbit fishes of the family Siganidae, sometimes kept in aquariums) is an example. It has a long whip-like tail, big staring eyes, large pectoral fins, and a large dorsal fin with a long spine behind the head. Its small mouth has lips and rabbit-like tooth plates. Around Iceland and Norway it migrates to shallow seas to breed, and is sometimes caught by fishermen.

Scientific name	*Chimaera monstrosa*
Classification	Subclass Holocephali. Order Chimaeriformes (chimaeras)
	Family Chimaeridae
Size	Up to about 1.5m (5ft) long
Distribution	Eastern and northern Atlantic; Mediterranean; possibly other oceans
Habitat	Deep temperate and cold waters, mainly at 300–500m (1000–1600ft)
Diet	Mainly bottom-living invertebrates
Reproduction	Spawns in spring; eggs in slender cases, 18cm (7in) long

Coelacanth

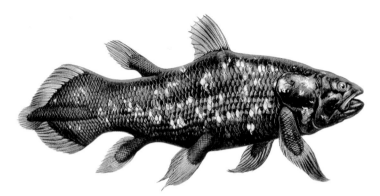

Coelacanths are sometimes called 'living fossils' because, until the first specimen was caught off South Africa and identified in 1938, they were thought to have been extinct for at least 65 million years. They are similar to the prehistoric fishes with fleshy, lobed fins that evolved into the amphibians, the first land animals. Their internal features show other ways in which fishes have evolved; the heart, for example, is much more primitive than that of other fishes. The coelacanth is a rare, endangered species; fortunately, only a few are caught each year – accidentally – by Comoran fishermen. Its eggs, which develop internally, are huge: 9cm (3½in) across.

Scientific name	*Latimeria chalumnae*
Classification	Subclass Sarcopterygii (fleshy-finned fishes)
	Order Coelacanthiformes; family Latimeriidae
Size	Up to about 1.8m (6ft) long; male smaller
Distribution	Indian Ocean, especially near Comoros Islands
Habitat	Cool water among rocky reefs and caves, at 150–750m (500–2500ft)
Diet	Fish; squid
Reproduction	Ovoviviparous; up to 25 young born after 13-month gestation

Atlantic sturgeon

Sturgeons are best known – and have long been caught – as the source of caviar, their immature eggs, which are salted to make a luxury food. These may be cut from the female's body, killing her, or may be stripped and the female returned to the water – much preferable for conservation reasons. Sturgeon flesh is also eaten, and it is now an endangered species. It is a primitive fish, and has rows of bony plates along its body. It breeds in fresh water; the young stay in the river of their birth for up to three years before returning to the sea. A similar (perhaps identical) species, *Acipenser oxyrhynchus*, lives along the North American Atlantic coast.

Scientific name	*Acipenser sturio*
Classification	Subclass Actinopterygii (ray-finned fishes; includes all fish species on following pages); order Acipenseriformes; family Acipenseridae
Size	Up to 3.5m (11½ft) long – female smaller; up to 400kg (880lb)
Distribution	Eastern Atlantic; Baltic; Mediterranean; *see also above*
Habitat	Shallow coastal waters (bottom-living); enters rivers to breed
Diet	Worms; crustaceans; molluscs; some fish
Reproduction	Spawns in spring and early summer on gravelly bottom of fast rivers

Atlantic tarpon

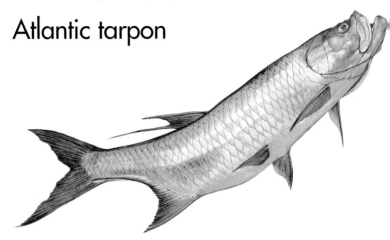

A huge silvery fish looking rather like a giant herring, the tarpon may weigh 125kg (275lb, or 19½ stones) or more. It fights fiercely and leaps when caught by game fishermen. It has a deep, compressed body (flattened from side to side) and large scales. The tarpon sometimes comes into brackish or fresh water. It is regarded by biologists as one of the most primitive of the Teleostei, the very large group that contains almost all bony fish species alive today. They all have a well-developed skull and spine, and a symmetrical tail fin (unlike the sturgeons, whose tail fin is bigger at the top). Most have a gas-filled swim bladder which helps the fish to float.

Scientific name	*Megalops* (or *Tarpon*) *atlanticus*
Classification	Order Elopiformes (tarpons and allies)
	Family Megalopidae
Size	Up to 2.4m (8ft) long
Distribution	Eastern and western Atlantic, to Argentina, Nova Scotia and Ireland
Habitat	Warm tropical, subtropical and Gulf Stream inshore waters
Diet	Mainly schooling fish; crabs
Reproduction	Spawns in spring and summer; vast numbers of eggs

Bonefish

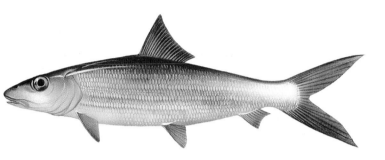

Even though they are so bony as to be almost inedible, bonefish are eagerly sought by game fishermen, using imitation shrimps and fish as bait, for the fierce fight they put up if hooked. Bonefish feed in the shallows, their body almost vertical as they forage for small, bottom-living creatures. They are slender and streamlined, with a deeply forked tail fin. It was long thought that a single species extended throughout the tropics, but biochemical and DNA studies have shown that the almost identical-looking fish in fact belong to at least five separate species. They all produce almost transparent larvae which eventually change into the adult form.

Scientific name	*Albula vulpes*
Classification	Order Albuliformes (bonefishes and allies)
	Family Albulidae
Size	Up to about 90cm (3ft) long
Distribution	Worldwide in tropical waters (*but see above*)
Habitat	Shallow inshore waters, especially over sand
Diet	Crabs; prawns; shellfish; small bottom-feeding fish
Reproduction	Spawns in shallow waters; eggs hatch to eel-like larvae

European eel

The life history of the European eel is extraordinary and still a mystery in some of its details. Eels spend most of their lives in rivers, where they have yellowish underparts. When mature and ready to breed, they turn silvery (as illustrated), develop enlarged eyes, and their gut shrivels; they will never eat again. They swim to the sea in autumn, and apparently in deep water to the Sargasso Sea, in the south western North Atlantic. There they spawn in spring at about 100–450m (300–1500ft) then die. The transparent larvae drift in the Gulf Stream and re-enter fresh water a small elvers. The American eel (*Anguilla rostrata*) has a similar life history.

Scientific name	*Anguilla anguilla*
Classification	Order Anguilliformes (true eels)
	Family Anguillidae
Size	Up to 1.4m (4½ft) long; female generally longer than male
Distribution	Breeds in Atlantic; spends most of life in fresh water
Habitat	Deep water during breeding migration; larvae drift near surface
Diet	Insects, crustaceans and fish in fresh water; adults do not eat at sea
Reproduction	Spawns in Sargasso Sea (western Atlantic); larvae drift in Gulf Stream

Mediterranean moray eel

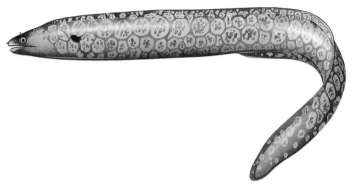

Moray eels live in rocky crevices and reefs throughout the tropical, subtropical and warm temperate seas of the world. They have a reputation for ferocity, and they are certainly voracious predators, lunging from their lair (where they lurk with only the head showing) to snatch their prey. They have a large mouth and sharp teeth, and if disturbed will give a diver a nasty bite. However, experts believe that the similarity of human fingers to an octopus's tentacles may often be to blame for attacks. Like most moray eels, the Mediterranean species has a boldly patterned body, mottled and banded. All morays lack scales and both pectoral and pelvic fins.

Scientific name	*Muraena helena*
Classification	Order Anguilliformes (true eels)
	Family Muraenidae
Size	Up to about 1.3m (4¼ft) long
Distribution	Mediterranean; adjacent areas of eastern Atlantic
Habitat	Rocky shores of warm temperate seas
Diet	Mainly fish, squid and cuttlefish
Reproduction	Spawns in summer; large (5mm; ⅕in) floating eggs

Conger eel

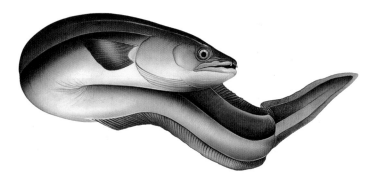

Huge fish with large jaws and extremely sharp teeth, conger eels are powerful predators that hunt mainly at night for fish such as pollack, hake, wrasse and sole, as well as squid, octopuses and crustaceans. By day, they hide among rocks or other 'cover', and divers exploring wrecks often find them inhabited by many large congers. They are distinguished from moray eels (*see p.203*) by their colouring and the fact that they have pectoral fins. They breed in deep water like the European eel (*see p.202*), and their larvae take one to two years to drift back to their coastal adult habitat. A similar species, *Conger oceanicus*, lives off eastern North America.

Scientific name	*Conger conger*
Classification	Order Anguilliformes (true eels)
	Family Congridae
Size	Up to 3m (10ft) long or more; weight up to 110kg (over 240lb)
Distribution	Eastern North Altantic; Mediterranean; breeds mid-Atlantic
Habitat	Rocky shores to 200m (650ft); old wrecks; breeds in deep water
Diet	Fish; crabs and other crustaceans; squid and octopuses. Feeds at night
Reproduction	Spawns at 3000-4000m (10 000–13 000ft) in summer

Brown garden eel

The name garden eel was coined in the early 20th century by American zoologist William Beebe, who first described the creatures living in dense colonies, half-buried, swaying in the current like a 'garden' or 'meadow' of sea-grass stems. The fish spend their whole life anchored in the sandy sea-bottom like this, eating passing food particles and always ready to retreat into their burrow if threatened by a predator. Males and females even mate part-buried, intertwining their bodies. The brown garden eel is the best-known Atlantic species. The spotted garden eel (*Heteroconger hassi*) lives in warm water from the Red Sea to Tahiti and California.

Scientific name	*Heteroconger* (or *Taenioconger*) *longissimus*, or *H.* (or *T.*) *halis*
Classification	Order Anguilliformes (true eels)
	Family Congridae; subfamily Heterocongrinae
Size	Up to about 60cm (24in) long
Distribution	Eastern and western Atlantic (Canaries to West Africa; Florida to Brazil)
Habitat	Sandy bottom of inshore warm waters, to 50m (165ft)
Diet	Small planktonic creatures and detritus
Reproduction	Mate while anchored in sand; juveniles burrow into sand

Atlantic or slender snipe eel

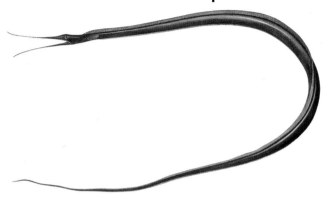

A very long, slender and fragile fish, with dorsal and anal fins that run almost the entire length of its body and a tail that ends in a whip-like filament, the snipe eel is named after its long, beak-like jaws. These resemble the beak of the snipe (a wading bird). The fish feeds by swimming along with its mouth open, using its sharp, backward-facing teeth to trap shrimps and other small creatures that swim into its gape. Only females and younger males have these long jaws; they degenerate in older males, which develop long, tubular front nostrils – possibly to help to locate females. Both males and females are believed to die after breeding.

Scientific name	*Nemichthys scolopaceus*
Classification	Order Anguilliformes (true eels), or separately in Saccopharyngiformes
	Family Nemichthyidae
Size	Up to about 1.3m (4¼ft) long
Distribution	Worldwide in tropical and temperate seas
Habitat	Mostly at 400–2000m (1300–6500ft); sometimes shallower in north
Diet	Mainly small crustaceans; some small fish
Reproduction	Eggs hatch into planktonic larvae

Pelican or gulper eel

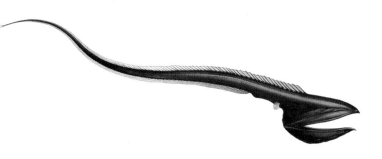

This is one of the most extraordinary-looking of all fish, with its slender body, long whip-like tail, and disproportionately large head and mouth (making up more than half of its true body length), with tiny teeth. Its gape is formed by a black elastic membrane, making the name 'pelican eel' an apt one. (Another name for it is umbrella-mouth gulper eel.) It has a flashing luminous organ at the tip of its tail – but no one knows whether the fish holds this in front of its mouth to lure prey, or whether it simply swims along with its mouth open, gulping whatever comes its way. The fish's stomach can expand to contain the rare catch of a large amount of prey.

Scientific name	*Eurypharynx pelecanoides*
Classification	Order Anguilliformes (true eels), or separately in Saccopharyngiformes
	Family Eurypharyngidae
Size	Up to about 75–100cm (30–39in) long
Distribution	Worldwide in temperate and tropical seas
Habitat	Deep water, mostly at 2000–7500m (6500–24 500ft)
Diet	Mainly small crustaceans; also fish, squid and other invertebrates
Reproduction	Eggs hatch into planktonic larvae which develop in shallower water

Atlantic herring

One of the most important commercial fishery species, herrings have been over-fished in recent years, and stocks are seriously depleted in many areas. Several distinct races live in different parts of their overall range, migrating to various spawning grounds – the floor of shallow bays and offshore banks – at different seasons. They swim in large schools, mostly near the surface at night, but deeper in daylight. Herrings breed rapidly, and can double their numbers in just over a year if not fished. The larvae swim to the surface in large schools, and reach maturity in three to nine years. The North Pacific herring, *Clupea pallasii*, is a similar species.

Scientific name	*Clupea harengus*
Classification	Order Clupeiformes (sardines, herrings and allies)
	Family Clupeidae
Size	Up to about 45cm (18in) long
Distribution	North Atlantic (Arctic to Carolinas and Bay of Biscay)
Habitat	Cool coastal waters and open sea; mainly near surface, to 200m (650ft)
Diet	Mainly small crustaceans; some small fish; sometimes filter-feeds
Reproduction	Races spawn at various seasons, on shallow sea-floor

waite shad

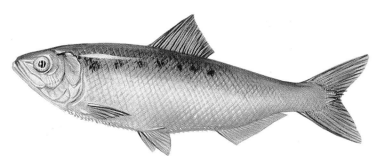

Shads are rather similar to herrings, but have a deeper body; it is silvery with a greenish or bluish cast. The twaite shad has a series of dark blotches – usually six to eight – along its sides, while the closely related but rarer allis shad (*Alosa alosa*) has only one or a few blotches. Twaite shads spend most of the year in shallow waters, feeding on shrimps, small fish and other floating organisms. They swim into the lower reaches of rivers in late spring to breed, before returning to the sea in late summer. The young return to the sea after about a year, when they are about 13cm (5in) long. Some shad populations live permanently in landlocked freshwater lakes.

Scientific name	*Allosa fallax*
Classification	Order Clupeiformes (sardines, herrings and allies)
	Family Clupeidae
Size	Up to about 60cm (24in) long
Distribution	Eastern North Atlantic; Baltic, Mediterranean and Black seas
Habitat	Inshore and coastal waters; estuaries; enters rivers to breed
Diet	Mainly crustaceans and small fish
Reproduction	Spawns spring and early summer on gravel beds in lower parts of rivers

European anchovy

More than 100 species of anchovies live in the world's oceans, and many are an important commercial catch. They are small fish that swim in large schools, and are also eaten by larger species, such as tunas. Anchovies are recognizable by their long snout and long, underslung mouth, which extends well behind the eyes. When feeding – mainly at night – they swim with their mouth agape to filter small floating creatures. One of the most important Pacific species is the anchoveta (*Engraulis ringens*), which feeds on plankton brought by the cold Peruvian current; when the *El Niño* phenomenon stops this current, the anchovies disappear.

Scientific name	*Engraulis encrasicholus*
Classification	Order Clupeiformes (sardines, herrings and allies)
	Family Engraulidae
Size	Up to about 20cm (8in) long
Distribution	Eastern Atlantic; Mediterranean and Black seas; western Indian Ocean
Habitat	Coastal and inshore temperate and tropical waters, usually near surface
Diet	Filter-feeds on small planktonic organisms
Reproduction	Spawns spring to autumn, often in estuaries and lagoons; eggs float

Dorab wolf-herring

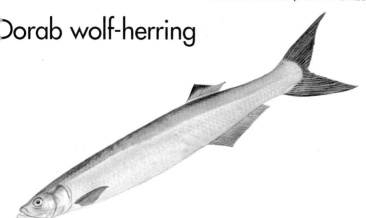

This large tropical and subtropical fish is very appropriately named, for it is a voracious predator of small schooling fish – such as its relatives the herrings and anchovies. Known also, particularly in south-western Asia and India, simply as the dorab, it has large, fang-like teeth and is very similar to its close relative the whitefin wolf-herring (*Chirocentrus nudus*). The only easy distinguishing features are the dorab's shorter pectoral fins and the black markings on the upper part of its dorsal fin. Both are very bony fish, but are important commercially in southern and south-east Asia, where more than 50 000 tonnes of the two are caught each year.

Scientific name	*Chirocentrus dorab*
Classification	Order Clupeiformes (sardines, herrings and allies)
	Family Chirocentridae
Size	Up to about 1m (3¼ft) long, possibly more
Distribution	Indian and western Pacific oceans (Red Sea to Australia and Japan)
Habitat	Warm coastal and shallow seas; also brackish waters
Diet	Mainly small schooling fish; probably also crustaceans
Reproduction	Details uncertain, but probably spawns in shallow water

Milkfish

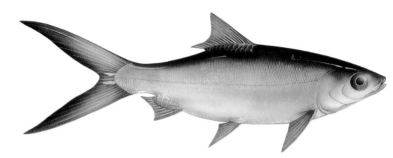

Big silvery fish of warm waters, milkfish have no teeth and are filter-feeders. They swim in schools near coasts and around island reefs, and lay their eggs in the sea – where the larvae remain for about two weeks. The larvae then move inshore, into river estuaries, brackish lagoons or mangrove swamps, where they grow into immature adults before returning to the sea. In many parts of south-east Asia and around the South China Sea, the larvae are collected and raised in village ponds until big enough to eat. Human manure may be used to encourage the growth of algae on which the fish feed. Milkfish thrive in water as warm as 32°C (90°F).

Scientific name	*Chanos chanos*
Classification	Order Gonorynchiformes (milkfish and allies)
	Family Chanidae
Size	Up to about 1.8m (6ft) long; weight up to 14kg (31lb)
Distribution	Indian and Pacific oceans (Red Sea and South Africa to California)
Habitat	Mainly shallow coastal waters (to 12m; 40ft) in tropics and subtropics
Diet	Filter-feeder on planktonic plants and animals
Reproduction	Spawns near surface at spring tides; larvae develop in brackish waters

Common jollytail, or inanga

Inanga is the Maori (native New Zealand) name for the fish called by European settlers the minnow (or, when young, whitebait); jollytail is the Australian name. The fish spends most of its adult life in the lower reaches of rivers, but migrates downstream to estuaries to breed. It spawns among vegetation at high spring tide (an extremely high tide), and the eggs hatch only after they have been covered again by the next spring tide. The larvae are washed into the sea and develop there for some months before returning to fresh water. Some fish have been caught far from land, and the species must have spread across the world by long sea journeys.

Scientific name	*Galaxias maculatus* or *G. attenuatus*
Classification	Order Osmeriformes (smelts and allies) or Galaxiiformes; sometimes included in Salmoniformes (salmon and allies); family Galaxiidae
Size	Up to about 18cm (7in) long
Distribution	Australia, New Zealand and South America, and adjacent islands
Habitat	Temperate fresh and salt water
Diet	Aquatic and surface-floating invertebrates
Reproduction	Spawns in estuaries at high tide; eggs hatch at next spring tide

European smelt

A curious feature of all smelts – the reason for which is unknown – is their cucumber-like smell. They are rather like slender trout, and (like trout) have a fleshy second dorsal fin near the tail. The European species, like most of its close relatives, spends most of its life at sea, rarely far from the shore, but it migrates to fresh water to breed. It swims upstream in early spring and sheds its eggs over the pebbles or gravel of the river bed, to which the eggs attach by a thin stalk. The adults return to the sea after some weeks or months, and the young fish later follow them. In some Scandinavian lakes there are permanent freshwater populations.

Scientific name	*Osmerus eperlanus*
Classification	Order Osmeriformes (smelts and allies); sometimes included in Salmoniformes (salmon and allies); family Osmeridae
Size	Up to about 46cm (18in) long
Distribution	North-east Atlantic; White Sea; Baltic
Habitat	Cool to cold coastal waters, to 50m (165ft); estuaries; rivers
Diet	Mainly shrimps and other small crustaceans; some small fish
Reproduction	Spawns in spring in gravel-bottomed rivers

Atlantic salmon

Often called the 'king of fish', the Atlantic salmon is one of the finest game and food fish. It spends its early years in a river, changing its appearance as it develops; the various stages have special names: alevin (larva), parr and smolt. The last, up to 25cm (10in) long, is the form that returns to the sea, where it lives and grows, usually for several years. Adults then find their way – probably by smell – back to the river where they were born, often leaping high waterfalls to reach their spawning grounds. These are the fish that anglers and fishermen catch. Many die after spawning, but emaciated 'kelts' may return to the sea and later breed again.

Scientific name	*Salmo salar*
Classification	Order Salmoniformes (salmon and allies)
	Family Salmonidae
Size	Up to about 1.5m (5ft) long; weight up to 47kg (104lb)
Distribution	Eastern and Western North Atlantic; some lakes; farmed in sea lochs
Habitat	Open sea, mostly in cold surface waters, to 10m (33ft); rivers and lakes
Diet	Squid; crustaceans; fish. Young feed on small invertebrates and fish
Reproduction	Spawns in river of birth; young return to sea after one to six years

Rainbow trout, or steelhead

Rainbow trout are named for the fine colouring of adult fish that spawn in fast-flowing streams; those living permanently in large inland lakes are a more uniform silver colour. The species originated in the American north-west, where populations that regularly migrate to the sea are known as steelheads; these are the biggest rainbow trout, and return to fresh water to breed after several years at sea. However, it seems to be naturally a freshwater species, and has been introduced to rivers and lakes worldwide – even to tropical lakes above about 1200m (4000ft). It is also widely farmed, and captive-bred fish are used to stock fishing lakes and ponds.

Scientific name	*Oncorhynchus mykiss* or *Salmo gairdneri*
Classification	Order Salmoniformes (salmon and allies)
	Family Salmonidae
Size	Up to about 1.2m (4ft) long; weight up to 26kg (57lb)
Distribution	Eastern Pacific (Alaska to Mexico); widely introduced around world
Habitat	Open temperate and warm waters, to 10m (33ft); rivers and lakes
Diet	Various invertebrates; small fish
Reproduction	In wild, spawns in spring in hollow in gravel of fast-flowing stream

Sockeye salmon; kokanee

These are two forms of the same Pacific salmon species, both of which breed in freshwater lakes and streams. The name sockeye applies to fish that return to the sea at the age of one to three years, and spend their adult life there before returning to their breeding grounds – a run of up to 2400km (1500 miles). Kokanee remain in fresh water; they never grow as big as sockeyes, which are among the most important commercially fished Pacific salmon (although endangered in some river systems). At spawning time, the fish turn from a silvery colour to red; the male is particularly brilliant, and develops a humped back, as illustrated. They then die.

Scientific name	*Oncorhynchus* (or *Salmo*) *nerka*
Classification	Order Salmoniformes (salmon and allies)
	Family Salmonidae
Size	Up to about 84cm (33in) long; weight up to 7.7kg (17lb)
Distribution	North Pacific rim, from Japan and Kamchatka to Alaska and California
Habitat	Open sea and coastal waters, to 240m (800ft); rivers and lakes
Diet	Mainly small invertebrates; sea-going adults also eat fish
Reproduction	Spawn in lakes or streams, then die; young sockeyes migrate to sea

Pacific hatchetfish

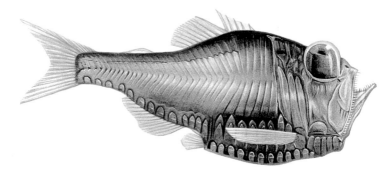

Despite its common name, this species is found in the Atlantic and Indian oceans as well as the Pacific. And despite their fearsome appearance, hatchetfishes are small, harmless fish that feed on small invertebrates. They, in turn, are important food for deep-swimming tuna in many areas. Hatchetfishes are middle to deep-water species with a short silvery body that is compressed from side to side. They have light-generating organs on their lower surface; the colour and pattern of light produced varies from species to species. The hatchetfish's mouth and the large, bulging eyes point upwards, suggesting that it usually attacks its prey from below.

Scientific name	*Argyroplectus affinis*
Classification	Order Stomiiformes (dragonfishes and relatives)
	Family Sternoptychidae
Size	Up to about 8.5cm (3⅓in) long
Distribution	Worldwide in tropical and warm temperate waters
Habitat	Mid-levels of deep ocean, but sometimes to 3870m (12 700ft)
Diet	Small crustaceans and other invertebrates
Reproduction	Eggs and larvae planktonic; swim to deep water as they mature

Sloane's viperfish

A fierce predator in spite of its relatively small size, the viperfish is armed with long, fang-like teeth set in jaws that are specially built to open extremely wide for attacking prey. Yet the lower teeth are so long that they extend above the fish's head when its jaws are closed. The frontmost ray of the viperfish's dorsal fin is lengthened into a filament like a fishing line; it is tipped with a light-generating organ to lure prey in the deep gloom in which it lives. The fish's ventral (lower) surface also has numerous light organs, which glow red, scattered all over it. Viperfish live at least 1000m (3300ft) deep, but are believed to move up to around 500m (1650ft) or even shallower levels at night, in order to feed.

Scientific name	*Chauliodus sloani*
Classification	Order Stomiiformes (dragonfishes and relatives)
	Family Stomiidae; sometimes classified separately in Chauliodontidae
Size	Up to about 35cm (14in) long
Distribution	Tropical and temperate parts of all oceans, but patchily distributed
Habitat	Deep oceanic waters, to 1000m (3300ft) or more
Diet	Fish; crustaceans
Reproduction	Spawns at any time of year; eggs and larvae planktonic

Scaly dragonfish, or boa fish

This dragonfish fully deserves its common names, with its fearsome fangs and its head and mouth wider than its body (and able to swallow prey larger than itself). It also has a strange-looking barbel (ending in a light-emitting organ and three short filaments) hanging from its mouth. There are several known subspecies with minor differences, living in various parts of the deep oceans, but all have rows of hexagonal (six-sided) scales along the sides of their body and a series of about 90 photophores (light-emitters) along the belly. Little is known of their habits, but they are believed to move from the deeps nearer the surface at night in order to feed.

Scientific name	*Stomias boa*
Classification	Order Stomiiformes (dragonfishes and relatives)
	Family Stomiidae or Stomiatidae
Size	Up to about 40cm (16in) long
Distribution	Atlantic, Indian and Pacific oceans, including Arctic and Antarctic waters
Habitat	Deep waters to 1000–4000m (3300–13 000ft); shallower at night
Diet	Fish; crustaceans
Reproduction	Eggs and larvae planktonic

Black dragonfish

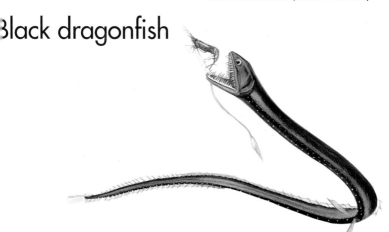

Known also as the ribbon sawtail fish, this deep-water species is rather like the scaly dragonfish (*opposite*), but has no scales on its body. However, its most extraordinary feature is its life history. Female larvae and adults are much bigger than their male counterparts. Female larvae grow to a length of about 7cm (2¾in), and have eyes at the end of stalks up to 2cm (¾in) long. As the larvae grow into the adult shape, the stalks are gradually absorbed until the eyes are in the normal position in the creature's head. Male larvae, on the other hand, remain small and, when adult, have no teeth and do not eat. They live only long enough to mate.

Scientific name	*Idiacanthus fasciola*
Classification	Order Stomiiformes (dragonfishes and relatives)
	Family Stomiidae; sometimes classified separately in Idiacanthidae
Size	Female up to about 50cm (19½in) long: male about 8cm (3in) long
Distribution	Probably worldwide in deep waters, except polar regions
Habitat	Deep waters, to 500–2000m (1650–6500ft); female shallower at night
Diet	Fish; crustaceans
Reproduction	Eggs planktonic; larvae differ according to sex (*see above*)

Showy bristlemouth

Some marine biologists believe that there are more individual creatures among bristlemouths of the genus *Cyclostoma* than of any other genus of vertebrates (animals with backbones) on Earth. This is because the oceanic plankton – the mass of floating organisms that drift around the world's oceans – contains billions upon billions of these tiny fish. Many species of bristlemouths (also sometimes known as lightfish) are even smaller than the so-called 'showy' species, and even the biggest is no more than 7.5cm (3in) long. Yet, like other members of their group, they have a relatively large mouth, and hunt small planktonic crustaceans and other creatures.

Scientific name	*Cyclothone signata*
Classification	Order Stomiiformes (dragonfishes and relatives)
	Family Gonostomatidae
Size	Up to about 3cm (just over 1in) long
Distribution	Pacific and Indian oceans
Habitat	Open ocean; mostly found near surface and at 400–500m (1300–1650ft)
Diet	Small planktonic creatures
Reproduction	Eggs and larvae planktonic

Spotted lantern fish

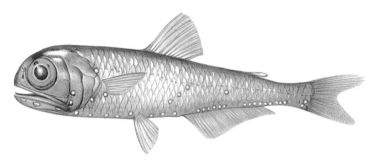

The species name *punctatum* means 'spotted', and this fish has light-producing photophores scattered over its head and body. There are many related species of lantern fish, all with a blunt, rounded head, long jaws and large eyes. Each has its own unique pattern of glowing spots, which is sometimes the only method of identification. In fact, the fish themselves may identify other members of their own species in the same way, for breeding. Lantern fishes are among the most abundant of all fish – more so even than sardines and anchovies. They are eaten by many larger species, and could become an important commercial catch in many areas.

Scientific name	*Myctophum punctatum*
Classification	Order Myctophiformes (lantern fishes and allies)
	Family Myctophidae
Size	Up to about 11cm (4½in) long
Distribution	North Atlantic (tropics to Arctic); Mediterranean
Habitat	Deep ocean, mostly at 700–800m (2300–2600ft); nearer surface at night
Diet	Small crustaceans; fish larvae
Reproduction	Female produces about 800–900 eggs in spring to summer; planktonic

Sea catfish

There are more than 2000 catfish species, easily recognized by the barbels around their mouth. Most live in fresh water, but this – the hardhead sea catfish – is one of more than 100 marine species. It swims in large shoals in shallow, often muddy water, and can make loud sounds by vibrating its swim-bladder. The most extraordinary feature of sea catfish of the family Ariidae is their breeding habits. Females produce relatively small numbers of large eggs, up to 2cm (¾in) across. The male holds as many as 55 eggs in his mouth until they hatch, and cannot eat during this time. After hatching, the young fish also hide from danger in his mouth.

Scientific name	*Arius felis*
Classification	Order Siluriformes (catfishes and allies); suborder Siluroidei
	Family Ariidae
Size	Up to about 70cm (27½in) long
Distribution	Western Atlantic (Massachusetts to Mexico)
Habitat	Coastal waters; estuaries
Diet	Bottom feeder, mostly at night, on crabs, shrimps and some small fish
Reproduction	Mouth-brooder (*see above*)

Red or diamond lizardfish

The lizardfishes have a large head, with a wide, toothy mouth, that looks very lizard-like especially when seen from the side. They are voracious carnivores, and habitually lie on the sea-bottom, propped on their lower fins, with their head raised in a lizard-like manner. Some species half-bury themselves in the sand, but the red lizardfish prefers a rocky bottom. The cryptic coloration of most species' back hides them well; whenever suitable prey passes, they suddenly lurch forward to catch it. A related species of the subfamily Harpadontinae, *Harpadon nehereus*, is widely caught in south and south-east Asia, dried and sold as the 'Bombay duck'.

Scientific name	*Synodus synodus*
Classification	Order Aulopiformes (lizardfishes, lancetfishes and allies)
	Family Synodontidae; subfamily Synodontinae
Size	Up to about 33cm (13in) long
Distribution	Tropical and subtropical parts of Atlantic; Caribbean
Habitat	Mainly shallow inshore waters; sometimes to 90m (300ft)
Diet	Mainly fish
Reproduction	Spawns near bottom in open waters

Atlantic cod

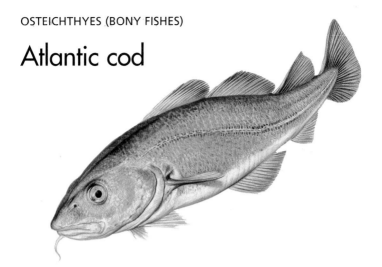

Many of the most important commerically caught fish, including hake, ling, whiting and haddock (*opposite*), belong to the cod family and its relatives, but few are as important as the common Atlantic cod. However, overfishing in recent decades has reduced stocks drastically, and some populations – particularly around Greenland and Newfoundland – are particularly endangered. Cod are prolific breeders, however, and with conservation measures their numbers could increase again quite rapidly. The similar but rather larger-headed Pacific cod (*Gadus macrocephalus*) is widely caught in the eastern and western far North Pacific.

Scientific name	*Gadus morrhua*
Classification	Order Gaddiformes (codfishes and allies)
	Family Gadidae
Size	Up to 2m (6½ft) long and 96kg (212lb); usually much less, to 1.3m (4¼ft)
Distribution	Atlantic north of Cape Hatteras and Bay of Biscay, to 78°N; Baltic
Habitat	Cool or cold inshore and offshore waters, to 600m (2000ft)
Diet	Fish; invertebrates
Reproduction	Spawns at various seasons, at or near sea-bottom

Haddock

Another important North Atlantic fishing catch, haddock do not grow as big as cod, and are easily distinguished by their sharply pointed first dorsal fin. (Both cod and haddock have three dorsal fins, but in cod the first one is rounded.) Haddock also have a much smaller chin barbel than cod. They have never been caught in quite such large quantities as cod, but they are less prolific breeders and are also threatened by overfishing. Young haddock live at first near the surface among the tentacles of a large jellyfish before heading for deep waters. Both cod and haddock may make long migrations between feeding and spawning grounds.

Scientific name	*Melanogrammus* (formerly *Gadus*) *aelgefinus*
Classification	Order Gaddiformes (codfishes and allies)
	Family Gadidae
Size	Rarely more than 1m (3¼ft) long; weight up to 17kg (37½lb)
Distribution	Atlantic north of Delaware Bay and Bay of Biscay, to 78°N; Barents Sea
Habitat	Cold inshore and offshore waters, mostly at 80–200m (260–650ft)
Diet	Mainly bottom-living crustaceans, molluscs, starfish, worms and fish
Reproduction	Spawns late winter to early summer at depth of 50–150m (165–500ft)

Alaska or walleye pollack

Since the 1980s, a greater tonnage of Alaska or walleye pollack has been caught than of any other fish species – almost 7 million tonnes a year at its peak in 1986, although less than 5 million tonnes by 1995. At one time, it was used only to make animal feed, but has become an important food for humans – mostly in the form of fish fingers and other frozen blocks, or as salted fish. Its roe is also eaten, and the flesh is processed to make surimi, or fish mince, for making 'crab sticks'. It is widely distributed, with at least 12 major populations. The fish swim near the sea-bottom in vast schools during the day, but often move to higher levels at night in order to feed.

Scientific name	*Theragra chalcogramma* (formerly *Gadus chalcogrammus*)
Classification	Order Gaddiformes (codfishes and allies)
	Family Gadidae
Size	Up to 80–90cm (31½–36in) long
Distribution	North Pacific, from Alaska to Sea of Japan and California
Habitat	Open sea, from near surface to almost 1000m (3300ft) deep
Diet	Fish; krill and other crustaceans
Reproduction	Spawns winter to summer in dense schools at 50–250m (165–800ft)

Rat-tail, or grenadier

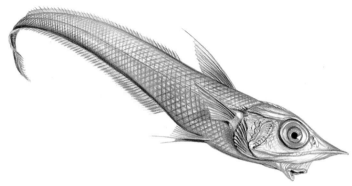

The various rat-tail or grenadier species are probably the most numerous and varied of all deep-water fish. They are curious-looking, with a large head and a long tapering tail; many, such as the species illustrated, have a pronounced snout. This and some others have a large light-producing organ on their belly, between the base of the pelvic fins. Rat-tails are rather slow swimmers with a characteristic head-down posture, eating more or less any small creatures or other food material – alive or dead – they encounter. Their large eyes suggest that they hunt luminescent deep-sea creatures, but they may also possibly feed nearer to the surface at night.

Scientific name	*Caelorinchus caelorhincus*; often spelt *Coelorinchus* or *Coelorhynchus*
Classification	Order Gaddiformes (codfishes and allies)
	Family Macrouridae
Size	Up to about 40cm (16in) long
Distribution	Warmer parts of Atlantic; Mediterranean
Habitat	Mainly deep water, to at least 1250m (4100ft); some species deeper
Diet	Bottom-living invertebrates and fish; also scavenges on dead creatures
Reproduction	Details uncertain, but eggs and larvae probably planktonic

Angler fish or goosefish

The angler fish makes excellent eating, but it looks so hideous that it is almost always displayed with the head and skin removed, and is sold as monkfish. The name angler fish is very apt, however, since it has a 'fishing rod' – a long, flexible spine properly called the *illicium* – growing from its head; this ends in a fleshy bait or lure called the *esca*. The fish lies on the bottom, well disguised by its colouring and fringe of skin flaps, and sometimes half-buried, until the lure attracts suitable prey. The slightest touch on the esca provokes a snapping reflex by its huge jaws. A similar but smaller species, *Lophius americanus*, lives off eastern North America.

Scientific name	*Lophius piscatorius*
Classification	Order Lophiiformes (angler fishes, frogfishes and allies)
	Family Lophiidae
Size	Up to about 2m (6½ft) long; weight up to 58kg (128lb)
Distribution	Eastern Atlantic (Barents Sea to West Africa); Mediterranean; Black Sea
Habitat	Inshore and offshore waters at 20–1000m (65–3300ft)
Diet	Fish; larger invertebrates
Reproduction	Spawns spring and summer; eggs planktonic, in gelatinous ribbon

Longnose batfish

Batfishes are bottom-living fish that lure their prey in a similar way to the related angler fishes (*opposite*). However, in batfishes, the lure and its 'fishing rod' are retractable into a tube just above the mouth when the fish is resting. When feeding, the lure is extended and vibrated to attract small prey – which the batfish snaps up. Batfishes are poor swimmers; they push themselves along the seabed with their fleshy pectoral fins. The longnose batfish was thought to be a single species, but fish found from the Bahamas north are now classified separately, as *Ogcocephalus corniger*, from those living from the West Indies to Uruguay, *O. vespertilio*.

Scientific name	*Ogcocephalus vespertilio* and *O. corniger*
Classification	Order Lophiiformes (angler fishes, frogfishes and allies)
	Family Ogcocephalidae
Size	Up to about 30cm (12in) long; *O. corniger* smaller
Distribution	Tropical and subtropical western Atlantic (*see above*)
Habitat	seabed, to depths of 30–240m (100–800ft)
Diet	Small crustaceans, worms and other invertebrates; small fish
Reproduction	Little is known of their biology

Garfish

The garfish, like all members of its family, has a long, slim body with the dorsal and anal fins set well back, and extremely elongated jaws armed with many needle-like teeth. It is a surface-swimming predator which chases its prey and often leaps well out of the water. Young garfish pass through a 'halfbeak' stage when the upper jaw is much shorter than the lower; these feed on plankton. (The upper jaw is permanently short in the halfbeaks of the related family Hemiramphidae.) Three distinct subspecies of garfish live in the northern Atlantic and Baltic, in the Mediterranean and warmer parts of the Atlantic, and in the Black Sea respectively.

Scientific name	*Belone belone*
Classification	Order Beloniformes (needlefishes, flying fishes and allies)
	Family Belonidae
Size	Up to about 94cm (37in) long
Distribution	Eastern North Atlantic; Baltic; Mediterranean; Black Sea
Habitat	Surface waters, mostly in open sea, but inshore and estuaries in summer
Diet	Small fish; some crustaceans
Reproduction	Spawns spring and summer; eggs have threads that attach to seaweed

Atlantic or Mediterranean flying fish

Flying fishes glide rather than actively fly through the air, and do so mainly to escape predators. They swim forwards at high speed in the water, boost themselves with a rapid vibration of the tail as they break the surface, then glide on their greatly enlarged fins. The Atlantic species is a so-called four-wing flying fish, using both its pectoral and smaller pelvic fins as wings. The tropical flying fish, *Exocoetus volitans*, which is widespread around the world, is a two-wing species, with only its pectoral fins enlarged. It is fished commercially in many areas, but the Atlantic species is also good to eat and could be exploited.

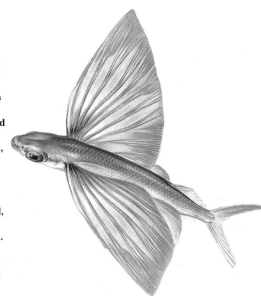

Scientific name	*Cheilopgon* (or *Cyselurus*) *heterurus*
Classification	Order Beloniformes (needlefishes, flying fishes and allies)
	Family Exocoetidae
Size	Up to about 40cm (16in) long
Distribution	Eastern and western Atlantic; western Mediterranean
Habitat	Temperate and subtropical inshore waters, near surface
Diet	Small crustaceans and other planktonic creatures
Reproduction	Spawns spring and summer; planktonic eggs with filaments

Mediterranean sand smelt

Several very similar species of sand smelts live along various parts of the African and European shoreline. *Atherina presbyter* (known simply as the sand smelt) is the most northerly, living along Atlantic coasts from Denmark to West Africa, while *A. hepsetus* and *A. boyeri* (the big-scaled sand smelt) are mostly Mediterranean species. All are silvery fish with a slender body and two widely spaced dorsal fins; some can be told apart only by small details such as the number of scales along the body and their relative eye size. They swim in large schools, and several species are widely caught for eating, sometimes as small 'whitebait'; they breed rapidly.

Scientific name	*Atherina hepsetus*
Classification	Order Atheriniformes (silversides and allies)
	Family Atherinidae
Size	Up to about 20cm (8in) long
Distribution	Mediterranean and adjacent parts of Atlantic; Black Sea
Habitat	Warm coastal waters; also brackish lagoons and estuaries
Diet	Bottom-living and floating crustaceans
Reproduction	Spawns near seabed; eggs anchored by filaments

California grunion

The 'grunion run' is a favourite spectacle for southern Californians. Every two weeks from March to August, for a few days after the new and full moons, female California grunion ride the surf at the spring (extra-high) tide, and swim and wriggle as far up the beach as they can. They half-bury themselves in the sand and lay their eggs, which are immediately fertilized by males. The eggs incubate in the damp sand until the next spring tide, about 10–12 days later; they then hatch within minutes of being immersed. Females may spawn several times at successive spring tides. Catching of spawning grunion is strictly controlled, using the hands only.

Scientific name	*Leuresthes tenuis*
Classification	Order Atheriniformes (silversides and allies)
	Family Atherinidae
Size	Up to about 19cm (7½in) long
Distribution	Eastern Pacific from Monterey Bay to southern Baja California
Habitat	Coastal waters; near surface, to 18m (60ft)
Diet	Small planktonic creatures
Reproduction	Spawns on beach at high spring tide; eggs hatch next high spring tide

Ribbon fish

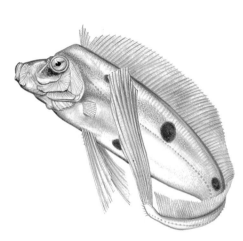

The ribbonfish and the closely related dealfishes (which belong to the same genus, *Trachipterus*) have an extraordinarily elongated but narrow body that tapers to the tail, with the dorsal fin extending along most of its length. They are rare ocean fish that mostly live at considerable depths, but they sometimes come to the surface and (together with the even larger oarfish, *Regalecus glesne*) are said to have originated many myths of sea-serpents. One curious feature of all fish of the order Lampridiformes is their extendible jaws; the upper jaw is not connected to the cheek bones, so it can protrude, expanding the mouth cavity by as much as 40 times.

Scientific name	*Trachipterus* (or *Trachypterus*) *trachypterus*
Classification	Order Lampridiformes or Lampriformes (oarfishes and allies)
	Family Trachipteridae
Size	Up to 3m (10ft) long
Distribution	Parts of Atlantic and Pacific oceans; Mediterranean
Habitat	Open ocean, mostly at mid-levels, to depth of 500m (1650ft)
Diet	Squid; midwater fish
Reproduction	Planktonic eggs incubate near surface; larvae feed on plankton

Opah or moonfish

Despite its very different shape – almost round when seen from the side, but highly compressed from side to side – the opah's extendible jaws show that it is a member of the same group as the ribbonfish and its allies (*see opposite*). It is a huge, brilliantly coloured fish that has been found in virtually every ocean, but only in very small numbers; it seems to spend most of its life alone. It is a strong swimmer, propelled by its large, muscular pectoral fins. It is occasionally caught by fishermen whose target is other fish – mainly tuna – and makes good eating. The smaller southern opah (*Lampris immaculatus*) of the Southern Ocean has no spots.

Scientific name	*Lampris guttatus* or *L. regius*
Classification	Order Lampridiformes or Lampriformes (oarfishes and allies)
	Family Lamprididae
Size	Up to 2m (6½ft) long; weight up to 270kg (600lb)
Distribution	Probably worldwide apart from Arctic and Antarctic
Habitat	Open ocean, mainly at depths of 100–400m (330–1300ft)
Diet	Squid, octopuses, crabs and other invertebrates; midwater fish
Reproduction	Planktonic eggs incubate near surface; larvae feed on plankton

Fifteen-spined or sea stickleback

Sticklebacks are found in fresh and salt water throughout much of the Northern Hemisphere, and have for generations fascinated young nature-explorers and aquarium-keepers. All are small fish with spines on their back (the origin of their common name), but this species (which may in fact have 14 to 17 spines) is much more slender and elongated than others. Unlike most sticklebacks, it lives only in seawater – in pools or among seaweeds and eel grasses along the shore. As with all species, the male makes a nest among the weeds and entices several females, one after the other, to spawn there. He fertilizes and guards the eggs until they hatch.

Scientific name	*Spinachia spinachia*
Classification	Order Gasterosteiformes (pipefishes and allies); Suborder Gasterosteoidei; Family Gasterosteidae
Size	Up to about 22cm (8½in) long
Distribution	North-eastern North Atlantic (northern Norway to Bay of Biscay); Baltic
Habitat	Shallow coastal waters; rock pools
Diet	Small bottom-living invertebrates
Reproduction	Male makes nest by gluing together seaweeds, and guards eggs

Tubesnout

The tubesnout's body and snout are even more elongated than the 15-spined stickleback's (*opposite*) and it has even more spines – up to 27 – on its back, but is very much the Pacific counterpart of the sea stickleback. (Another similar species, *Aulichthys japonicus*, lives along seashores of northern Japan and Korea.) A major difference is that, while sticklebacks tend to be solitary or live in pairs, the tubesnout is found in large shoals of hundreds or even thousands of fish – mostly among kelp beds and patches of eel grass, or in rocky areas where the seabed is sandy. But schools also sometimes swim much farther offshore, near the surface.

Scientific name	*Aulorhynchus flavidus*
Classification	Order Gasterosteiformes (pipefishes and allies); Suborder Gasterosteoidei; Family Aulorhynchidae
Size	Up to about 18cm (7in) long
Distribution	Eastern North Pacific (Alaska to Baja California)
Habitat	Usually shallow coastal waters; sometimes in schools well offshore
Diet	Small crustaceans; fish larvae
Reproduction	Male makes nest by gluing together seaweeds, and guards eggs

Lesser pipefish

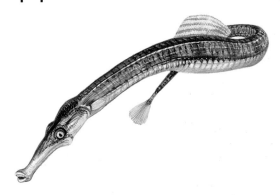

Sometimes called Nilsson's pipefish after the naturalist who identified and named it in the 1850s, this is one of numerous similar pipefish species, all of which have a long tubular snout and an elongated tubular body encased in bony, jointed rings. The fish vary in size and coloration, but some species may be distinguishable only by the number of rings; in the lesser pipefish there are 13 to 17 body rings and a further 37 to 42 along the tail. In all pipefishes, the male has a pouch of skin on his abdomen in which the female deposits about 100 eggs; he fertilizes and keeps them in the pouch until they hatch and the young are about 14mm (just over ½in) long.

Scientific name	*Syngnathus rostellatus*
Classification	Order Gasterosteiformes (pipefishes and allies); suborder Syngnathoidei
	(but sometimes classified as Syngnathiformes); Family Syngnathidae
Size	Up to about 17cm (6½in) long
Distribution	Eastern North Atlantic (Norway to Bay of Biscay); Kattegat
Habitat	Shallow coastal waters and estuaries
Diet	Small invertebrates; fish larvae
Reproduction	Male carries eggs and larvae in pouch on rear of abdomen

Long-snouted seahorse

Seahorses are unmistakeable, and are among the most fascinating of all bizarre-shaped fishes. They were first described scientifically by Aristotle over 2300 years ago, and are widely displayed in aquariums. They are, in fact, quite closely related to pipefishes, but have their body bent into an S-shape; the tail is prehensile – it can be wound around a seaweed stem or a branch of coral to anchor the fish. When swimming, the dorsal and pectoral fins oscillate, but the body hardly moves. As with pipefishes, the male seahorse has a brood pouch in which the eggs incubate and the larvae develop until they are 'pumped' out through a small opening.

Scientific name	*Hippocampus guttulatus* or *H. ramulosus*
Classification	Order Gasterosteiformes (pipefishes and allies); suborder Syngnathoidei (but sometimes classified as Syngnathiformes); Family Syngnathidae
Size	About 16–18cm (6¼–7in) long
Distribution	Eastern Atlantic (British Isles to Canaries); Mediterranean; Black Sea
Habitat	Shallow inshore waters among seaweeds and eel grass; lagoons
Diet	Small bottom-living invertebrates
Reproduction	Male carries eggs and larvae in abdominal pouch for 3–5 weeks

John dory

The John dory is a rather ugly-looking fish with long dorsal spines, but it makes excellent eating and is caught by fishermen in many parts of the world. However, it is a largely solitary species and is not suitable for large-scale fishing. It is not a strong swimmer, but its body is highly compressed from side to side, making it inconspicuous from the front so that it can approach prey unnoticed. It has greatly extendible jaws which it then shoots forward to seize the prey. According to legend, the round black mark on each side of the John dory's body is the thumb-print of St Peter, and its name in several European languages translates as St Peter's fish.

Scientific name	*Zeus faber*
Classification	Order Zeiformes (dories and allies);
	Family Zeidae
Size	Up to about 90cm (3ft) long and 8kg (17½lb); usually 30–40cm (12–16in)
Distribution	Almost worldwide in temperate waters
Habitat	Offshore waters, to 400m (1300ft); closer inshore in summer
Diet	Mainly schooling fish such as herrings, anchovies, etc.; some crustaceans
Reproduction	Spawns in shallow water inshore; large (2mm [½in]) floating eggs

Flying gurnard

In spite of its common name and its extremely large, wing-like or fan-like pelvic fins, the flying gurnard cannot fly. (For this reason, some experts prefer the name helmet gurnard – its head is bony.) The fish seems to use its 'wings' to make itself look bigger to potential predators – to ward them off long enough to make an escape. The front part of each pelvic fin forms a separate lobe, and the fish uses this to 'walk' over the sandy, muddy or rocky sea floor, searching for crustaceans, clams and other food items. A rather similar but smaller related species, the oriental flying gurnard (*Dactyloptena orientalis*), lives in the Indian and western Pacific oceans.

Scientific name	*Dactylopterus* (formerly *Trigla*) *volitans*
Classification	Order Scorpaeniformes (scorpionfishes and allies); sometimes classified separately in Dactylopteriformes; Family Dactylopteridae
Size	Up to about 50cm (20in) long
Distribution	Tropical and warm temperate east and west Atlantic; Mediterranean
Habitat	Bottom-living in shallow waters; reefs
Diet	Bottom-living crustaceans (especially crabs); molluscs; small fish
Reproduction	Spawns on seabed

Lionfish

The lionfishes and their close relatives the turkeyfishes (*Dendrochirus* species, also often called lionfishes) are extraordinary-looking, brightly coloured fish that are well described by the alternative name butterfly cod. Their colouring and form are a warning to other reef-dwellers, for the lionfish is venomous: it has poison glands at the base of the extremely sharp dorsal spines. It hides in an inconspicuous place in daytime, and hunts its prey by night. Anyone disturbing the fish with hands or feet is liable to get an extremely painful, but rarely fatal sting. (Heat on the wound relieves the pain.) Yet, despite the risk, it is caught and makes good eating.

Scientific name	*Pterois volitans*
Classification	Order Scorpaeniformes (scorpionfishes and allies);
	Family Scorpaenidae
Size	Up to about 38cm (15in) long
Distribution	Indian; Pacific oceans (Malaysia and Australia to Japan and Polynesia)
Habitat	Tropical and subtropical lagoons and reefs, to 50m (165ft)
Diet	Small fish and crustaceans
Reproduction	Spawns on sea-bottom

Stonefish

If the lionfish (*opposite*) advertises its nature colourfully, the stonefish disguises itself – and is far more dangerous. It and its equally venomous estuarine relative, *Synanceia horrida* or *S. trachynis*, lie half-hidden in sand or mud, motionless for long periods and sometimes encrusted with algae. When prey approaches, suddenly opening jaws suck in the victim. If disturbed, the stonefish raises its dorsal spines, which are needle-sharp and have grooves that inject venom into a wound like a hypodermic needle. It is the most toxic venom in the fish world, causing severe pain and sometimes death. Fortunately, an antidote has been developed in Australia.

Scientific name	*Synanceia verrucosa*
Classification	Order Scorpaeniformes (scorpionfishes and allies);
	Family Scorpaenidae or Synanceiidae
Size	Up to about 40cm (16in) long
Distribution	Indian; Pacific oceans (East Africa to Australia, Ryukyu Is and Polynesia)
Habitat	Bottom-living in warm; shallow lagoons; pools and reefs; to 30m (100ft)
Diet	Bottom-living fish and crustaceans
Reproduction	Details uncertain, but presumably spawns on sea-bottom

Redfish, or ocean perch

One of the most commercially important species of the scorpionfish group, the redfish, ocean perch or rosefish is sometimes sold as 'Norway haddock'. It is a gregarious fish that swims in schools, and is caught by deep-water trawlers in the northern seas. It is a heavy-bodied, slow-growing species, and most of those caught are much smaller than the maximum size given below. It has spines on the dorsal and anal fins and on the cheeks, and the large head has a protruding lower jaw. The redfish's eggs are fertilized internally; males and females mate in late summer or autumn but fertilization is delayed and the young born six to nine months later.

Scientific name	*Sebastes marinus*
Classification	Order Scorpaeniformes (scorpionfishes and allies)
	Family Scorpaenidae or Sebastidae
Size	Up to about 1m (3¼ft) long; maximum weight 15kg (33lb)
Distribution	Northern Atlantic and Barents Sea, south to North Sea and New Jersey
Habitat	Offshore, at 100–1000m (330–3300ft); nearer surface at night to feed
Diet	Schooling fish such as herring; crustaceans and other invertebrates
Reproduction	Ovoviviparous; young, about 6mm (¼in) long, born spring and summer

Tub gurnard

The true gurnards or sea robins of the family Triglidae look quite similar to the flying gurnard (*see p.243*), but they are classified in a separate order as well as a separate family. The tub gurnard is typical of the family, with a large polygonal head armoured with bony plates, a pointed snout and a steep brow. The front three rays of the pectoral fins are separated to form finger-like feelers, which the fish uses to 'walk' on the sea-bottom and to sense the presence of prey. Like many gurnards, it can make a grunting noise by using muscular contractions to pump air through its swim bladder. This may be the origin of its name, from the Latin *grunnire* (to grunt).

Scientific name	*Chelidonichthys* (or *Trigla*) *lucerna*
Classification	Order Scorpaeniformes (scorpionfishes and allies);
	Family Triglidae
Size	Up to about 75cm (30in) long; maximum weight 6kg (13lb)
Distribution	Eastern Atlantic (Norway to Mauritania); Mediterranean; Black Sea
Habitat	Bottom-living, to 150–300m (500–1000ft), on mud, sand or gravel
Diet	Bottom-living crustaceans; molluscs and fish
Reproduction	Eggs and larvae planktonic

Lumpsucker or lumpfish

The name lumpsucker comes from the warty bumps that cover the fish's scaleless body, and from the round sucker disc – actually modified pelvic fins – on its underside. The male fish uses this sucker to attach itself to a rock while guarding the eggs that the female lays in shallows in the summer. The male develops reddish underparts (as illustrated) at this time; the female remains bluish-green. While on guard, the male does not feed; he constantly fans fresh water over the eggs with his fins. Lumpsuckers, or lumpfish, are caught for their flesh but particularly for their roe, which is dyed black or orange to make an economical caviar substitute.

Scientific name	*Cyclopterus lumpus*
Classification	Order Scorpaeniformes (scorpionfishes and allies)
	Family Cyclopteridae
Size	Female up to about 60cm (24in) long; male up to about 50cm (20in)
Distribution	North Atlantic (Hudson Bay and Barents Sea to Maryland and Spain)
Habitat	Mainly on rocky seabed, to 50–400m (165–1300ft); breeds inshore
Diet	Jellyfish, worms, small crustaceans and other invertebrates; some fish
Reproduction	Large numbers of sticky eggs, laid in mass on rocks; guarded by male

Dusky grouper or dusky perch

Perch and their allies (order Perciformes) make up the biggest and most varied group of all vertebrates (animals with backbones), although DNA studies result in their classification constantly being reorganized. The biggest species in the group is the 4.6m (15ft) black marlin (*Makaira indica*), but the dusky grouper is one of the biggest inshore fishes in many parts of its range, which includes the Mediterranean. It is a deep-bodied, massive fish that is highly territorial and generally lives alone among rocks. Very little is known about its breeding habits except that all young fish are female and some later become male – but the age of the change is uncertain.

Scientific name	*Epinephelus marginatus* or (incorrectly) *E. guaza*
Classification	Order Perciformes (perches and allies); Suborder Percoidei
	Family Serranidae
Size	Up to about 1.5m (5ft) long; maximum weight 60kg (132lb)
Distribution	Tropical and subtropical parts of Atlantic and southwest Indian oceans
Habitat	Reefs and rocky coasts, at depth of 8–300m (25–1000ft)
Diet	Crabs and other crustaceans; squid and octopuses; fish
Reproduction	Changes sex (*see above*); other details uncertain

Cardinal fish

The brightly coloured cardinal fish looks at first glance rather like a goldfish (*Carassius auratus*), but the eyes and mouth are much bigger and it has two dorsal fins, not one. It is one of some 200 similar species found in warm waters worldwide; freshwater species live in Australia, New Guinea and some Pacific islands. Most, if not all (including this species), are mouth-brooders; the male takes the fertilized eggs into his mouth and keeps them there until the larvae hatch. The normal flow of water through the parent's mouth and gills keeps the eggs supplied with oxygen. Some cardinal fish form schools, but others are found singly.

Scientific name	*Apogon imberbis*
Classification	Order Perciformes (perches and allies); Suborder Percoidei
	Family Apogonidae
Size	Up to about 15cm (6in) long
Distribution	Eastern Atlantic (Portugal to Gulf of Guinea); Mediterranean
Habitat	Rocks, underwater caves and muddy seabeds, at 10–200m (30–650ft)
Diet	Small invertebrates and fish
Reproduction	Breeds in summer; mouth-brooder (*see above*)

Gilthead seabream

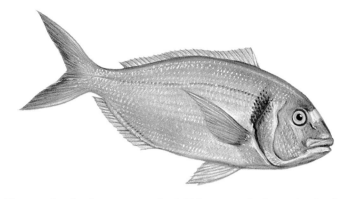

Many species of seabreams or porgies (which are completely unrelated to freshwater bream species) are fished commercially, and the gilthead is also an important farmed fish. Its common name comes from the golden mark between its eyes, which fades after death. Like most other members of its family, it has strong teeth for eating shellfish, and is regarded as a nuisance on commercial mussel and oyster beds. All giltheads are born males but become females when about three years old. A similar, related species, *Chrysophrys* (or *Pagrus*) *auratus*, lives in Australian and New Zealand waters, where it is called the snapper or red bream.

Scientific name	*Sparus auratus*
Classification	Order Perciformes (perches and allies); Suborder Percoidei
	Family Sparidae
Size	Up to about 1.4m (4½ft) long and 18kg (40lb); usually less than half this
Distribution	Eastern Atlantic (British Isles to Cape Verde); Mediterranean
Habitat	Bottom-living in coastal waters, to 150m (500ft); also brackish lagoons
Diet	Mostly shellfish and crustaceans; also squid and fish
Reproduction	Changes sex (*see above*); spawns in winter in Mediterranean

Striped red mullet

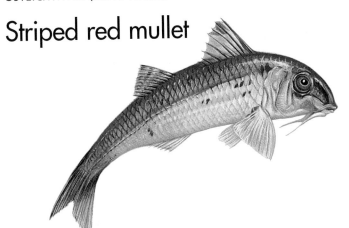

This species is distinguished from the otherwise very similar plain or common red mullet (*Mullus barbatus*) by the yellow stripes along its body and the very long barbels below its chin. (In the common species, these are no longer than the pectoral fins.) Red mullets use their barbels to locate their prey (they have taste buds at the tips), then often dig it from the sand or mud of the seabed. Both species have been widely caught for food, especially in warm regions, since ancient times; they are rather bony but make excellent eating. The goatfishes of North America, especially the red goatfish (*M. auratus*), are more or less closely related species.

Scientific name	*Mullus surmuletus*
Classification	Order Perciformes (perches and allies); Suborder Percoidei; Family Mullidae
Size	Up to about 40cm (16in) long; weight up to 1kg (2¼lb)
Distribution	Eastern Atlantic (Norway to Senegal); Mediterranean and Black Seas
Habitat	Bottom-living in coastal waters, to 60m (200ft)
Diet	Bottom-living small crustaceans, worms, molluscs and fish
Reproduction	Spawns in summer; eggs and larvae float

Long-nosed
or copper-banded butterfly fish

Butterfly fishes are small, beautifully patterned and colourful disc-shaped reef fishes that are favourite specimens for salt water aquariums. Many, including the copper-banded species, have a false 'eye' spot on the upper rear part of the body, while the true eye is camouflaged by the body markings; this pattern presumably confuses would-be predators. The closely related margined butterfly fish (*Chelmon marginalis*) is similar but with a slightly different colour pattern. The long nose of these and some other species is adapted to picking small invertebrates from coral crevices; other butterfly fishes have short jaws for biting off coral polyps (*see p.25*).

Scientific name	*Chelmon* (or *Chaetodon*) *rostratus*
Classification	Order Perciformes (perches and allies); Suborder Percoidei;
	Family Chaetodontidae
Size	Up to about 20cm (8in) long
Distribution	North-eastern Indian Ocean and western Pacific (Ryukyu Is to Australia)
Habitat	Warm, shallow waters along rocky shores and reefs; lagoons
Diet	Small bottom-living invertebrates
Reproduction	Prolonged larval stage; larvae have characteristic bony head plates

Common remora, or sharksucker

Sharksuckers are not strictly parasites, but they attach themselves to a host with a large sucker on the back of their head. The host is usually a shark (especially, in the case of the common remora, the blue shark; *see p.184*), but may be another large fish, turtle or whale, or even a ship. Young fish swim freely, and the sucker disc develops as they grow. It is in fact a modified dorsal fin with a series of slat-like ridges which the remora moves to create suction. It may damage the host's skin, but does not draw nourishment from the host; instead remoras eat small parasitic crustaceans found on the host's skin and sometimes swim free to catch other prey.

Scientific name	*Remora remora*
Classification	Order Perciformes (perches and allies); Suborder Carangoidei
	Family Echeneidae
Size	Up to about 85cm (33in) long
Distribution	Worldwide in tropical and warm temperate waters
Habitat	Mainly offshore, but anywhere carried by host
Diet	Mainly parasitic copepods (small crustaceans) on host
Reproduction	Eggs float, but other details uncertain

Crevalle jack

Jacks vary widely in body form, but most have two, small, separate spines just in front of the anal fin, and many have a distinct narrowing of the body just in front of the tail. The Mediterranean and Atlantic crevalle jack (*Caranx hippos*) is found over a wide range, from Portugal to Angola and from Nova Scotia to Uruguay. The Pacific crevalle jack (*C. caninus*), found from California to the Galápagos Islands and Peru, is probably the same species, but Indian Ocean specimens probably belong to a separate species, the giant trevally (*C. ignobilis*). Crevalle jacks form fast-swimming schools, and are caught commercially. They often grunt when captured.

Scientific name	*Caranx hippos* and *C. caninus*
Classification	Order Perciformes (perches and allies); Suborder Carangoidei; Family Carangidae
Size	Up to about 1.2m (4ft) long; maximum weight 32kg (70lb)
Distribution	Warm temperate and tropical Atlantic and east Pacific; Mediterranean
Habitat	Mainly offshore waters, surface to 350m (1150ft); sometimes in rivers
Diet	Mainly fish; also crustaceans and other invertebrates
Reproduction	Spawns in coastal waters and brackish estuaries

Common dolphinfish

With their beautiful metallic colouring (which fades on death) and distinctive long dorsal fin, the two species of dolphinfish are unmistakeable. Both look similar, but the pompano dolphinfish (*Coryphaena equiselis*) is smaller and its anal fin has a convex (outward-curving) edge. They swim in small schools, hunting surface-swimming fish such as flying fish (*see p.233*), which rarely escape even by leaping from the water. Both commercial and sports fishermen catch dolphinfish, and they are farmed in some regions; they are excellent to eat. They are sometimes called simply dolphins, risking confusion with the small whales (*see pp.299–300*).

Scientific name	*Coryphaena hippurus*
Classification	Order Perciformes (perches and allies); Suborder Carangoidei; Family Coryphaenidae
Size	Up to about 2.1m (7ft) long; maximum weight 40kg (88lb)
Distribution	Worldwide in tropical and subtropical waters
Habitat	Open waters and near coast, to 85m (275ft)
Diet	Wide range of fish; also squid and crustaceans
Reproduction	Spawns in open sea and inshore

Striped or flathead mullet

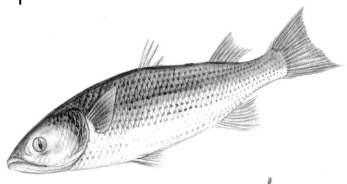

Only distantly related to the red mullets of the family Mullidae (*see p.252*), fishes of this family are generally known as grey mullets or simply mullets. There are a number of similar species, all quite elongated, cylindrical, silvery fish, which can be difficult to distinguish despite belonging to several different genera. The striped or flathead mullet is an important commercial catch, and is widely farmed in freshwater ponds in south-east Asia. Like other members of the group, it swims in large schools and is adapted to feeding on the detritus of the sea-bottom, sucking up mud and sand and digesting food from it with its muscular gizzard and long gut.

Scientific name	*Mugil cephalus*
Classification	Order Perciformes (perches and allies); Suborder Mugiloidei; Family Mugilidae
Size	Up to about 1.2m (4ft) long; maximum weight 8kg (17½lb)
Distribution	Worldwide in tropical and warm temperate waters
Habitat	Mainly bottom-living in coastal and offshore waters; estuaries and rivers
Diet	Small algae and planktonic animals in bottom detritus; algae in rivers
Reproduction	Spawns at sea, on bottom

Clown fish

More correctly called the orange clownfish or clown anemone fish, this is one of
27 *Amphiprion* species that live among the tentacles of sea anemones on reefs.
The fish are protected from the anemone's sting by a thick layer of mucus and by
their undulating swimming motion, which does not trigger the anemone's stinging
response. The anemone is preened by the fish and cleaned of debris, but the fish
gets the main advantage by being protected from its predators. Almost all fish of the
family Pomacentridae (known as damselfishes) are brightly coloured, but this and
the very similar *A. ocellaris* (often kept in aquariums) are among the best known.

Scientific name	*Amphiprion percula*
Classification	Order Perciformes (perches and allies); Suborder Labroidei;
	Family Pomacentridae
Size	Up to about 11cm (4¼in) long
Distribution	South-western Pacific (north Queensland to New Guinea and Vanuatu)
Habitat	Reefs and lagoons, to 15m (50ft)
Diet	Very small crustaceans and other reef organisms
Reproduction	Eggs laid in prepared nest on coral or rock; guarded by both parents

Rainbow wrasse

Many wrasses are colourful but in this species, like some others, mature males are much more so than females and young fish, which are coppery-brown with a broad white stripe. However, to confuse matters, some old females change their sex. Like other members of the family Labridae, it has a modified throat structure that acts like a second set of jaws to help to crush and grind its hard-shelled prey. It is sometimes called the Mediterranean rainbow wrasse to distinguish it from the rainbow or painted wrasse of the West Indies, *Halichoeres pictus*. Those found south of Cape Verde, West Africa, may belong to a separate species, *Coris atlantica*.

Scientific name	*Coris julis*
Classification	Order Perciformes (perches and allies); Suborder Labroidei; Family Labridae
Size	Up to about 30cm (12in) long
Distribution	Eastern Atlantic (Norway to Gabon; *but see above*); Mediterranean
Habitat	Shallow water along rocky shores and weed beds; mature males deeper
Diet	Small crustaceans, sea urchins and other invertebrates
Reproduction	Eggs planktonic; some old females become males

Queen parrotfish

Parrotfishes get their name from their beak-like mouth. This is formed by their fused front teeth, and is well adapted to grazing algae from coral and rocks, at the same time biting off coral chunks. They grind it all into paste with plate-like teeth in their throat, releasing nutrients in the algae (like a cow chewing the cud), and converting the coral into sand. Until they realized that male and female parrotfish look very different, scientists used to think there were many more than the 84 species now recognized. Female queen parrotfish are brown and white. It is one of several species that excretes a mucus cocoon at night; it probably repels predators.

Scientific name	*Scarus vetula*
Classification	Order Perciformes (perches and allies); Suborder Labroidei; Family Scaridae
Size	Up to about 60cm (24in) long
Distribution	Western Atlantic (Bermuda to northern South America); Caribbean
Habitat	Warm waters on and near coral reefs
Diet	Algae scraped from rocks and dead coral
Reproduction	Eggs scattered; mature females may change into males

Atlantic wolf fish

Wolf fishes used to be grouped with blennies – small fishes mostly found in rock pools and along rocky shores – but are now classified in a separate suborder with the eelpouts and their allies, many of which are deep-water species. They are big fish with a big head and sharp fang-like teeth, which they use to break open clams and other hard-shelled invertebrates; they then crush these with their broad molar teeth. They are an important commercial catch. The larger, similar-shaped but browner spotted wolf fish or spotted cat (*Anarhichas minor*) lives in the same waters, while the Bering wolf fish (*A. orientalis*) is found in the northern Pacific.

Scientific name	*Anarhichas lupus*
Classification	Order Perciformes (perches and allies); Suborder Zoarcoidei; Family Anarhichadidae
Size	Up to about 1.5m (5ft) long; maximum weight 24kg (53lb)
Distribution	Atlantic (Spitsbergen and Greenland to British Isles and New Jersey)
Habitat	Bottom-living (mostly on rocks) in offshore waters, to 500m (1650ft)
Diet	Crabs, lobsters, shellfish, sea urchins and other invertebrates
Reproduction	Spawns in winter, laying clusters of sticky eggs on seabed

Greater weever

Weevers are wedge-shaped fish that have poison glands along the spines of their front dorsal fin and also on the spine on their gill-covers. They often bury themselves in the sand with only their eyes and back showing, and spread the venomous spines in defence if they are disturbed. A person who accidentally treads on the fish is liable to get an extremely painful wound that causes swelling, although rarely death. The greater weever is one of the largest species, but the lesser weever (*Echiichthys vipera*), which is less than half the length, is more common. Most weevers live off Europe and Africa, but one possible species is known from Chile.

Scientific name	*Trachinus draco*
Classification	Order Perciformes (perches and allies); Suborder Trachinoidei; Family Trachinidae
Size	Up to about 53cm (21in) long
Distribution	Eastern North Atlantic (Norway to Canary Is); Mediterranean; Black Sea
Habitat	Bottom-living, on sand, mud or gravel; shallows to about 150m (500ft)
Diet	Small fish and bottom-living crustaceans, mainly at night
Reproduction	Eggs planktonic, laid in summer

Sand eel

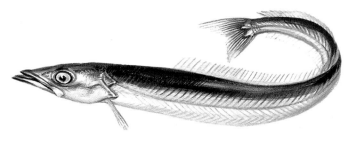

Various species of sand eels (which are unrelated to true eels; *see p.202*) live in temperate and tropical waters in most parts of the world. They vary in size, but are all similar to the lesser (or small) sand eel illustrated. They are sometimes called sand lances, because of their habit of diving into sand or fine gravel; they alternate between lying buried in this way and swimming freely in schools. Sand eels are an important part of the marine food chain because they eat plankton and are in turn food for larger creatures – fish, mammals and birds. In areas such as the North Sea their numbers have been greatly reduced by industrial fishing to make animal feed.

Scientific name	*Ammodytes tobianus* and other species
Classification	Order Perciformes (perches and allies); Suborder Trachinoidei; Family Ammodytidae
Size	Up to about 20cm (8in) long; other species up to 40cm (16in)
Distribution	Eastern North Alantic (Spitsbergen to Spain); Mediterranean
Habitat	Sandy-bottomed inshore waters; hibernates in winter buried in sand
Diet	Very small planktonic plants and animals
Reproduction	Spawns on sea-bottom, but larvae planktonic

Atlantic mudskipper

Mudskippers are among the most bizarre of all fish. They live on and in shallow mudflats of the mangrove swamps that fringe many tropical shores, and spend as much or more time out of the water as in it. When the tide falls, they emerge from a burrow and walk or skip across the mud and mangrove roots using their muscular pectoral fins. Their skin is richly supplied with blood vessels and can absorb oxygen directly from the air, making them truly amphibious. Mudskippers' eyes protude from the top of their head, enabling them to spot insects and other prey easily and to watch for predators such as birds. As the tide rises, they retreat into their burrow.

Scientific name	*Periophthalmus barbarus* or *P. koelreuteri*
Classification	Order Perciformes (perches and allies); Suborder Goboidei; Family Gobidae
Size	Up to about 25cm (10in) long
Distribution	Coasts of tropical West Africa (Senegal to Angola) and offshore islands
Habitat	Shallows and mudflats of estuarine mangrove swamps (*see above*)
Diet	Crabs, insects and other invertebrates of mud surface; plant material
Reproduction	Spawns in burrow in mud

Blue tang

The blue tang is one of the most distinctively coloured members of its family – the surgeonfishes – found in the tropical and subtropical western Atlantic. Juveniles are a beautiful clear yellow colour, becoming blue with a yellow tail as they grow, and finally maturing all-blue. Surgeonfishes are so called because they have a sharp scalpel-like spine projecting from each side of their body, just in front of the tail. The blue tang is a deep-bodied fish with a prominent 'scalpel'; this can inflict a painful wound on a diver who ventures too close, for the fish is not afraid of humans. It is rarely seen north of Florida, but is also found as far east as Ascension Island.

Scientific name	*Acanthurus coeruleus*
Classification	Order Perciformes (perches and allies); Suborder Acanthuroidei; Family Acanthuridae
Size	Up to about 39cm (15½in) long
Distribution	Atlantic (New York [rarely] and Bermuda to Gulf of Mexico and Brazil)
Habitat	Warm waters of coral reefs; inshore weed beds and rocky areas
Diet	Algae
Reproduction	Scatters eggs on and near reef

Great barracuda

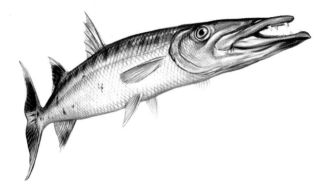

The biggest of the 20 or more barracuda species, the great barracuda is a slender, swift and powerful swimming machine. It swims mostly at or near the surface, usually on its own or in small groups. (Some other barracuda species form dense schools.) Their pointed jaws and prominent sharp teeth give barracudas a fierce appearance, and they show great curiosity, approaching boats, divers and other objects. They do sometimes attack humans – usually with a single fierce strike – but this is rare unless they are provoked. They can cause a severe wound, but barracuda attacks are rarely fatal. They are a popular prey of game fishermen.

Scientific name	*Sphyraena barracuda*
Classification	Order Perciformes (perches and allies); Suborder Scombroidei; Family Sphyraenidae
Size	Up to about 2m (6½ft) long; maximum weight about 50kg (110lb)
Distribution	Worldwide in tropical and subtropical waters, except eastern Pacific
Habitat	Muddy estuaries to open sea; young in lagoons, estuaries, mangroves
Diet	Mainly fish; also squid, cuttlefish and crustaceans
Reproduction	Spawns in inshore waters

Swordfish

With a huge, flattened 'sword' making up as much as one-third of its total length and a tall, sickle-shaped dorsal fin, the swordfish is one of the biggest and most spectacular of all the so-called billfishes. (This group includes the marlins and sailfishes of the family Istiophoridae. The black marlin, *Makaira indica*, is about the same size as the swordfish, but is heavier, with a smaller bill.) Swordfish use their sword to slash at and kill their prey; they have no teeth. They are widely sought by game and commercial fishermen, some of whom use harpoons, but their numbers are threatened by overfishing. Large swordfish may be toxic because of high levels of mercury in their flesh.

Scientific name	*Xipias gladius*
Classification	Order Perciformes (perches and allies); Suborder Scombroidei; Family Xiphiidae
Size	Up to about 4.5m (15ft) long; maximum weight 650kg (over 1400lb)
Distribution	Virtually worldwide, including cold waters in summer; migratory
Habitat	Mainly open sea, from surface to 800m (2600ft); also coastal waters
Diet	Wide range of fish; squid and octopuses; crustaceans
Reproduction	Spawns near surface in tropical and warm temperate spawning grounds

Northern or Atlantic bluefin tuna

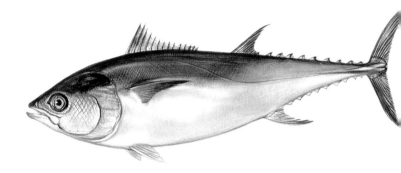

Sailfish (*Istiophorus* species) are the fastest fish measured in short bursts, at over 110km/h (68mph), but the northern bluefin tuna is unmatched for endurance; one tagged specimen covered 7700km (4785 miles) in 119 days. The reason is that fish of the Scombridae family have a high proportion of red muscle, which is well adapted to prolonged effort. This muscle is also what makes tuna so 'meaty' and valuable for eating cooked or raw (as sashimi). However, despite producing millions of eggs, bluefin numbers increase only slowly, and all bluefin species around the world – as well as many other tuna species – are severely threatened by overfishing

Scientific name	*Thunnus thynnus*
Classification	Order Perciformes (perches and allies); Suborder Scombroidei; Family Scombridae
Size	Up to about 4.6m (15ft) long; maximum weight 684kg (1507lb)
Distribution	Throughout most of Atlantic; Mediterranean; southern Black Sea
Habitat	Mainly open ocean; also inshore
Diet	Small schooling fish; squid; crabs
Reproduction	Spawns at various seasons, in open water; very large numbers of eggs

Black-finned icefish

Members of the suborder Notothenioidei are found only in Antarctic waters, and some have a special blood protein that acts like an antifreeze, enabling them to survive at the temperature of freezing seawater, –1.9°C (28.6°F). Icefish are strange, semi-transparent fish that completely lack the oxygen-carrying red pigment haemoglobin in their blood – which, like their gills and flesh, is white. Apart from one species of larval eel (which develops haemoglobin as it matures), no other fish is known to lack the pigment; the reason is unknown. Icefish have to rely on oxygen simply dissolving in their blood plasma, and as a result can move only sluggishly.

Scientific name	*Chaenocephalus aceratus*
Classification	Order Perciformes (perches and allies); suborder Notothenioidei; Family Channichthyidae
Size	Up to about 72cm (28in) long
Distribution	Atlantic part of Southern Ocean, south of Falklands and South Georgia
Habitat	Mainly bottom-living, to about 750m (2500ft)
Diet	Small fish; krill and other crustaceans
Reproduction	Details uncertain

Atlantic halibut

Unlike other fishes of similar shape, true flatfishes live on their side, one of their eyes having migrated, or moved, close to the other during larval development; the fins on either side are in fact the dorsal and anal fins. Flatfishes are classified as right- or left-eyed, depending on which side of the fish has both eyes and is usually uppermost. The Atlantic halibut – one of the biggest and best to eat – belongs to the right-eyed flounder family. It has a relatively elongated body. Females live longer and grow larger than males. They shed up to 2 million eggs, but halibut grow slowly and start to breed only at 10 to 14 years old, so are very vulnerable to overfishing.

Scientific name	*Hippoglossus hippoglossus*
Classification	Order Pleuronectiformes (flatfishes); Suborder Pleuronectoidei; Family Pleuronectidae
Size	Up to about 2.4–3m (8–10ft) long; maximum weight 320kg (700lb)
Distribution	Atlantic (Virginia and Bay of Biscay to Greenland and Barents Sea)
Habitat	Mainly bottom-living, to 2000m (6500ft); sometimes in midwater
Diet	Mainly fish; also squid, large crustaceans and other invertebrates
Reproduction	Spawns in winter and early spring; eggs float in midwater

Turbot

Despite its relatively small size, the turbot is probably the most prized of all the flatfishes for its fine flavour and texture; some people claim that it is the finest-tasting of all sea fish. It is almost circular in outline and, like other members of its family, is left-eyed. (In the illustration, the dorsal fin is to the lower left.) However, as with all flatfishes, a few individuals are found with the eyes on the 'wrong' side. Its brown left (upper) side is scaleless but has large bony tubercles. The pale right (lower) side has no tubercles except in the Black Sea subspecies, *Scophthalmus maximus maeoticus*. Hybrids occur between turbot and the brill (*S. rhombus*).

Scientific name	*Scophthalmus maximus* or *Psetta maxima*
Classification	Order Pleuronectiformes (flatfishes); Suborder Pleuronectoidei;
	Family Scophthalmidae; sometimes included in Bothidae
Size	Up to about 1m (3¼ft) long; maximum weight 25kg (55lb)
Distribution	Eastern Atlantic (Arctic to Morocco); Mediterranean; Black Sea; farmed
Habitat	Bottom-living in relatively shallow waters, to 70m (230ft); estuaries
Diet	Mainly bottom-living fish; also crustaceans, shellfish and worms
Reproduction	Spawns in spring and summer; eggs and larvae float

Common or Dover sole

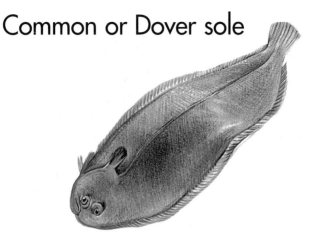

Soles form a separate suborder of flatfishes, having a small twisted mouth, small teeth and often rather small and asymmetrical pelvic and pectoral fins. Most – including members of the true sole family, Soleidae, and American soles, Achiridae – are right-eyed. The common or Dover sole is the commonest true sole in European waters, and is a highly regarded food fish. However, in North America the name Dover sole is used for a less valued flounder, *Microstomus pacificus*, also known as the slippery or slime sole. The common sole is a migratory species, wintering in deeper waters but moving to shallows in spring to spawn. It feeds mainly at night.

Scientific name	*Solea solea* or *S. vulgaris*
Classification	Order Pleuronectiformes (flatfishes); Suborder Soleoidei; Family Soleidae
Size	Up to about 70cm (27½in) long
Distribution	Eastern Atlantic (Norway to Senegal); Kattegat; Mediterranean
Habitat	Bottom-living, on sand or mud of continental shelf, to 150m (500ft)
Diet	Crustaceans; molluscs; worms; some small fish
Reproduction	Spawns in spring and early summer in shallow water; eggs float

Red-toothed triggerfish

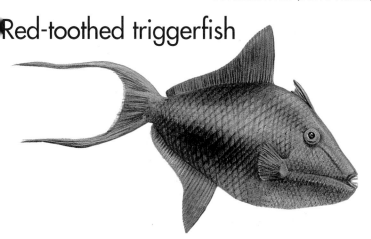

Triggerfishes get their name from the way the long front spine of their dorsal fin can be locked in position when it is raised as a defensive response to a predator. The small second spine moves forward, locking the first with a ball-and-socket mechanism; the long spine can be flattened again only by releasing this 'trigger'. Triggerfishes can also enlarge their pelvis by rotating the pelvic bone, making themselves look bigger to an attacker. By using both mechanisms, a fish can wedge itself in a rock or coral crevice and is very difficult to dislodge. The tropical red-toothed species is a popular marine aquarium fish, and is also eaten fresh or salted.

Scientific name	*Odonus niger*
Classification	Order Tetraodontiformes (triggerfishes and allies); Suborder Balistoidei; Family Balistidae
Size	Up to about 50cm (20in) long
Distribution	Indian and Pacific oceans (Africa to Japan, Polynesia and Australia)
Habitat	Mainly seaward side of coral reefs, at 5–40m (16–130ft)
Diet	Sponges; small planktonic animals
Reproduction	Spawns in sandy depression, guarded by male

Long-nosed or harlequin filefish

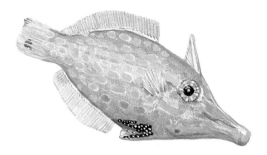

Filefishes are closely related to triggerfishes (*see p.273*), and also have a locking mechanism to keep the long first dorsal spine in position. However, they are much slimmer-bodied, and their scales have bristles that give the skin a file-like texture. Filefishes are among the few creatures that feed directly on corals, because these have powerful stinging cells. The long-nosed filefish is even more unusual in eating the polyps of only one coral genus – the branching or plate coral *Acropora*. The Red Sea long-nosed filefish, *Oxymonacanthus halli*, is very similar to the Indo-Pacific species, but is smaller and has a black bar on the tail in place of a black spot.

Scientific name	*Oxymonacanthus longirostris*
Classification	Order Tetraodontiformes (triggerfishes and allies); Suborder Balistoidei; Family Monacanthidae; sometimes included in Balistidae
Size	Up to about 12cm (5in) long
Distribution	Indian and Pacific oceans (East Africa to Ryukyu Is, Samoa and Australia)
Habitat	Clear lagoons and seaward side of coral reefs, to 30m (100ft)
Diet	Polyps of *Acropora* corals
Reproduction	Spawns on bottom, on site prepared by male

Long-horned cowfish

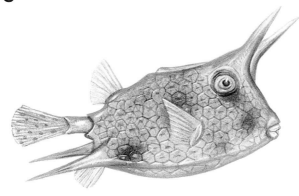

This box-like fish has such a curious appearance that it is sometimes dried to make an ornament. Like other members of its family – which includes boxfishes and trunkfishes – its body is encased in bony, mostly six-sided plates, with gaps for the eyes, small mouth, fins and other openings. The purpose of the 'horns' is not known. As if the hard shell and cryptic coloration were not enough protection from predators, most trunkfish secrete a highly toxic substance called ostracitoxin from their skin. The long-horned cowfish also has a curious method of feeding: it blows away sand from the lagoon floor with a water jet to expose small invertebrates.

Scientific name	*Lactoria cornuta* or *Ostracion cornutus*
Classification	Order Teraodontiformes (triggerfishes and allies); Suborder Balistoidei; Family Ostraciidae (or Ostraciontidae)
Size	Up to about 46cm (18in) long
Distribution	Indian and Pacific oceans (Red Sea and East Africa to Korea, Polynesia)
Habitat	Bottom-living on sand or gravel of shallow lagoons and coastal reefs
Diet	Bottom-living invertebrates
Reproduction	Spawns at dusk; eggs float

Long-spined porcupine fish

Porcupine fishes, like their less spiny relatives the puffer fishes of the family Tetraodontidae, are able when threatened to rapidly take large amounts of water into their stomach and puff themselves up like a ball. Few predators are big or brave enough to swallow such a large, spiny object. Many species of both families also have highly toxic skin, flesh or internal organs. (Puffers include the fugu [*Takifugu* species], whose flesh is highly prized by the Japanese despite the organs containing deadly poisons.) The long-spined porcupine fish is not toxic, but the similar and closely related common or spotted porcupine fish (*Diodon histrix*) is.

Scientific name	*Diodon holocanthus*
Classification	Order Tetraodontiformes (triggerfishes and allies); Suborder
	Tetraodontoidei; Family Diodontidae
Size	Up to about 50cm (20in) long
Distribution	Almost worldwide in tropical and subtropical waters
Habitat	Shallow reefs, lagoons and open sandy sea-bottoms, to 100m (330ft)
Diet	Shellfish; sea urchins; crabs; hermit crabs. Feeds at night
Reproduction	Spawns on bottom, in nest

Ocean sunfish

One of the most curious-looking of all fishes, the ocean sunfish looks as if the rear part of its body has been cut off, for it stops immediately after the tall, narrow dorsal and anal fins, and it has virtually no tail. The size and bulk of what 'remains' is shown by the fact that it is the heaviest of all bony fishes (that is, excluding sharks and rays). Despite its size and ungainly appearance, it can swim at a good speed, but it also often drifts or swims slowly at the surface, its dorsal fin breaking out of the water. Its mouth is small, with beak-like teeth; it feeds mainly on soft-bodied invertebrates. Another record it holds is for the number of eggs it produces: over 300 million.

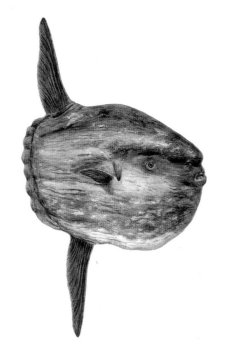

Scientific name	*Mola mola*
Classification	Order Tetraodontiformes (triggerfishes and allies); Suborder Tetraodontoidei; Family Molidae
Size	Up to 3.3m (11ft) long; maximum weight 2 tonnes or more
Distribution	Worldwide in warm and temperate waters
Habitat	Open sea, often near surface, but sometimes to 300m (1000ft)
Diet	Small squid; fish; starfish; jellyfish and other invertebrates; also plants
Reproduction	Mature female may produce more than 300 million floating eggs

Leatherback turtle

The leatherback is the biggest turtle, weighing up to 910kg (2000lb), but averages less than half this weight. Its carapace (shell) has no horny external plates but is like hard rubber, with seven ridges; unlike in other turtles, it is not joined to the spine or ribs. Leatherbacks go ashore only to lay eggs, which take up to four months to hatch, depending on temperature. The hatchlings are about 65mm (2½in) long, and must dig themselves free and scramble to the sea. Adults swim long distances – even into subpolar regions – with flippers spanning up to 2.7m (9ft). They have soft jaws and eat mainly jellyfish, but may die if they mistake plastic debris for food.

Scientific name	*Dermochelys coriacea*
Classification	Order Chelonia; suborder Cryptodira (hidden-necked turtles)
	Family Dermochelyidae
Size	1.5–2m (5–6½ft) long
Distribution	Worldwide but rare
Habitat	Mainly warm oceans; breeds on tropical and subtropical beaches
Diet	Jellyfish; sea squirts
Reproduction	Buries several clutches of 80–100 eggs in sand, at 10-day intervals

Green turtle

Once widely hunted for their carapace (shell) and meat, and used to make turtle soup, green turtles are now an endangered species. They are the largest hard-shelled turtles, and have extraordinary migratory habits. They always return to breed on the beach where they were born, often travelling 1300km (800 miles) or more from their shallow, warm-water feeding grounds. There are only a relatively few breeding beaches, where hundreds of turtles go each year. Important breeding sites include Ascension Island in the Atlantic, the Galápagos Islands, and certain islands and beaches in Hawaii, south-east Asia, the West Indies, Mexico and Florida.

Scientific name	*Chelonia mydas*
Classification	Order Chelonia; suborder Cryptodira (hidden-necked turtles)
	Family Cheloniidae
Size	1–1.2m (3¼–4ft) long
Distribution	Worldwide in tropics and subtropics
Habitat	Seas no colder than 20°C (68°F); breeds on beaches
Diet	Mainly seaweed and other sea plants; some crustaceans and jellyfish
Reproduction	Buries one or more clutches of about 100 eggs in sand of beach of birth

Salt-water or estuarine crocodile

The biggest of all crocodiles – and the most dangerous to humans – the salt-water or estuarine crocodile is slightly misnamed. Although it does live in salt or brackish (weakly salty) water near coasts and in estuaries and mangrove swamps, it is also seen in freshwater swamps and rivers – where it breeds. Breeding coincides with the wet season. A female scrapes together a mound of vegetation on land, in which she lays her eggs and which she defends. Decaying vegetation keeps the eggs warm, but in hot weather she may cool them with water. Salt-water crocodiles feed in and out of the water, sometimes leaping into the air or rushing onto land for prey.

Scientific name	*Crocodylus porosus*
Classification	Order Crocodilia (crocodiles and alligators)
	Family Crocodylidae; subfamily Crocodylinae
Size	Up to 7.5m (about 25ft) long or even more; usually 4–5m (13–16½ft)
Distribution	Southern India, through south-east Asia to northern Australia and Fiji
Habitat	Brackish estuaries, swamps and coastal waters; breeds near fresh water
Diet	All kinds of aquatic and land (waterside) animals – including humans
Reproduction	Buries 50 or more eggs in mound of vegetation beside river or swamp

Marine iguana

With the appearance of prehistoric monsters, marine iguanas are unique in several ways. They live only on the Galápagos Islands of the Pacific and are the world's only truly marine lizards. They are usually dark grey or blackish, but on some islands are blotched red and greenish, as shown. They live and breed close to rocky shores, and brave waves and spray to get the seaweed that is their main food, which grows between high- and low-water levels. They also swim and dive – large adults as deep as 15m (50ft) – for food. When they haul themselves out on a rock, water sprays from their nostrils, excreting excess salt from a special nasal gland.

Scientific name	*Amblyrhynchus cristatus*
Classification	Order Squamata; sub-order Lacertilia or Sauria (lizards)
	Family Iguanidae
Size	Up to 1.2–1.5m (4–5ft) long
Distribution	Galápagos Islands
Habitat	Shoreline; splash and intertidal zone; inshore waters
Diet	Mainly seaweed; sometimes small invertebrates
Reproduction	Up to six eggs, laid in burrow dug in sand or volcanic ash near shore

Sea-krait, or common sea snake

Unlike most sea snakes (such as the banded species; *opposite*), the sea-krait is not fully adapted to life in the water. Both it and its close and somewhat bigger relative the yellow-lipped sea-krait (*Laticauda semifasciata*) come ashore to lay their eggs. Using the wide scales on their belly – which are more like those of a land snake than of a true sea snake – for grip, they clamber over rocks and even climb low cliffs to find a suitable cave or rock crevice. For this reason, herpetologists (reptile experts) regard them as intermediate types between land and true aquatic snakes. Like all sea snakes, however, they have powerful venom with which they very quickly immobilize their prey.

Scientific name	*Laticauda laticaudata*
Classification	Order Squamata; suborder Ophidia or Serpentes (snakes)
	Family Elapidae (subfamily Hydrophiinae), Hydrophiidae or Laticaudidae
Size	Averages about 80cm (32in) long
Distribution	Eastern Indian Ocean to western Pacific
Habitat	Tropical inshore waters; breeds on shore
Diet	Fish
Reproduction	Lays eggs in cave or rock crevice

Banded sea snake

One of the true sea snakes, this species is so well adapted to life in water that it cannot survive if washed up on land. Unlike the sea-krait (*see opposite*), it has no wide belly scales with which to grip the ground and return itself to the sea. Adaptations to aquatic life include a laterally flattened body, a paddle-like tail with which to propel itself, and nostrils situated on the top of its head that can be closed by flaps when it dives. It does not even come ashore to breed; it is ovoviviparous – that is, the female produces eggs that are retained within her body until they hatch and are born, but unlike a mammal she does not nourish the developing young.

Scientific name	*Hydrophis cyanocinctus*
Classification	Order Squamata; suborder Ophidia or Serpentes (snakes)
	Family Elapidae (subfamily Hydrophiinae) or Hydrophiidae
Size	Up to 2m (6½ft) long
Distribution	Indian and Pacific oceans from Persian Gulf to Japanese waters
Habitat	Mostly coastal waters
Diet	Fish
Reproduction	Ovoviviparous; gives birth to 2–6 live young

Polar bear

One of the biggest of all bears, the polar bear is the only truly semi-aquatic species, equally at home on the tundra, pack-ice and ice floes, and swimming in the ice-cold Arctic seas. Its thick oily coat is impermeable to water and gives good insulation. Even the soles of its feet are fur-covered for good grip on the ice and snow and to retain warmth. A membrane between the toes helps swimming. Except during the mating season, adult polar bears are mostly solitary animals. Pregnant females spend much of the winter sleeping in a den dug under the snow, where they give birth, but others travel in the winter, moving over the ice in search of prey.

Scientific name	*Ursa maritimus*
Classification	Order Carnivora; suborder Fissipedia (terrestrial carnivores)
	Family Ursidae
Size	Up to 2.5m (8¼ft) long, sometimes more, including 10cm (4in) tail
Distribution	Arctic
Habitat	Coasts, ice floes and pack-ice
Diet	Seals; fish; seabirds; hares; reindeer; musk-oxen; plant matter in summer
Reproduction	One to four cubs, born in winter after about 9-month gestation

Sea otter

Sea otters – the most fully marine otters – are loved for their playfulness and their skill in diving for food, then floating on their back and cracking open an abalone shell by hitting it on a stone balanced on their chest. They sleep in the same way, entwined in a frond of kelp (a large seaweed) as an anchor. Sea otters have thick double-layered fur – among the finest of any animal – that keeps them warm by trapping air near the skin. They were widely hunted until protected internationally in 1911. Only about 2000 remain in California, but they are recovering elsewhere. Among their greatest dangers are oil spills, which ruin their fur's water-resistance.

Scientific name	*Enhydra lutris*
Classification	Order Carnivora; suborder Fissipedia (terrestrial carnivores)
	Family Mustelidae; subfamily Lutrinae
Size	Up to 1.4m (4½ft) long, excluding tail
Distribution	Pacific coasts of Russia, Alaska and parts of US West Coast
Habitat	Shallow inshore waters, especially where kelp grows
Diet	Mainly shellfish such as abalones; sea urchins; crabs; octopuses; fish
Reproduction	Single pup may be born in any season, after 4–5 month gestation

South American fur seal

Fur seals and sealions are distinguished from true seals (*see pp.291–297*) by their external ears and long hind flippers, which they can use to help to propel themselves on land. They probably evolved from bear-like animals. Various species of fur seals of the genus *Arctocephalus* live in many seas and coasts in the Southern Hemisphere and as far north as California. South American (or southern) fur seals have long been hunted for their fur, leather and oil, but are now protected in all countries. Males defend territories during the breeding season. Females give birth soon after coming ashore, and then mate with the male in whose territory they are.

Scientific name	*Arctocephalus australis*
Classification	Order Carnivora; suborder Pinnipedia (marine carnivores)
	Family Otariidae; subfamily Arctocephalinae
Size	Varies over range; males up to 2m (6½ft) long; females up to 1.5m (5ft)
Distribution	Southern Atlantic and Pacific coasts, southern Brazil to Peru; Falkland Is
Habitat	Inshore and offshore seas; breeds on rocky coasts
Diet	Fish such as anchovies, sardines ; crustaceans; mussels; squid; penguins
Reproduction	Polygynous. Single pup, born in spring after 11–12 month gestation

Northern fur seal

Northern fur seals, like other related species, have long been hunted for their fur, but since 1985 they have been hunted only by the native peoples of the Aleutian and other islands where they breed. Almost three-quarters of the million or so northern fur seals breed on the Pribilof Islands of the southern Bering Sea, but other breeding colonies (called rookeries) are scattered from California to the Kuril Islands, north of Japan. Mature males arrive in late spring, and establish breeding territories. Females arrive, give birth within two days, and then mate with a male – who may have a harem of up to 100 females. All return to the sea by early winter.

Scientific name	*Callorhinus ursinus*
Classification	Order Carnivora; suborder Pinnipedia (marine carnivores)
	Family Otariidae; subfamily Arctocephalinaé
Size	Male up to 2.1m (7ft) long; female up to 1.4–1.5m (4–5ft)
Distribution	Northern Pacific Ocean, from Japan to Alaska and California
Habitat	Cold seas, to 160km (100 miles) offshore; breeds on rocky islands
Diet	Mainly fish and squid
Reproduction	Polygynous. Single pup, born in summer after 11–12 month gestation

California sealion

Named for their noisy, roaring bark, sealions are closely related to fur seals (*see pp.286–287*) and are similarly able to use both front and hind flippers to manoeuvre on land. In fact, most trained 'performing seals' in zoos and marine parks are California sealions. Very closely related animals – which may form a subspecies of the California sealion or a separate species, *Zalophus wollebaeki* – live in the Galápagos Islands. Another, now extinct group used to live in the Sea of Japan. Like fur seals, male California sealions keep a breeding harem of females – usually about 15-strong. They set up a territory on a beach or rocky coast and mate with the females soon after these give birth.

Scientific name	*Zalophus californianus*
Classification	Order Carnivora; suborder Pinnipedia (marine carnivores)
	Family Otariidae; subfamily Otariinae
Size	Male up to 2.5m (8¼ft) long; female about 1.8m (6ft)
Distribution	Southern British Columbia to Mexico; *see also above*
Habitat	Inshore seas; breeds on rocky, gravel or sandy islands and coasts
Diet	Fish; squid; octopuses; crabs; lobsters; shellfish
Reproduction	Polygynous. Single pup born early summer after 11–12 month gestation

Steller's or northern sealion

This species is the biggest of the sealions, although females reach only about a third of the weight of males (which can weigh as much as 1 tonne). It sometimes kills other, smaller seals. It is an endangered species, numbers having dropped severely – in Alaska especially – since the 1970s, and is protected in most areas. A major reason for its decline is thought to be the impact of commerical fishing. Pollack, salmon and other commercially caught fish form an important part of its diet, and these have been caught in huge numbers in the Bering Sea and Gulf of Alaska. Like other species, females give birth and then mate with males, who defend their territory and mate with many females.

Scientific name	*Eumetopias jubatus*
Classification	Order Carnivora; suborder Pinnipedia (marine carnivores)
	Family Otariidae; subfamily Otariinae
Size	Male up to 3.3m (11ft) long; female up to 2m (6½ft)
Distribution	North Pacific rim, from Japan, Russia and Alaska to California
Habitat	Shallow coastal and offshore waters; breeds on rocky sloping beaches
Diet	Fish; squid; octopuses
Reproduction	Polygynous. Single pup born early summer after 11–12 month gestation

Walrus

Unmistakable with their huge tusks in both sexes, walruses are otherwise rather like very large sealions (*see pp.288–289*). There are two subspecies, in separate populations in the northern Atlantic, from Canada to northern Russia (the Atlantic walrus), and in the Bering and Chukchi seas (the larger and much more numerous Pacific walrus). A third population, in the Laptev Sea north of Siberia, may be a third subspecies or a form of Pacific walrus. Walruses spend most of their life at sea in groups, diving to forage for food on the bottom. They haul themselves out on ice floes, or sometimes shores, mainly to breed in winter and spring, but mate at sea.

Scientific name	*Odobenus rosmarus*
Classification	Order Carnivora; suborder Pinnipedia (marine carnivores)
	Family Odobenidae
Size	Male up to 3.6m (12ft) long; female up to 2.6m (8½ft)
Distribution	Far North Atlantic, North Pacific and Arctic
Habitat	Cold shallow seas; ice floes; shores
Diet	Bottom-living molluscs (mainly clams); crustaceans; other invertebrates
Reproduction	Single calf, born on pack-ice in spring after 15–16 month gestation

Mediterranean monk seal

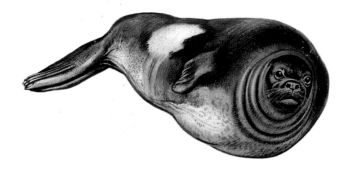

This species – once found widely on the Mediterranean, Black Sea and African Atlantic coasts – is the most critically endangered of all seals, with fewer than 500 survivors. The closely related Hawaiian monk seal (*Monachus schauinslandi*) is also rare, and the Caribbean species (*M. tropicalis*) became extinct in the 1950s. The name comes from their dark coat, said to be like a monk's robes. Mediterranean monk seals often have a pale patch – especially marked in young animals, and variable in shape – on their underparts. They used to live on beaches, but tourism, building, fishing and the use of motor boats has driven them to isolated caves.

Scientific name	*Monachus monachus*
Classification	Order Carnivora; suborder Pinnipedia (marine carnivores)
	Family Phocidae; subfamily Monachinae
Size	Male about 2.4m (8ft) long; female slightly smaller
Distribution	Small areas of Mediterranean and north-west African Atlantic coast
Habitat	Rocky coasts, sea-caves and shallow coastal waters
Diet	Wide variety of fish; squid; octopuses
Reproduction	Single pup, born most often in autumn after 10–11 month gestation

Leopard seal

With their spotted coat, large head and mouth, relatively slim body and long front flippers, leopard seals are distinctive and aptly named. The name also fits their hunting habits. In the Antarctic summer – December to February – they kill many young of other seal species, especially crab-eaters (*Lobodon carcinophagus*), and also small penguins. These are caught mainly in the water, where leopard seals' long flippers give them great speed and manoeuvrability. As with other seals, males and females mate (in the water in this case) soon after the young are born, but the fertilized embryo is believed not to start growing until several months later.

Scientific name	*Hydrurga leptonyx*
Classification	Order Carnivora; suborder Pinnipedia (marine carnivores)
	Family Phocidae; subfamily Lobodontinae
Size	Male 2.4–3.2m (8–10½ft) long; female slightly larger
Distribution	Around Antarctica; strays north to Australasia, South America, Africa
Habitat	Cold Antarctic waters; edges of ice shelf; pack-ice; ice floes; islands
Diet	Squid; fish; other seals and penguins in summer; krill in winter
Reproduction	Single pup, born on ice floe in summer after 11–12-month gestation

Southern elephant seal

Male elephant seals are highly territorial and aggressive in the breeding season, engaging in fierce – though rarely fatal – head-swinging battles with rival males to control patches of beach. This lets them collect a harem of up to 100 females, with which they will mate soon after the females give birth. Because of this competition, the males have evolved to a huge size; most are smaller than the maximum figures below, but males average four times the weight of females – the biggest difference among all mammals. The main breeding colonies are on South Georgia and other subantarctic islands, and in southern Argentina. The rest of the year is spent at sea.

Scientific name	*Mirounga leonina*
Classification	Order Carnivora; suborder Pinnipedia (marine carnivores)
	Family Phocidae; subfamily Cystophorinae
Size	Male up to 6.2m (20½ft) long and 4 tonnes; female up to 3.7m (12ft)
Distribution	South polar, subpolar regions; South Atlantic; southern South America
Habitat	Cold seas; beaches
Diet	Squid and fish
Reproduction	Polygynous. Single cub, born in spring after 11–12-month gestation

Hooded seal

Male hooded seals have an inflatable crest, or hood, on their forehead. They breed on drifting ice floes, out of the reach of predatory polar bears. When a female leaves the water to give birth in spring, a male accompanies and guards her, waiting for the chance to mate. He wards off rivals by inflating his hood. His septum – the red membrane separating the nostrils – is also inflatable, making a balloon the size of a football, and he uses this both to court the female and to threaten rivals. As soon as the pup is weaned and abandoned by its mother – only a few days after birth, the shortest time of any mammal – the male and female mate in the water.

Scientific name	*Cystophora cristata*
Classification	Order Carnivora; suborder Pinnipedia (marine carnivores)
	Family Phocidae; subfamily Cystophorinae
Size	Male about 2.6m (8½ft) long; female about 2m (6½ft)
Distribution	Far North Atlantic Ocean; parts of Arctic Ocean
Habitat	Deep waters; ice floes; edge of pack-ice
Diet	Fish; squid and other invertebrates
Reproduction	Single pup, born in spring after 12-month gestation

Ringed seal

Named for the markings on its coat, the ringed seal is the smallest of all seals and is the most common and widespread Arctic species. Isolated groups live in the Baltic and in Lakes Ladoga and Saimaa, in Russia and Finland. Ringed seals breed and live much of their life on the pack-ice, making and keeping open holes in the ice through which they dive for food. As the ice breaks up in summer, they move north to follow it. Baltic and lake populations frequently land on shore in summer, when the ice has melted. Like all true seals – which evolved from otter-like animals – their short flippers are little use on land or ice except to haul themselves clumsily along.

Scientific name	*Phoca* (or *Pusa*) *hispida*
Classification	Order Carnivora; suborder Pinnipedia (marine carnivores)
	Family Phocidae; subfamily Phocinae
Size	Varies over range; 1.1–1.75m (3½–5¾ft) long; male longer than female
Distribution	Arctic Ocean; Bering Sea; Sea of Okhotsk; Baltic Sea; subarctic lakes
Habitat	Pack-ice; cold seas (including under ice)
Diet	Fish; crustaceans; other invertebrates
Reproduction	Single pup, born in den on ice in spring, after 11-month gestation

Ribbon seal

It is difficult to understand why the ribbon seal evolved its distinctive coat pattern, but it may be a form of cryptic coloration that helps to disguise it against the broken pack-ice where it spends the winter and spring, and where it breeds. Apart from the four yellowish to white 'ribbons', each 10–12cm (4–5in) wide, around the neck, fore-flippers and hind part of the body, adult males are almost black. Females have the same pattern, but it is less distinct against their lighter brown coat. The pattern takes about four years to develop; pups are born silvery-white, but this coat is shed after about a month, and young seals are blue-black with silvery underparts.

Scientific name	*Phoca* (or *Histriophoca*) *fasciata*
Classification	Order Carnivora; suborder Pinnipedia (marine carnivores)
	Family Phocidae; subfamily Phocinae
Size	About 1.6m (5¼ft) long, sometimes up to 1.9m (6¼ft)
Distribution	North Pacific, from Sea of Okhotsk to Bering and Chukchi seas
Habitat	Edge of pack-ice in winter and spring; open sea when ice melts
Diet	Mainly fish; also squid, shrimps and crabs
Reproduction	Single pup, born on ice in spring after about 12-month gestation

Common or harbour seal

Known in North America as the harbour seal, the common seal is not, in fact, as common in many areas (including the British Isles) as its cousin the grey seal (*Halichoerus grypus*). But it is certainly the most widespread species, found in the north of both Eastern and Western hemispheres. As might be expected from such a wide range, it varies considerably in size and colouring, and at least five separate subspecies are recognized. Common seals rarely venture very far from coasts, and haul themselves out of the water to breed, moult and rest in favourite sheltered spots, forming colonies of 1000 or more. Their breeding season varies with location.

Scientific name	*Phoca vitulina*
Classification	Order Carnivora; suborder Pinnipedia (marine carnivores)
	Family Phocidae; subfamily Phocinae
Size	Male 1.4–1.8m (4½–6ft) long (largest in northern Pacific); female smaller
Distribution	Eastern and western North Atlantic and North Pacific oceans
Habitat	Temperate and subpolar coastal areas
Diet	Variety of fish, crustaceans and molluscs
Reproduction	Single pup, born in spring or early summer after 11-month gestation

Common or harbour porpoise

Porpoises are sociable marine mammals that travel in schools of 12 to 16 or more, migrating with the seasons between colder and warmer waters. They have acute hearing, and communicate by means of clicks and squeaks. They are distinguished from dolphins by their rounded head (without a beak) and short, triangular dorsal fin. When swimming, they break the surface but, unlike dolphins, rarely leap. They feed mainly on fish and crabs, consuming about 50 mackerel or herring-sized fish each day. They dive for several minutes at a time, using echo-location to find their prey. Pairs mate in late summer after long courtship rituals swimming side by side.

Scientific name	*Phocoena* (or *Phocaena*) *phocoena*
Classification	Order Cetacea; suborder Odontoceti (toothed whales)
.	Family Phocoenidae
Size	About 1.5–2m (5–6½ft) long
Distribution	North Atlantic and Pacific oceans; Mediterranean; Black Sea
Habitat	Mainly estuaries and other shallow waters
Diet	Mainly fish; some crabs and other invertebrates
Reproduction	Single calf, born in summer after 10–11 month gestation

Common dolphin

The beauty and playfulness of dolphins has endeared them to artists and others for thousands of years. They travel in large schools, often of several hundred and sometimes a thousand or more animals, that leap and splash noisily; they often swim beside or just ahead of a ship's bows. They have intricate but very variable markings, with a yellowish or tan patch on each side. A number of geographical forms have been identified, divided into two basic types: the long-beaked and short-beaked forms. Common dolphins can dive to depths of nearly 300m (1000ft) as they hunt for fish, guided by echo-location. Many are caught in fishing nets and drown.

Scientific name	*Delphinus delphis*
Classification	Order Cetacea; suborder Odontoceti (toothed whales)
	Family Delphinidae
Size	Up to about 2.5m (8¼ft) long
Distribution	Worldwide in tropical and warm temperate waters
Habitat	Open ocean and shallow coastal seas
Diet	Mainly fish and squid
Reproduction	Single calf, born in summer after 10–11 month gestation

299

Bottlenose dolphin

Growing much bigger than common dolphins (*see p.299*), bottlenose dolphins are highly intelligent mammals that are often kept in captivity in zoos and marine parks, and taught to perform tricks to entertain visitors. They are just as playful (and longer-lived) in the wild, sometimes swimming alongside boats and swimmers, surfing and lobtailing (slapping their tail against the water). They travel in small schools, and often leap out of the water, sometimes as high as 3m (10ft). Bottlenose dolphins show many geographical variations of colour and shape, but the two main forms are the smaller inshore and more robust offshore types.

Scientific name	*Tursiops truncatus*
Classification	Order Cetacea; suborder Odontoceti (toothed whales)
	Family Delphinidae
Size	Up to about 3.6m (12ft) long
Distribution	Worldwide in tropical and temperate waters
Habitat	Open ocean (especially in tropics), shallow coastal waters and estuaries
Diet	Mainly deep-swimming fish inshore; other fish; cuttlefish; crustaceans
Reproduction	Single calf, born in summer after 12–13 month gestation

Orca, or killer whale

Despite their name, size and fierce appearance, killer whales are intelligent and playful, and tame in captivity. But in the wild they prey on almost anything that swims, from fish, penguins and seals up to sharks and even large whales; they have been known to attack boats. They are bulky, distinctively patterned whales with a dorsal fin up to 1.8m (6ft) tall in males. Despite their size they can swim at almost 55km/h (35mph) and can leap clear of the water. They sometimes 'skyhop' – raise their head vertically from the water – to scan the horizon for prey. They travel in coordinated pods (family groups) in search of food, but do not migrate regularly.

Scientific name	*Orcinus orca*
Classification	Order Cetacea; suborder Odontoceti (toothed whales)
	Family Delphinidae
Size	Male about 7–10m (23–33ft), female about 4.5–6m (15–20ft), long
Distribution	Worldwide, especially in colder waters
Habitat	Deep and shallow waters, usually within 800km (500 miles) of land
Diet	Very varied: fish; squid; birds; turtles; seals; dolphins; other whales
Reproduction	Single calf born after about 12-month gestation

Long-finned pilot whale

There are two distinct populations of the long-finned pilot whale: one in the northern North Atlantic, the other around the globe in the Southern Ocean and southern parts of the Atlantic, Indian and Pacific oceans. They may, in fact, form separate species – just as the pilot whale itself was only in the 1970s realized to consist of very similar but separate long-finned and short-finned species (the latter, *Globicephala macrorhynchus*, living in warmer waters). Both have a markedly bulbous forehead, which is largest in older males. The long-finned species has a strong blow, more than 1m (3¼ft) high, and can dive to at least 600m (2000ft).

Scientific name	*Globicephala melas* or *G. malaena*
Classification	Order Cetacea; suborder Odontoceti (toothed whales)
	Family Delphinidae
Size	Male may exceed 6m (20ft) long; female slightly smaller
Distribution	Cold temperate waters of all oceans except North Pacific
Habitat	Mainly deep offshore waters; often feeds in coastal and shallow waters
Diet	Mainly squid; also fish
Reproduction	Single calf, usually born in late summer after 12–16 month gestation

Narwhal

Mature male narwhals have a single long tusk – very rarely two – protruding from the upper lip; it is usually about 2m (6½ft) long but may grow to 3m (10ft). It is in fact an elongated tooth. Narwhals are born with only two teeth, both in the upper jaw. Usually, neither erupt (enlarge and grow from the gum) in females, and only one in males; it penetrates the lip and grows in a 'barley-sugar' twist. Some females also grow a tusk, but it is much shorter than the male's. The tusk's purpose is uncertain, but it is probably a sexual characteristic; male narwhals have been seen jousting with their tusks, and the winner can probably mate with more females.

Scientific name	*Monodon monoceros*
Classification	Order Cetacea; suborder Odontoceti (toothed whales)
	Family Monodontidae
Size	Up to about 6m (20ft) long, including male's tusk; female 4.5m (12ft)
Distribution	Far northern North Atlantic and Arctic oceans
Habitat	Cold Arctic waters, near pack-ice
Diet	Fish; squid; crustaceans
Reproduction	Single calf, born in summer after 15-month gestation

White whale, or beluga

The white whale is aptly named, although it is dark brownish-grey at birth and only gradually becomes paler up to the age of about six years. Like the related narwhal (*see p.303*), it has no dorsal fin, but it does have a ridge on its back. It is sociable, usually swimming slowly in groups ranging from 10 or 20 up to many hundreds in estuaries when food is plentiful in summer. White whales, or belugas, are among the most vocal of all whales, making a wide variety of clicks, squeaks, trills and lowing sounds; whalers used to call them 'sea canaries'. They manoeuvre well in shallow water and among ice floes, but are often attacked by polar bears.

Scientific name	*Delphinapterus leucas*
Classification	Order Cetacea; suborder Odontoceti (toothed whales)
	Family Monodontidae
Size	Up to about 4–5m (13–16½ft) long; sometimes up to 6m (20ft)
Distribution	Arctic and northern oceans, to Japan, southern Alaska and Nova Scotia
Habitat	Offshore waters; coasts; bays; rivers; not open ocean
Diet	Mainly crustaceans; some fish
Reproduction	Single calf, born in summer after 14-month gestation

Northern bottlenose whale

The bottlenosed whales – the southern species (*Hyperoodon planifrons*) lives south of 30°S – belong to the beaked-whale family, so called because of their beak-like snout. They have a bulbous forehead and usually only two teeth, both in the lower jaw. All members of the family dive deeply, but the northern bottlenose is believed to be one of the deepest and longest-diving of all whales. It often dives for more than an hour, and may remain submerged for as long as two hours. It swims in small groups of up to ten individuals. It is attracted by the sound of ships, and many thousands were killed by whalers until the species was given protection in 1977.

Scientific name	*Hyperoodon ampullatus*
Classification	Order Cetacea; suborder Odontoceti (toothed whales)
	Family Ziphiidae
Size	Up to about 7–10m (23–33ft) long
Distribution	North Atlantic and Arctic oceans, north of about 30°N to pack-ice
Habitat	Mainly in ocean waters deeper than about 1000m (3300ft)
Diet	Mainly squid
Reproduction	Single calf, born in spring or summer after 12-month gestation

Sperm whale

Unjustly immortalized in Herman Melville's book *Moby Dick* as an aggressive killer, the sperm whale was for centuries hunted for its spermaceti, long used as a fine lubricating oil; for ambergris, a fatty substance from its intestines used as a 'fixative' in perfume making; for its fatty blubber; and for its 20cm (8in) teeth, a kind of ivory. It is now largely protected, but long-term hunting of large males reduced numbers greatly. Spermaceti is found in cavities in the whale's head, and may aid buoyancy during deep dives to 1000m (3300ft) or more; it may also help to focus echo-location sounds. Adult males are solitary but breeding females live in groups.

Scientific name	*Physeter macrocephalus* or *P. catodon*
Classification	Order Cetacea; suborder Odontoceti (toothed whales)
	Family Physeteridae
Size	Male up to about 18m (about 60ft) long; female much smaller
Distribution	Worldwide except coldest polar regions
Habitat	Open ocean and offshore waters in water more than 200m (650ft) deep
Diet	Giant deepwater squid; fish; other squid and octopuses
Reproduction	Usually single calf, born after 14–16 month gestation

Grey whale

Sometimes called the California grey whale because it is often seen off the coasts of that state, where it winters, this species is a filter-feeding baleen whale. It has no teeth, but up to 180 fringed, horny plates of baleen, or whalebone, hang from the roof of its mouth. It feeds on the bottom, sucking in sediments; small food organisms trapped by the baleen are retrieved by the whale's tongue. Atlantic grey whales were hunted to extinction. A few may survive in the western Pacific, wintering off Korea and migrating to the Sea of Okhotsk in summer. California grey whales migrate farther than any other mammal, 10 000km (6000 miles) to the Bering Sea.

Scientific name	*Eschrichtius robustus*
Classification	Order Cetacea; suborder Mysticeti (baleen whales)
	Family Eschrichtidae
Size	Up to about 14m (46ft) long; male slightly smaller than female
Distribution	Summer in northern North Pacific; winters farther south (*see above*)
Habitat	Coastal waters; farther offshore in summer
Diet	Small crustaceans, worms and other bottom-living invertebrates
Reproduction	Single calf, born in winter after 12-month gestation

Blue whale

The blue whale is the biggest animal on Earth – heavier, if not longer, than the biggest dinosaur. (The species name *musculus* – Latin for 'mouse' – is a joke.) Its average of 24m (80ft) and 100 tonnes is, however, much less than the record size (*see below*). Blue whales can swim at up to 50km/h (almost 30mph), and were safe from whalers until steam power and exploding harpoons enabled large-scale killing. Despite protection since 1966, only about 5000 blue whales survive. They feed only in summer, in cold regions, on krill (*see p.133*). Their baleen plates (*see p.307*) are up to 1m (3¼ft) long, and sieve up to 3.5 tonnes of food – 40 million krill – each day.

Scientific name	*Balaenoptera musculus*
Classification	Order Cetacea; suborder Mysticeti (baleen whales)
	Family Balaenopteridae
Size	Up to 33m (110ft) long, weighing 190 tonnes; usually smaller
Distribution	Worldwide, but biggest population in Southern Hemisphere
Habitat	Open ocean, mainly in cold waters in summer, warmer parts in winter
Diet	Krill – small planktonic cold-water crustaceans
Reproduction	Single calf, born in winter in warm waters after 12-month gestation

Minke whale or lesser rorqual

The minke is the smallest and most plentiful of the rorquals (members of the family Balaenopteridae), and was hunted on a large scale only after larger, more profitable whales were given protection in the 1970s. Since 1986, there has been a voluntary restriction on killing it, and only a few countries still do so, but it is still regarded as a threatened species. Like other baleen whales, it has a series of pleats or grooves – up to 70 in this species – in its throat. These allow the throat to expand as it takes in huge mouthfuls of water and floating food; the whale then expels the water through the baleen plates (*see p.307*), which filter it and trap food organisms.

Scientific name	*Balaenoptera acutorostrata*
Classification	Order Cetacea; suborder Mysticeti (baleen whales)
	Family Balaenopteridae
Size	Up to about 10m (33ft) long
Distribution	Virtually worldwide
Habitat	Deep, cool water, nearer Equator in winter; sometimes inshore
Diet	Planktonic krill; small fish; small squid
Reproduction	Single calf, born in winter in warm waters after 10-month gestation

Humpback whale

The hump of this whale is not very noticeable, but it has a stocky body, numerous rounded knobs on the head, lower jaw and flippers, and often many attached barnacles (*see p.125*). The flippers are long and black in Pacific humpbacks, mostly white in Atlantic animals. They are slow swimmers and often move inshore, so they were easy prey to whalers until they were given protection in 1966. They number 15 to 20 thousand, only one-fifth their original population. Humpbacks often isolate a school of fish with a ring-'net' of bubbles from their blow-hole before feeding. Males perform complex 'songs', up to 20 minutes long, as part of their mating ritual.

Scientific name	*Megaptera novaeangliae*
Classification	Order Cetacea; suborder Mysticeti (baleen whales)
	Family Balaenopteridae
Size	Up to about 15m (50ft) long; female slightly larger than male
Distribution	Worldwide; migrates
Habitat	Winters in tropical waters; spends summer in temperate and polar areas
Diet	Planktonic krill; small fish
Reproduction	Single calf, born in winter in warm waters after 12-month gestation

Bowhead or Greenland right whale

One of the most endangered whale species – with perhaps fewer than 10 000 remaining – the bowhead whale was long killed for its blubber and baleen. The blubber is up to 70cm (28in) thick, and yielded large quantities of oil. The baleen, or whalebone, is the long, thin, horny plates in the whale's mouth that filter tiny food organisms from seawater. The bowhead has longer baleen – up to 4.5m (14½ft) – than any other whale; it is springy and was widely used until the early 20th century for women's corsets, umbrella ribs, fishing rods and many other things. Today only native Arctic peoples are allowed to kill a few bowhead whales, a traditional food.

Scientific name	*Balaena mysticetus*
Classification	Order Cetacea; suborder Mysticeti (baleen whales)
	Family Balaenidae
Size	Up to 18–20m (60–65ft) long
Distribution	Arctic Ocean and adjacent seas and bays
Habitat	Mostly fairly shallow cold seas
Diet	Very small planktonic krill, copepods and similar creatures
Reproduction	Single calf, born in spring or early summer after 12-month gestation

North American manatee

The sea cows are also known as sirenians because they are said to be the origin of the legend of sirens, or mermaids, that lured seafarers to their death. They do not look much like the popular image of a mermaid, but females are said to help their new-born young to rise out of the water to breathe rather like a human mother holding her baby. The manatee is the bigger of the two best-known species (*see also opposite*), weighing up to 600kg (1300lb). It uses its flippers to gather food, and plays an important part in keeping waterways from becoming choked with weed. When not grazing, manatees rest under water, rising every few minutes to breathe.

Scientific name	*Trichecus manatus*
Classification	Order Sirenia (sea cows)
	Family Trichechidae
Size	Up to about 3–4m (10–13ft) long
Distribution	Atlantic and Caribbean from Florida and Gulf coast to Guyana
Habitat	Coastal waters, lagoons and river mouths with water at least 20°C (68°F)
Diet	Wide range of aquatic plants; some invertebrates
Reproduction	Single young, born after about 12-month gestation

Dugong

Dugongs are slightly smaller than manatees and have more streamlined flippers and a divided tail rather like a whale's or dolphin's; this enables them to swim at more than 20km/h (12mph). Males have two short tusks that are usually almost hidden by the upper lips. Dugongs live in several separate areas of tropical and subtropical waters – along the coast of eastern Africa and Madagascar, around Sri Lanka and southern India, from southern China to the Philippines, and off northern Australia and New Guinea. They live in family groups, browsing on sea plants – each eating about 40kg (90lb) a day – and only coming to the surface to breathe.

Scientific name	*Dugong dugon*
Classification	Order Sirenia (sea cows)
	Family Dugongidae
Size	Up to about 3m (10ft) long
Distribution	Parts of Indian Ocean, western Pacific Ocean and adjacent seas
Habitat	Shallow coastal waters
Diet	Seaweeds and other aquatic plants
Reproduction	Single young, born after about 12-month gestation

Glossary

Adaptation Changes in characteristics or behaviour that help a creature to survive and thrive in a particular environment.

Adipose fin Small, fleshy fin without *rays*.

Amphibious Describes a creature that can live parts of its life permanently in both water and air.

Antarctic Zone around the South Pole with cold water where surface ice exists for at least part of the year.

Aquatic Living in or related to water.

Arctic Zone around the North Pole with cold water where surface ice exists for at least part of the year.

Baleen Whalebone; the hard but flexible horn-like material found in plates in the mouth of certain whales, used for *filter-feeding*.

Barbel Long, slender sense organ on the head of some fish.

Benthic Bottom-living.

Blow Expulsion of air, water and vapour by a whale from the nostrils on the top of its head.

Blubber Thick layer of fat below the skin of a whale or other marine creature.

Brackish Describes water that is less salty than normal seawater, as in a river estuary or swamp.

Capsule, egg Tough case enclosing an egg, notably of many sharks and related fishes.

Carapace A protective shell or membrane encasing much of the body of a crustacean.

Caudal On or related to the tail; *see p.11*.

Chelipeds The adapted first walking legs of crustaceans, armed with pincers.

Chordate Animal that, at some time in its life history, has a stiffening rod (notochord) in its body; includes *vertebrates* and some others (*see pp.172–313*).

Cnidarian Group of *invertebrates* with a hollow body cavity; includes jellyfish and corals (*see pp.18–27*).

Commensal Living within or on another creature (its host), but not feeding on or damaging the host.

Compressed (of fish) Narrow, or flattened from side to side.

Continental shelf Relatively shallow and flat area of sea-bed extending from the coast around a continent.

Convergent evolution Occurs when two quite different fish or other creatures evolve into a similar form because they live in similar environments.

Crustacean Group of *invertebrates* with a body in segments, jointed limbs and a hard, horny outer skeleton; includes crabs, lobsters and prawns (*see pp.122–149*).

Cryptic colouring Confusing pattern that gives camouflage.

Detritus Broken sandy or gravelly material containing small food organisms or particles.

Dimorphism Having two distinctly different appearances, generally as between male and female.

Dorsal On or related to the back or upper side; *see p.11*.

Echinoderm Group of *invertebrates* with a skeleton of chalky material just below the skin and often with spines; includes sea urchins and starfishes (*see pp.152–171*).

Echo-location Detecting objects and other creatures by sending out sounds and detecting the echoes; highly developed in many whales, dolphins and porpoises.

Ecological niche A particular set of environmental and other conditions.

Endangered Existing in such small numbers that the species is in danger of becoming extinct.

Environment The physical and biological surroundings a creature inhabits.

Family See p.12.

Fin Usually flat structure used by fish to propel themselves; *see p.11*.

Filter-feeding Feeding by filtering small organisms (plant and/or animal) from a constant flow of water.

Fish Major group of cold-blooded *vertebrates*, almost all of which are wholly aquatic, having gills and fins, and a skeleton of either cartilage (*see pp.175–197*) or bone (*see pp.198–277*).

Genus See p.12.

Gestation Period of egg and embryo development within the parent's body.

Gill rakers Horny growths in some fishes' gills, used for *filter-feeding*.

Gill Respiratory organ of fish and many marine *invertebrates*, by which they obtain oxygen from the water and get rid of carbon dioxide.

Habitat A particular type of *environment*, such as the water around a coral reef.

Hermaphrodite Having both

male and female reproductive organs.

Host The organism on or in which a *parasite* lives.

Ichthyologist Scientist who studies fish.

Incubation Period during which eggs develop into *larvae* or young individuals.

Invertebrate Animal without a backbone.

Kelp Type of very large *sessile* seaweed.

Larva (plural larvae) Newly hatched creature, generally retaining the egg's yolk-sac and often having a different form from the adult.

Lateral line Series of sensory receptors, able to detect sound and vibration, along a fish's body; often visible.

Mammal Major group of warm-blooded *vertebrates* that bear live young, which they feed with their mother's milk; marine mammals include seals, whales and their relatives (*see pp.284–313*).

Metabolic rate The speed at which a creature's internal processes – particularly its use of food energy – take place.

Metamorphosis Change from larval or immature into different-shaped adult form.

Mollusc Group of aquatic invertebrates having a soft body with an internal or external shell; includes snails, squids, octopuses and shellfish such as oysters and clams.

Mouth-brooder A fish in which an adult (usually the male) holds the incubating eggs (and sometimes the young larvae) in its mouth.

Nekton Any free-swimming creature of the water column.

Oceanic Describes deep waters beyond the *continental shelf*.

Order *See p.12.*

Organism Any living thing, plant or animal.

Oviparous Egg-laying.

Ovoviviparous Producing eggs that incubate within the female's body without being nourished by her; see also *viviparous*.

Parasite Creature that lives in or on another creature (the *host*) and gets nourishment from the host.

Pectoral (especially of fins) Relating to the front sides of the body; *see p.11.*

Pelagic Living in the water column.

Pelvic (especially of fins) Relating to the rear sides of the body; *see p.11.*

Photophore Light-emitting organ, as in many deep-sea creatures.

Plankton Any plant or animal (referred to as planktonic) that floats in the water, carried by currents rather than its own swimming action.

Pod School or family group of whales, dolphins or seals.

Polar *Arctic* or *Antarctic*.

Range The parts of the world a creature normally inhabits.

Ray Stiff, bony structure supporting a *fin*.

Reptile Major group of cold-blooded, air-breathing vertebrates with external scales or horny plates; includes turtles, crocodiles, lizards and snakes (*see pp.278–283*).

Scales Small bony or horny plates on the skin of many fish, reptiles and other creatures.

School Group of fish or other marine creatures that swim together.

Sessile Attached to rock or other solid material.

Shoal Same as *school*.

Species *See p.12.*

Spine (1) Backbone. (2) Sharp, thorn-like extended *ray* on a fish, or similar sharp structure on another creature, such as a sea urchin.

Subtropical Zone just outside (to north or south) of the tropics, with warm water even in winter.

Swim bladder Gas-filled organ in many fish that gives buoyancy; *see p.12.*

Temperate Zone with cool but ice-free water between the subtropical and polar zones.

Territorial Describes a creature whose individuals or pairs live and breed in their own specific area, usually defending it against others, especially of their own species.

Tropical Permanently warm zone either side of the Equator, in which the sun is directly overhead at some time(s) of the year.

Tubercle Small warty or bony protruberance.

Ventral On or related to the belly or lower side; *see p.11.*

Vertebrate Animal with a backbone.

Viviparous Giving birth to live young that have been nourished by the female during development.

Water column The main mass of water between the bottom and the surface.

Yolk sac Part of an egg containing nutritious yolk to feed the developing egg and *larva*.

Index

NEW TESTAMENT
GREEK

An Introductory Grammar

by

ERIC G. JAY, M.A., Ph.D.

(Professor of Systematic Theology,
McGill University)

LONDON
S·P·C·K
1961

First published in *1958*
Reprinted, *1961*
S.P.C.K.
Holy Trinity Church
Marylebone Road
London N.W. *1*

Made and printed in Great Britain by
William Clowes and Sons, Limited, London and Beccles

τῇ ἰδίᾳ γυναικὶ
πολυτλήμονι καὶ μακροθυμούσῃ
ὁ συγγραφεὺς
ἐν εὐχαριστίᾳ

CONTENTS

CONTENTS

INTRODUCTION

Those who wish to begin a study of New Testament Greek have not, at present, a very wide choice of Grammars. It has appeared to the compiler of this one that there is a gap between those which are so elementary that they do not adequately equip the student to deal with the text of the New Testament itself, and those which take too much for granted in the reader. Most of those who come to the study of New Testament Greek are men in training for the Ministry who have left school some years before, and who have probably forgotten the details and technical terms even of English grammar. This book is an attempt to bridge this gap, by providing full explanations even of the simplest constructions, by introducing the student at the earliest possible stage to sentences from the New Testament both in exercises and in examples, and by dealing, at least briefly, towards the end of the book with some of the more difficult grammatical points which in some Grammars are reserved for a second volume.

Even the most modern methods of learning a language do not enable the student to avoid the task of learning a great many things by heart. It is hoped that the way in which the declensions of nouns, adjectives, and pronouns, and the paradigms of verbs are set out will lighten this labour.

St Mark's Gospel is frequently chosen by teachers as the first New Testament book to set the beginner to read. Examination syllabuses often dictate this policy. For this reason examples of constructions, etc., dealt with in the text are given from St Mark wherever possible.

It is easy for the beginner to delude himself into a false optimism when he acquires a certain ease in translating Greek into English. Facility in translating English into Greek, however, is a better sign of progress. For this reason in the exercises more sentences are given for translation into Greek than into English.

To teachers of the language my debt to other works will be obvious. I have consulted H. P. V. Nunn's *Elements of New Testament Greek* (C.U.P.); J. H. Moulton's *An Introduction to the Study of New Testament Greek* (Epworth Press); the same author's *A Grammar of New Testament Greek* (T. & T. Clark), especially vol. II, *Accidence and Word-formation*; and C. F. D. Moule's *An Idiom-Book of New Testament Greek* (C.U.P.). These two last mentioned will be indispensable to the more advanced student. *A Primer of Greek Grammar*, by E. Abbott and E. D. Mansfield (Rivingtons), from which I myself as a beginner in classical Greek derived much help, suggested the method of setting out the paradigms of verbs used in Appendix 4.

Chapter 1

THE GREEK LANGUAGE

Greek is the language spoken and written in ancient times by the Hellenes, the inhabitants of that part of the Balkan peninsula which is the modern Greece, and by Greek settlers in other parts of the Mediterranean area. It had several dialects: Ionic, Aeolic, and Doric. Ionic, in which Homer wrote, became the literary dialect, and developed into the Attic used by the great playwrights, orators, and philosophers of Athens in Attica. The Attic dialect, owing to the importance of Athens and of its literary men, established itself in the fifth and fourth centuries B.C. as the chief dialect of Greece.

After the time of Alexander the Great, who died in 323 B.C., Attic Greek became to a large extent the language of literature and commerce over the whole area of his conquests. But it was an Attic which, partly, no doubt, because Alexander's soldiers and the merchants who followed on their heels were drawn from all parts of the Hellenic world, had absorbed elements from other Greek dialects, and which was capable of absorbing elements from the native languages of the peoples of the Near and Middle East who now began to use it. This flexible, developing language is sometimes called Hellenistic Greek, and sometimes *Koinē* (i.e. Common) Greek, from ἡ κοινὴ διάλεκτος *the common language*. The New Testament writers used the *Koinē*. See p. 265, where the question of the influence upon them of Semitic languages is discussed.

THE GREEK ALPHABET

The word *alphabet* is derived from the first two of the twenty-four letters commonly used by the Greeks. These were adapted from the Phoenician alphabet. The letters given in the first column below are those which are found on ancient Greek inscriptions and are used in modern printed books as capitals. The student will notice that nearly half of them are used also as English capitals. The letters in the second column were evolved from the capitals for the purpose of ordinary writing. They are known as cursives, or "running" letters, and are used for the small letters in printed books.

		Name of letter	English sound
A	α	alpha	a
B	β	bēta	b
Γ	γ	gamma	g (hard)
Δ	δ	delta	d
E	ε	ĕpsīlon	e (short)
Z	ζ	zēta	z
H	η	ēta	e (long)
Θ	θ	thēta	th
I	ι	iōta	i
K	κ	kappa	k
Λ	λ	lambda	l
M	μ	mū	m
N	ν	nū	n
Ξ	ξ	xī	x
O	ο	ŏmīkron	o (short)
Π	π	pī	p
P	ρ	rhō	r
Σ	σ, ς	sigma	s
T	τ	tau	t
Y	υ	upsīlon	u
Φ	φ	phī	ph
X	χ	chī	ch (i.e. kh)
Ψ	ψ	psī	ps
Ω	ω	ōmĕga	o (long)

NOTICE: 1. The cursive alpha, α, is in some books printed *a*, but the student should write it as a figure 8 laid on its side and opened on the right, similar to the α used in this book.

2. Γ, γ, *gamma*, is pronounced hard as in *gate*, never soft as in *page*. But before another guttural it is pronounced *n* (the gutturals are γ, κ, ξ, χ; see below, p. 5). Thus ἄγγελος a *messenger*, is pronounced *angelos* (*g* hard).

Notice that γ is written through the line, whilst ν, *nu*, is written on the line.

3. *Epsilon* means literally "unaspirated *e*". Prior to about 400 B.C. the letter had sometimes been used to denote the aspirate. *Psilon* is from the Greek ψιλός *bare* or, as a term in grammar, "without the aspirate." Cf. *upsilon*.

4. H is the capital of the long *e*, *ēta*, not an aspirate as in English.

5. ν (the capital is N) is the English *n*, and is not to be confused with the English *v*, which has no Greek equivalent.

6. *Ŏmīkron* means "little *o*". *Mikron* is from the Greek μικρός *little*.

7. P is the capital *r*, not *p* as in English.

8. *Sigma* is written σ at the beginning or in the middle of a word, ς at the end.

9. Χ, χ, *chī*, is not the English *x* sound (this is ξ, *xī*). Nor is it pronounced *ch* as in *cherry*, but hard as in *chasm*.

10. *Ōmĕga* means "great *o*". *Mega* is from the Greek μέγας *great*.

11. ζ, ξ, and ψ are double consonants, being respectively *dz*, *ks*, and *ps*.

BREATHINGS

When a word begins with a vowel or diphthong there must be a slight preliminary emission of breath, as a few moments of experiment will make plain. This breathing may be "rough" (i.e. an aspirate, the English *h*) or "smooth". Greek marks both the rough breathing and the almost imperceptible smooth breathing by a comma placed over the initial vowel, ' for the rough breathing, ' for the smooth. If the vowel is a capital letter the breathing is placed to the left of it, and if the word begins with a diphthong the breathing is placed over the second of its vowels. These points may be illustrated by a few words which occur early in St Mark's Gospel:

ἀποστέλλω	(Mark 1.2)	apŏstellō
ὁδόν	(1.2)	hodon
Ἀρχή	(1.1)	arkhē
εὐαγγελίου	(1.1)	euangeliou
οὖ	(1.7)	hou

When the consonant ρ begins a word it is given a rough breathing, and is pronounced *rh*:

ῥήξει (Mark 2.22) rhēxei

When ρ occurs twice together in a word the first is sometimes given a smooth breathing, and the second a rough breathing; thus, ἔῤῥηξεν. (But Nestle's text does not print these breathings; see Luke 9.42, ἔρρηξεν, errēxen.)

The vowel υ always has a rough breathing when it begins a word. But if the υ is the first letter of a diphthong the breathing is written over the second:

ὑποδημάτων	(Mark 1.7)	hupŏdēmatōn
ὑμᾶς	(1.8)	humās
υἱοῦ	(1.1)	huiou

THE VOWELS

H (η) and ω are long vowels; ε and o are short by nature, but may become long by position (see below). A (α), ι, and υ are sometimes long and sometimes short.

A short vowel becomes long by position when followed by two or more consonants or by a double consonant. Thus the α in εὐαγγελίου is long. No difference is made in pronouncing the α, but in poetry the syllable would be reckoned as a long beat. If, however, of the two consonants following a short vowel the first is a mute and the second a liquid or nasal (see below, p. 5, for these terms) then the vowel may be regarded as long or short.

DIPHTHONGS

Two different vowel sounds may combine to make one syllable, known as a diphthong (Greek δίφθογγος *with two sounds*; δι = δίς *twice*, φθόγγος *a sound*).

There are eight Greek diphthongs, in four of which ι is the second vowel, υ being the second vowel in the other four:

αι   ει   οι   υι
αυ   ευ   ηυ   ου

When the first vowel is long and the second is ι, the ι is written underneath the long vowel (*iota subscript*); thus, ᾳ, ῃ, ῳ. This ι is not pronounced.

PRONUNCIATION

There is no clear evidence about the way in which Greek was pronounced in ancient times. Modern Greek, in which a number of the vowels and diphthongs are given the long *e* sound, is no guide. In England several different systems of pronouncing the vowels and diphthongs are in use. The student will do well to follow the guidance of his teacher. But those who are attempting to learn the language without a teacher may safely adopt the following scheme:

Vowels: α, when short, as *a* in *cat*
        α, when long, as *a* in *pass*
        ε   as *e* in *peg*
        η   as *ee* in *deep*
        ι, when short, as *i* in *pin*
        ι, when long, as *i* in *pile*

o   as *o* in *pot*
υ, when short, as *u* in *put*
υ, when long, as *u* in *cute*
ω   as *o* in *pole*

Diphthongs: αι as *ai* in *aisle*⎫indistinguishable
   ει as *ei* in *eider*⎭
   οι as *oi* in *boil*
   υι as *wi* in *wine*
   αυ as *au* in *caught*
   ευ as *eu* in *euphony*⎫indistinguishable
   ηυ as *eu* in *euphony*⎭
   ου as *ou* in *count*

CLASSIFICATION OF CONSONANTS

Consonants are classified in two different ways:

A. According to whether or not they can be pronounced by themselves and without a vowel sound:

1. *Mutes* cannot be pronounced without the assistance of a slight vowel sound. They can be sub-divided into:

(*a*) *Hard Mutes*, κ, τ, π.
(*b*) *Soft Mutes*, γ, δ, β.
(*c*) *Aspirated Mutes*, χ, θ, φ.

Trial will show that the above consonants cannot be pronounced without the utterance of a slight vowel sound.

2. *Semivowels* are consonants which can be sounded without the assistance of even a slight vowel sound. They can be subdivided into:

(*a*) *Nasals*, so called because the breath passes through the nose; γ when pronounced as *n*, μ, and ν.
(*b*) *Spirants* (from Latin *spiro*, "breathe"), σ.
(*c*) *Liquids*, so-called to express the rippling nature of the sound, λ, ρ.

B. According to the organ used in pronunciation:

1. *Gutturals*, i.e. sounds made by the throat, γ, κ, χ.
2. *Labials*, i.e. sounds made by the lips, β, π, φ, μ.
3. *Dentals*, i.e. sounds made by the teeth, δ, τ, θ, ν, σ.

In addition there are the *double consonants*, ζ (*dz*), ξ (*ks*), and ψ (*ps*).

Obsolete Consonants

Greek originally possessed two other spirants which fell out of use in very early times:

1. Ϝ (ϝ), the digamma, so called because it was written like one gamma, Γ, placed over another. Its sound was like the English *w*. Its original presence accounts for peculiarities in certain words (see, for example, pp. 108 and 118).

2. The *y* sound, called "consonantal iota". No sign for it has survived; but its original presence accounts for certain formations (see pp. 125, 132, 153, 186).

MODIFICATION OF CONSONANTS IN INFLECTION

Changes in a consonant are sometimes brought about when a word inflects (i.e. as a word changes its form in the course of the declension of a noun or the conjugation of a verb—see pp. 17, 19), especially if it is brought together with another consonant. The student is advised not to attempt to learn the following notes at this stage. They will be more easily understood when he has come across actual examples in his study of the noun and the verb. As he does so he will be referred back to this section.

Gutturals (γ, κ, χ)

Before τ a guttural becomes κ by assimilation. τ is the hard dental, and it assimilates a soft or an aspirated guttural (γ or χ) to its own hardness.

Thus: πέπρακται for πέπραγται.
δέδεκται for δέδεχται.

Before δ a guttural becomes γ. δ is the soft dental, and it assimilates a hard or an aspirated guttural (κ or χ) to its own softness.

Thus: ὄγδοος *eighth* (but ὄκτω, *eight*).

Before θ a guttural becomes χ by assimilation.

Thus: ἀχθῆναι for ἀγθῆναι.

Before μ a guttural becomes γ by assimilation, the nasal μ being a soft labial.

Thus πέπλεγμαι for πέπλεκμαι.

Before σ a guttural coalesces with the σ to form the double consonant ξ.

Thus: γυναιξίν for γυναικσίν.

When a guttural follows ν, the ν may become γ (see below under *Nasals*).

Labials (β, π, φ)

Before τ a labial becomes π by assimilation. τ is the hard dental, and it assimilates a soft or an aspirated labial (β or φ) to its own hardness.

Thus: βλάπτω for βλάβτω.
γέγραπται for γέγραφται.

Before δ a labial becomes β by assimilation.

Thus: ἕβδομος, *seventh* (but ἕπτα, *seven*).

Before θ a labial becomes φ by assimilation.

Thus: ἐλείφθην for ἐλείπθην.

Before μ a labial becomes μ (complete assimilation).

Thus: τέτριμμαι for τέτριβμαι.

Before σ a labial coalesces with the σ to form the double consonant ψ.

Thus: φλεψίν for φλεβσίν.
ἀλείψω for ἀλείφσω.

When a labial follows ν, the ν may become μ (see below, under *Nasals*).

Dentals (δ, τ, θ)

Before μ a dental becomes σ.

Thus: πέπεισμαι for πέπειθμαι.

Before σ a dental disappears.

Thus: παισίν for παιδσίν.
ἤλπισα for ἤλπιδσα.

When two dentals are brought together the first may become σ.

Thus: ἐπείσθην for ἐπείθθην.
πέπεισται for πέπειθται.

When in inflection ντ is followed by σ, the ντ is dropped, but compensation is usually made by lengthening the previous vowel.

Thus: ἄρχουσιν for ἄρχοντσιν (dat. plur. of ἄρχων).
λυθεῖσιν for λυθέντσιν.

Nasals (ν)

The nasal ν may be assimilated before gutturals, labials, and liquids.

Before a guttural ν may become γ.

Thus: συγχαίρω for συν-χαίρω.
ἐγκαινίζω for ἐν-καινίζω.

Before a labial ν may become μ.

Thus: συμβάλλω for συν-βάλλω.
ἐμβαίνω for ἐν-βαίνω.

Before a liquid ν may be completely assimilated.

Thus: συλλαλῶ for συν-λαλῶ.

In the New Testament the manuscripts are not uniform in respect of verbs compounded with συν and ἐν, and the ν is often retained without change. Nestle's text does not assimilate this ν.

Before σ, ν may effect different changes:

(*a*) The ν may be dropped, as frequently in the dative plural of 3rd decl. nouns. Thus ποιμέσιν for ποιμένσιν (see p. 94).

(*b*) The σ may be dropped, and compensation made by lengthening the previous vowel. This accounts for certain aor. indic. forms, e.g. ἔμεινα for ἔμενσα (see p. 185).

(*c*) The vowel ε may be inserted between the ν and the σ. This initiates a process by which the σ then drops out as coming between two vowels (see p. 9), and the vowels thus brought together contract. Thus the fut. indic. act. of μένω is μένσω which becomes μενέσω, μενέω, and so μενῶ (see p. 184).

Liquids (λ, ρ)

A liquid may assimilate a preceding ν.

Thus: συλλαλῶ for συν-λαλῶ (see above).

Liquids tend not to tolerate a following σ. The conjunction is avoided in one of two ways:

(*a*) The σ may be dropped, and compensation made by lengthening the previous vowel. This occurs in the aorist of some verbs with liquid stems (see p. 153).

Thus: ἤγειρα for ἤγερσα.
ἤγγειλα for ἤγγελσα.

(*b*) ε may be inserted between the liquid and the σ, and the same process is then followed as described on p. 8 in the case of the future of μένω (see p. 153).

Thus: ἐγέρσω becomes ἐγερέσω, ἐγερέω, and so ἐγερῶ.
ἀγγέλσω becomes ἀγγελέσω, ἀγγελέω and so ἀγγελῶ.

Spirants (σ)

When σ follows a guttural or a labial, the double consonants ξ and ψ are produced (see above under *Gutturals* and *Labials*).

A dental before σ drops out (see above under *Dentals*).

When inflexion brings σ between two vowels the σ may be dropped, and the vowels thus brought together contract:

Thus: γένους, the genitive of γένος (stem γενεσ-), is for γένε(σ)ος.

It is to be noticed that the σ by no means always drops out between two vowels, e.g. ἔλυσα, τίθησιν. Where it does drop out attention will be drawn to it in the text.

For the combination νς see above under *Nasals*.

Aspirates

Before a rough breathing a hard mute (κ, τ, π) is aspirated:

Thus: οὐχ ἵστημι for οὐκ ἵστημι.
ἀφίστημι for ἀπ-ἵστημι.

When two consecutive syllables begin with an aspirate, the first aspirate usually becomes the corresponding unaspirated letter:

Thus: τριχός for θριχός, the genitive of θρίξ *hair*.
τίθημι for θίθημι, which shows a reduplication of the stem θε- (see p. 230).
ἐτέθην for ἐθέθην (stem θε-).

CONSONANTS AT THE END OF WORDS

The only consonants which can stand at the end of a Greek word are ν, ρ, and ς (whether as a single consonant or in the double consonants ξ and ψ). The only exceptions to this rule are the words ἐκ *out of, from*, and οὐκ *not*. But foreign words incorporated into Greek (e.g. the large number of Hebrew and Aramaic nouns used in the New Testament) are immune from the rule. Thus we get Ἀβραάμ, Δαυείδ, Ἰσραήλ.

Movable ν

ν is often added to the final ι of the dat. plur. of 3rd decl. nouns, adjectives and pronouns; thus γυναιξί(ν); and to the final ε or ι (but not ει and αι) of 3rd pers. sing. and plur. verb forms; thus ἔλυε(ν), λύουσι(ν). Strictly it is only added to facilitate pronunciation when the next word begins with a vowel, but in the New Testament this movable ν is found sometimes even when a consonant follows.

Movable ς

A few words have a movable ς, namely οὕτω(ς) *thus*; ἄχρι(ς) *until*; μέχρι(ς) *until*; ἐκ (ἐξ) *out of, from*.

CONTRACTION OF VOWELS

When certain vowels are brought together in a word they coalesce or contract. This is of importance in the pres. and imperf. tenses of *verbs* whose stems end in α, ε, or ο (see p. 109).

αε and αη become α (long).
αο and αω become ω.
ι in the second syllable is written *subscript*. Thus αει becomes ᾳ.
υ disappears. Thus αου becomes ω.
εε becomes ει.
εο becomes ου.
ε followed by any long vowel or diphthong disappears. Thus εω becomes ω; εου becomes ου.
ο and a long vowel become ω. Thus οη becomes ω.
ο and a short vowel become ου. Thus οο becomes ου; οε becomes ου.
ο before ου disappears. Thus οου becomes ου.

Any combination of ο with an ι becomes οι. Thus οει, οῃ, οοι all become οι.

The following contractions should be noted in *nouns* and *adjectives*:

εα may become η. Thus γένη for γένεα (see p. 120). Cf. also ἀληθῆ for ἀληθέα (see p. 152).
οα may become ω. Thus αἰδῶ for αἰδόα, accusative of αἰδώς *modesty*; but such nouns are rare. Cf. also μείζω for μείζοα (see p. 152).

Stops and other Signs

Greek manuscripts often omit punctuation signs altogether, but those used in printed Greek are:

Comma and full stop as in English.

Colon or semi-colon, a dot above the line · (see Mark 1.2, τὴν ὁδόν σου·).

The note of interrogation, expressed by ; (see Mark 1.24, Ναζαρηνέ;).

Other signs to be noted are:

Apostrophe (Greek ἀποστροφή *a turning away*)

The same sign as for a smooth breathing, ', is used to mark the cutting off or *elision*, as it is usually called, of a vowel. This sometimes happens to the last vowel in a word when the next word begins with a vowel (see Mark 1.5, ὑπ' αὐτοῦ for ὑπὸ αὐτοῦ).

Diaeresis (Greek διαίρεσις *a taking apart*)

Two dots over the second of two vowels occurring together indicates that they are to be pronounced separately and not as a diphthong. Thus ἀΐδιος, Rom. 1.20, *a-idios*, not *aidios*. Notice that where the diaeresis occurs at the beginning of a word, as in ἀΐδιος, the breathing is written over the first vowel, and not over the second as it would be if the syllable were a diphthong.

Accents

The system of Greek accents was invented by Aristophanes of Byzantium about 200 B.C. as an aid to the correct pronunciation of the Greek language at a time when its use was being widely extended amongst foreign peoples. The accents indicated not a stress on a syllable but the pitch of the voice as the syllable was pronounced. The *acute* accent marked a rise in the voice, and the *circumflex* a rise followed by a fall. Syllables with neither an acute nor a circumflex accent were regarded as possessing a *grave* accent, though it was not usually marked. The syllable which had a grave accent, whether marked or not, was pronounced at the ordinary pitch of the voice.

Acute Accent	μαρτυρία, ἐγένετο.
Circumflex Accent	φῶς, εὗρεν
Grave Accent	καὶ, αὐτὸς.

NOTICE: 1. If an accent occurs on a diphthong it is written over the second of the vowels: τοῦτο, πιστεύω.

2. When a word has an acute accent on the last syllable (e.g. ἀρχή in Mark 1.1), the acute is turned into a grave accent unless a full stop or a colon immediately follows. Cf. αὐτὸν in Mark 1.5 with αὐτόν in Mark 1.10.

3. An accent on the initial syllable of a word beginning with a capital vowel is written to the left of it with the breathing: Ἤγγιζεν, Ἦλθον. But if the syllable is a diphthong the accent is placed above the second vowel: Οὕτως, Οὗτος.

The Use of the Accents

As we have already said, it is impossible to be certain how the Greeks themselves pronounced their language. Nor is it possible to be certain in what way or to what extent the pitch of the voice rose and fell with the accents. The accents, therefore, although they are employed in printed Greek, do not now serve their original purpose. Their main practical use is that they serve to distinguish between certain words which differ only in their accentuation.

The main rules of accentuation are given in Appendix 1. Their application to verbs is quite regular, and the student should not find it difficult to accustom himself to placing the correct accent on all verb forms. The accents of nouns will be found more difficult. There is no way in which the student can discover by rule that, for instance, the accent of λόγος falls on the first syllable, and that of ὁδός on the last. It is simply a matter of learning the position of the accent on the nominative singular of the nouns.

Anybody with pretensions to Greek scholarship will pride himself on making very few mistakes with accents, though few, perhaps, are infallible. The beginner will be wise to do no more than note carefully such words as differ only in their accents (see Appendix 2), and to learn the accentuation of the verb. He may leave the accents of the other parts of speech to be learned by constant use as time goes on.

EXERCISE I

Put the following into Greek letters. Insert the proper breathings and stops. Only use capital letters where indicated. A stroke over the letters *e* and *o* indicates that they are long; otherwise they are short. When the letter *i* is in brackets it is to be written in Greek underneath the preceding vowel (*iota subscript*).

1. Archē tou euangeliou Iēsou Christou Huiou Theou. 2. Kai eiselthōn palin eis Kapharnaoum di' hēmerōn ēkousthē hoti en oikō(i) estin.

3. Kai eisēlthen palin eis sunagōgēn, kai ēn ekei anthrōpos exērammenēn echōn tēn cheira. 4. Kai palin ērxato didaskein para tēn thalassan. kai sunagetai pros auton ochlos pleistos, hōste auton eis ploion embanta kathēsthai en tē(i) thalassē(i), kai pas ho ochlos pros tēn thalassan epi tēs gēs ēsan. 5. Kai ēlthon eis to peran tēs thalassēs eis tēn chōran tōn Gerasēnōn. 6. Kai exēlthen ekeithen, kai erchetai eis tēn patrida autou, kai akolouthousin autō(i) hoi mathētai autou. 7. Kai sunagontai pros auton hoi Pharisaioi kai tines tōn grammateōn elthontes apo Hierosolumōn. 8. En ekeinais tais hēmerais palin pollou ochlou ontos kai mē echontōn ti phagōsin, proskalesamenos tous mathētas legei autois. 9. kai elegen autois Amēn legō humin hoti eisin tines hōde tōn hestēkotōn hoitines ou mē geusōntai thanatou heōs an idōsin tēn basileian tou Theou elēluthuian en dunamei. 10. Kai ekeithen anastas erchetai eis ta horia tēs Ioudaias kai peran tou Iordanou, kai sunporeuontai palin ochloi pros auton, kai hōs eiōthei palin edidasken autous.

The above ten sentences are the first verses of the first ten chapters of St Mark, by reference to which in Nestle's text (British and Foreign Bible Society) you may verify your work.

Chapter 2

THE VERB

There can be no complete sentence without a verb. Hence it may be said that the verb is the most important part of speech in any language. It will be well, therefore, for the student to learn something of the Greek Verb at the outset of his studies, for he will thus be able from the beginning to express complete ideas in Greek, even if they are simple ones.

We shall first present the Present Indicative Active (i.e. the Present Tense of the Indicative Mood in the Active Voice) of the simplest type of Greek verb. But it will be necessary as a preliminary to understand what is meant by the terms *Voice, Mood,* and *Tense.*

VOICE

In Greek the verb may have three Voices: Active, Middle, and Passive. The Voices are forms of the verb used to denote different relations between the action described by the verb and its subject.

1. *The Active Voice* is used to express the idea that the subject of the verb is the actual doer of the action described, e.g. *he looses, they killed.* It is the Active Voice which will concern us first.

2. *The Middle Voice* expresses the idea that the action described is of special advantage or significance to the subject. There is no Middle Voice in English, and a Greek example must be given: λύεται τὸν ὄνον may be translated *he looses the ass,* but the verb, λύεται, being in the Middle Voice, implies that he does so for himself, perhaps in order to ride it. The beginner is apt to jump to the conclusion that the Greek Middle Voice is reflexive. This is not so. It denotes that the subject performs the action *for* himself, but not *to* himself.

3. *The Passive Voice* expresses the idea that the subject of the verb is passive in respect of the action described; the subject is the recipient of the action: e.g. *he is being loosed, they were killed.*

MOOD

The moods are different forms of the verb denoting the modes in which the idea of the verb is employed.

1. *The Indicative Mood* (i.e. the "pointing out" mood) is used to make a statement whether in present, future, or past time. It is also generally used in asking direct questions. Thus, the italic words in the following examples are all verbs in the Indicative Mood: he *goes*, they *will go*, you *were going*, *did* we *go*? It is the Indicative Mood which will concern us first.

2. *The Imperative Mood* (Latin, *impero*, "command") is used when an order is given, e.g. please *go*, *run* quickly.

3. *The Conjunctive Mood* takes two forms:

(a) *The Subjunctive*, employed when the idea of the verb is subjoined to some other verb and expresses contingency (i.e. that a thing may or may not happen). In the sentences "I am coming, that I *may see* for myself", "If they *should invade*, we shall fight", the verbs in italics would be put into the Greek Subjunctive.

(b) *The Optative*, used principally to express a wish: "O that I *might die*."

4. *The Infinitive Mood*, described in some grammars as the Verb Infinite, expresses the idea of the verb without limiting it by specifying person or number, e.g. *to loose, to be about to loose, to have loosed*.

Belonging also to the Verb Infinite are the Participle and the Verbal Adjective, which will be described later.

TENSE

The tenses (from Latin *tempus*, "time") are forms of the verb used to denote the time at which the action is regarded as taking place. The tenses in New Testament Greek are Present, Future, Imperfect, Aorist, Perfect, and Pluperfect, the usual English equivalents of which are as follows:

Present	I kill (*Active*)	I am (being) killed (*Passive*)
Future	I shall kill	I shall be killed
Imperfect	I was killing	I was being killed
Aorist	I killed	I was killed
Perfect	I have killed	I have been killed
Pluperfect	I had killed	I had been killed

The Present Tense will concern us first.

We can, of course, only deal with the Greek verb piecemeal, taking one voice, mood, and tense at a time. But it is important even in these early stages for the student to gain an idea of the full extent of the Greek verb.

He will then be less apt to be overwhelmed a little later on when he is faced one by one with the tenses other than the present, with all these tenses in the middle and passive voices as well as the active, and with the moods other than the indicative. Constant reference should, therefore, be made to Appendix 4 where the verb λύω in all its voices, moods, and tenses is set out. By so doing whenever a new part of the verb is reached in the text of the book, the student will gradually acquire a valuable familiarity with the whole paradigm.[1]

The Stems of Verbs

We shall have occasion very frequently to speak of the *stem* of a verb. By this term is meant the part of the word (it may be one syllable or more) which expresses the root idea, and to which various additions may be made, such as the *augment* or the *reduplication* in front of it used in certain parts of the verb; or the endings which may be added (suffixes), which help both to denote the tense and the person(s) who perform(s) the action. In the case of the English verb "to know" we may regard the syllable *know* as the stem. Personal suffixes can be added as in *knows, knoweth*, and *knowest*. Other additions appear in the participles *knowing* and *known*. In the Greek verb λύω the stem is λυ-.

Many verbs have both a *verb stem* and a *present stem*: the simple form of the stem is found in the second (or strong) aorist indicative active (verbs whose stems end in a vowel do not possess this tense), while the present indicative shows a lengthened form of it. This is true of βάλλω (see p. 44). The verb stem is βαλ- which is found in the 2nd aor. indic. act. ἔ-βαλ-ον, while the present indicative has the lengthened βαλλ-. It is convenient, therefore, in such cases to speak of *verb stem* (from which all tenses are formed except the pres. and imperf.) and *present stem*. The verb stem of βάλλω is βαλ-. Its present stem is βαλλ-. But in the case of λύω (and of many other verbs) there is no such lengthening of the stem in the pres. indic., and λυ- is its stem throughout the paradigm.

We shall have occasion later on to notice that in some verbs a change of vowel in the stem is found in some of the tenses: e.g. πείθω *persuade, counsel* has a 2nd perf. act. πέ-ποιθ-α (πε- is not part of the stem: it is the reduplication—see above). But as a similar phenomenon is found in English, e.g. *know, knew; run, ran*, the student should not find this a great difficulty. Instances of it in the Greek verb will be noted as we come to them.

[1] *Paradigm*, Greek παράδειγμα *a model*, is a term used in Grammar for the complete "model" of a verb, in all its voices, moods, and tenses.

THE PRESENT INDICATIVE ACTIVE

We are now in a position to present the Present Indicative Active of the verb. As our model we take λύω, one of the simplest verbs whose 1st person singular in the pres. indic. act. has the suffix -ω added to the stem.[1] (The more difficult -μι verbs we leave until later.)

	Singular		Plural	
1st person	λύ-ω	I loose,	λύ-ομεν	we loose
2nd „	λύ-εις	thou loosest, you loose	λύ-ετε	you loose
3rd „	λύ-ει	he, she, or it looses	λύ-ουσι(ν)	they loose

NOTICE: 1. Other permissible translations of the Present Tense are the continuous *I am loosing* and the emphatic *I do loose*.

2. The stem is λυ-, which is unchanged throughout. The endings vary according to the person and number of the subject. This change of ending in the tenses is known as *Inflexion* or *Flexion*.

3. The personal pronoun is not necessarily written in Greek. The ending of the verb clearly indicates the person: λύομεν means *we loose*: there is no need of the pronoun "we" (ἡμεῖς), but it may be written if emphasis is required. In the case of the 3rd person singular the context usually decides whether λύει means *he looses*, *she looses*, or *it looses*. The verb may, however, have a pronoun as its subject when needed: thus, οὗτος λύει *this man is loosing*.

4. The Greek verb may have a noun as its subject, just as in English. Thus ἄνθρωπος λύει *a man looses*; ἄνθρωποι λύουσιν *men loose*.

5. In the 3rd person plural, λύουσι is usually written if the next word in

---

[1] Verbs whose 1st persons sing. pres. indic. end in -ω are sometimes called *thematic* verbs. The -μι verbs are unthematic. The term "thematic" applied to a verb denotes that it has tenses in which the vowels o and ε, known in this connection as "thematic vowels", are employed between the stem (or theme) and the ending.

The original personal suffixes of the pres. indic. act. were -μι, -σι, -τι, -μεν, -τε, -ντι. The use of the thematic vowel gives us λυ-ο-μι, λυ-ε-σι, λυ-ε-τι (which by the classical period were modified into λύω, λύεις, λύει), λυ-ο-μεν, λυ-ε-τε, λυ-ο-ντι (modified into λύουσι). Compare the unthematic verb τίθημι (p. 230) in which the thematic vowels do not appear.

Thematic verbs employ the thematic vowels in the pres. and imperf. indic., and pres. imper. in all three voices.

the sentence begins with a consonant; λύουσιν is usually written if the next word begins with a vowel, and at the end of a sentence (see p. 10).

6. The pres. indic. act. of other verbs in -ω may easily be formed on the model of λύω by adding the personal endings as above to the present stem. Thus the pres. indic. act. of λέγω *speak* (stem λεγ-) is as follows:

λέγω	*I speak*	λέγομεν	*we speak*
λέγεις	*thou speakest*	λέγετε	*you speak*
λέγει	*he, she, it speaks*	λέγουσι(ν)	*they speak*

Practise this with the other verbs in the list at the head of Exercise 2.

7. The Greek question mark is ;. Thus, λύομεν *we loose*, but λύομεν; *do we loose?* or *are we loosing?*

Vocabulary

ἄγω *I lead*	θύω *I sacrifice*
ἀκούω *I hear*	κελεύω *I order*
βάλλω *I throw, put*	λέγω *I speak*
γράφω *I write*	πέμπω *I send*
ἐσθίω *I eat*	πιστεύω *I believe*
ἔχω *I have, hold*	φέρω *I bring*
θεραπεύω *I heal*	φυτεύω *I plant*

Note that a Greek word may often be used rightly to translate different English words of similar meaning: e.g. κελεύω translates *I command* or *I bid*; λέγω *I say*; φέρω *I bear* or *I carry*.

EXERCISE 2

Translate into English:

1. θύουσιν. 2. φέρομεν. 3. ἐσθίετε. 4. γράφει. 5. λέγω; 6. πιστεύεις. 7. θεραπεύετε. 8. ἔχομεν. 9. φυτεύει. 10. κελεύουσιν.

EXERCISE 3

Translate into Greek:

1. He hears. 2. They speak. 3. Thou eatest. 4. We are throwing. 5. Is she leading? 6. You (plur.) believe. 7. We sacrifice. 8. It brings. 9. They are planting. 10. We lead. 11. Thou art writing. 12. Am I sending? 13. We hear. 14. You (plur.) throw. 15. It eats. 16. Dost thou bid? 17. They are carrying. 18. We are leading. 19. She possesses. 20. You (plur.) say.

Chapter 3

THE NOUN

Greek nouns differ from English in several important respects, of which the most important are *Declension, Gender,* and *Case.*

DECLENSION

Greek nouns are grouped together in classes known as *Declensions.* This word (from Latin *declino*) refers to the declining, bending, or inflexion of the endings of the noun in its different cases. Notice the change in the endings of ἀρχή given below. Nouns which decline in the same way, or in much the same way, are said to be of the same declension. There are three declensions in Greek.

GENDER

English nouns which are the names of males are masculine, those which are the names of females are feminine, and all others are regarded as neuter.

The gender of Greek nouns is not quite so simple as this. Names of men and male animals are indeed masculine, and those of females are feminine. But the student must accustom himself to the fact that a great many nouns which would be called neuter in English are masculine or feminine in Greek. Thus:

1st decl. nouns whose nom. sing. ends in -η or -α are feminine.
 ” ” ” ” ” ” ” ” -ης or -ας are masculine.
There are no neuter nouns of the 1st decl.

2nd decl. nouns whose nom. sing. ends in -ος are masculine, with a very few, but important, exceptions which are feminine (see p. 33).

2nd decl. nouns whose nom. sing. ends in -ον are neuter.

For the gender of 3rd decl. nouns see pp. 74f., 94f., 120ff.

Since any adjective which qualifies a noun must agree with that noun in gender (as well as in number[1] and case), it is important to pay close attention to the genders of nouns.

---

[1] A noun may be singular or plural. Classical Greek occasionally employed a Dual Number when two things were referred to; thus τὼ οἰκία *the two houses* (nom. and acc.), τοῖν οἰκίαιν (gen. and dat.). But the Dual Number is not used in the New Testament.

CASE

English has only a rudimentary system of cases. English grammars speak of subjective, objective, and possessive cases of the noun or pronoun. The noun used as the subject of a sentence is in the subjective case, used as object to a verb it is in the objective case, and when it indicates possession it is in the possessive case. Thus in the sentences: "The book bores me"; "I like the book"; "the book's cover is red", the noun *book* is respectively in the subjective, objective, and possessive cases. No change in the spelling of the noun is involved in the English cases, but when the noun is in the possessive case *'s* is usually added in the singular, and *s'* usually in the plural.[1]

In ancient languages the noun has more cases. The Latin noun has six, and the Greek five. There is evidence that the common stock language from which they were derived had even more. The different ways in which a noun may be used in a sentence are thus more clearly indicated in these ancient languages by the modification of the noun's ending which takes place according as the noun is used.

FIRST DECLENSION NOUN, ἀρχή (*f.*) *beginning*

Case	Singular	Plural
Nominative	ἀρχή	ἀρχαί
Vocative	ἀρχή	ἀρχαί
Accusative	ἀρχήν	ἀρχάς
Genitive	ἀρχῆς	ἀρχῶν
Dative	ἀρχῇ	ἀρχαῖς

NOTICE: 1. The vocative of nouns declined like ἀρχή is the same as the nominative in both the singular and plural.

2. *Iota* is written under the final -η of the dative singular (*iota subscript*). It is not pronounced.

3. The declension of other nouns of this class may be easily found by changing the endings in accordance with the model of ἀρχή. Thus, ἐντολή *command*:

[1] When the English plural involves a change other than the addition of *s* (e.g. man, *men*), the possessive plural has *'s*. Thus *men's*, *women's*.

	Singular	Plural
N., V.	ἐντολή	ἐντολαί
A.	ἐντολήν	ἐντολάς
G.	ἐντολῆς	ἐντολῶν
D.	ἐντολῇ	ἐντολαῖς

Practise this with the nouns in the list at the head of Exercise 4.

THE MEANING OF THE CASES

We give here only the simplest uses of the cases. Others will be noted later.

The Nominative Case is used when the noun is the subject of a sentence: παιδίσκη λέγει *a young girl speaks.*

The Vocative Case is used when a person or thing is addressed. Sometimes ὦ is written in front: ἀδελφή or ὦ ἀδελφή *sister, O sister.*

The Accusative Case is used when the noun is the direct object of a verb: βάλλει λόγχην *he throws a spear.*

The Genitive Case is used to denote possession: ἀδελφῆς means *of a sister, belonging to a sister, sister's*; νύμφης ἑορτή *a feast of a bride, a bride's feast.* The genitive may be written either before or after the noun on which it depends: ἑορτή νύμφης would also mean *a bride's feast.*

The Dative Case is used for the indirect object, and may be translated *to* or *for.* The direct object of a sentence is that which the verb acts upon directly. Thus in the sentence, "the boy kicks the ball", the noun *ball* is the direct object, and in Greek would be put into the accusative case. But a verb may have an indirect object, as in the sentences, "the boy gives the book to the girl", "the master does the work for the children", where *girl* and *children* are indirect objects of the verbs, and in Greek would be put in the dative case.

THE ARTICLE

English has both the Definite Article *the*, and the Indefinite Article *a* or *an.* Greek has no indefinite article. Thus ἐντολή may be translated either *commandment* or *a commandment.*

The Greek definite article is ὁ (masculine), ἡ (feminine), τό (neuter). The Article, being an adjective, must agree with the noun which it qualifies (see p. 36), and hence it declines. We give here the declension of the feminine:

	Singular	Plural
N.	ἡ	αἱ
A.	τήν	τάς
G.	τῆς	τῶν
D.	τῇ	ταῖς

NOTICE: 1. The endings are similar to the endings of the 1st decl. nouns ending in -η.

2. There is no vocative case, since the definite article cannot be used in a direct address.

3. The article must agree with its noun not only in gender, but also in its case and number (singular or plural). Thus, τὴν ἀρχήν *the beginning* (object of verb); τῆς ἀδελφῆς *of the sister*; αἱ λόγχαι *the spears* (subject of a verb); τῶν ἐντολῶν *of the commandments*.

4. Generally speaking, the Greek article is used where English would use *the* before a noun. But Greek commonly has the article with an abstract noun; thus ἡ ἀγαπή *love*; ἡ δικαιοσύνη *righteousness*. The article also is very often employed with a proper noun; ἡ Μάρθα *Martha*.

5. An article may sometimes be separated from its noun, especially by an adjective or a phrase which is equivalent to an adjective. This may occur when the possessive genitive is attached to a noun: ἡ τῆς παιδίσκης λύπη *the girl's grief*; ἡ τῆς βροντῆς φωνή *the voice of thunder*. However, ἡ λύπη τῆς παιδίσκης and ἡ φωνὴ τῆς βροντῆς may be written.

Vocabulary

ἀγαπή *love*
ἀδελφή *sister*
βροντή *thunder*
γῆ *earth, land, country*
γραφή *writing, scripture*
διδαχή *teaching*
δικαιοσύνη *righteousness*
εἰρήνη *peace*
ἐντολή *commandment*
ἑορτή *feast*

ζώνη *girdle, belt*
κεφαλή *head*
λόγχη *spear*
λύπη *grief*
νύμφη *bride*
παιδίσκη *young girl*
συναγωγή *synagogue, assembly*
τιμή *honour*
φωνή *voice*
ψυχή *life, soul*

The student should be on the watch for Greek words which have passed into the English language. To note these will help him to remember the meanings of the Greek words. Thus in the above list γραφή is incorporated in a number of English words, e.g. *graphite, photography*; φωνή in *microphone, symphony*; ψυχή in *psychology*.

EXERCISE 4

Translate into English:

1. αἱ ἀδελφαὶ ἐσθίουσιν. 2. ἀκούεις τὴν τῆς νύμφης φωνήν. 3. πιστεύομεν τὰς διδαχὰς τῆς συναγωγῆς. 4. λύπην φέρει τῇ παιδίσκῃ. 5. ἀδελφὴ ἀδελφῇ λέγει. 6. ἡ ἐντολὴ κελεύει τὴν ἑορτήν.

EXERCISE 5

Translate into Greek:

1. Of a sister. 2. For the brides. 3. The voices (acc.). 4. Of girdles. 5. To a young girl. 6. Peace (acc.). 7. Of righteousness. 8. For the land. 9. The feasts (acc.). 10. Of a spear. 11. The soul of the young girl. 12. For the love of the bride. 13. The thundering of the voices. 14. The writing (acc.) of the commandments. 15. The synagogues.

EXERCISE 6

Translate into Greek:

1. The sister sends. 2. He sends the sister. 3. Thou art leading the bride. 4. The bride hears the voice. 5. The earth has peace. 6. We write the commandments. 7. You (plur.) hear the teaching. 8. He brings the scriptures for the young girl. 9. The sisters speak to the bride. 10. The bride is speaking to the sisters.

Chapter 4

THE IMPERFECT AND FUTURE INDICATIVE ACTIVE OF λύω

IMPERFECT INDICATIVE ACTIVE

	Singular		Plural	
1st person	ἔ-λυ-ον	I was loosing	ἐ-λύ-ομεν	we were loosing
2nd ,,	ἔ-λυ-ες	thou wert loosing, you were loosing	ἐ-λύ-ετε	you were loosing
3rd ,,	ἔ-λυ-ε(ν)	he (she, it) was loosing	ἔ-λυ-ον	they were loosing

NOTICE: 1. The Imperfect is a past tense, denoting continuous action, repeated action or attempted action. *I was loosing, I used to loose, I tried to loose*, are all permissible translations of ἔλυον.

2. Greek marks past time in the historic tenses of the indicative mood by the *augment* (see p. 25). This is normally the vowel ἐ placed in front of the present stem.

3. The endings of the Imperfect Indicative Active are -ον, -ες, -ε(ν), -ομεν, -ετε, -ον. These endings are added directly to the stem.[1]
There are thus three constituent parts in the Imperfect, the augment, stem, and personal ending.

4. As with λύουσι (3rd pers. plur. pres. indic.) so with ἔλυε, ν may be added to make pronunciation easier. It is nearly always added when a vowel follows, and often at the end of a sentence.

5. The 3rd pers. plur. is spelled as the 1st pers. sing. The context will nearly always make clear which is intended.

6. The imperf. indic. act. of other verbs in -ω may easily be found on

---

[1] It is more accurate to say that the personal endings of the imperf. indic. act. are ν (originally μ which Greek does not tolerate at the end of a word), ς, τ (dropped because τ cannot conclude a Greek word), μεν, τε, ν(τ). These endings are added to the stems of verbs of the -ω class by means of the helping thematic vowel, ε or ο (see note on p. 17). But for the beginner it is sufficient to know that the imperf. indic. act. endings are as above.

the model of ἔλυον by prefixing the augment to the present stem and adding the personal endings.   Thus:

$$
\begin{array}{ll}
\text{ἔ-λεγ-ον} \quad I \; was \; speaking & \text{ἐ-λέγ-ομεν} \\
\text{ἔ-λεγ-ες} & \text{ἐ-λέγ-ετε} \\
\text{ἔ-λεγ-ε(ν)} & \text{ἔ-λεγ-ον}
\end{array}
$$

Practise this with θύω, κελεύω, πιστεύω, γράφω, πέμπω, βάλλω, φέρω.

THE AUGMENT

The augment is the ἐ prefixed in the historic tenses of the indicative mood, and it signifies past time.   But when the stem of a verb begins with a vowel the augment usually causes a lengthening of the initial vowel. Thus:

ἀκούω *hear*;	imperf. indic. act.:	ἤκουον	(α lengthening to η).
ἄγω *lead*;	,, ,, ,,	ἦγον	(α ,, ,, η).
ἐσθίω *eat*;	,, ,, ,,	ἤσθιον	(ε ,, ,, η).
ἐγείρω *arouse*;	,, ,, ,,	ἤγειρον	(ε ,, ,, η).
ὀνομάζω *name*;	,, ,, ,,	ὠνόμαζον	(ο ,, ,, ω).
ἰσχύω *am strong*;	,, ,, ,,	ἴσχυον	(ῐ ,, ,, ῑ).
ὑβρίζω *insult*;	,, ,, ,,	ὕβριζον	(ῠ ,, ,, ῡ).

In the last two instances the augment is not discernible in the written word, but when pronounced the ι and the υ are lengthened.

Some stems beginning (or appearing to begin) with ε have augment ει instead of η.   Of these verbs the commonest is ἔχω *have*; imperf. εἶχον. The stem of ἔχω is in reality σεχ-.   For further notes on the augment see pp. 66, 262.

THE FUTURE INDICATIVE ACTIVE

	Singular		Plural	
1st person	λύ-σ-ω	*I shall loose*	λύ-σ-ομεν	*we shall loose*
2nd ,,	λύ-σ-εις	*thou wilt loose, you will loose*	λύ-σ-ετε	*you will loose*
3rd ,,	λύ-σ-ει	*he (she, it) will loose*	λύ-σ-ουσι(ν)	*they will loose*

NOTICE: 1. The Greek future is usually translated by the English *shall* (1st pers.) and *will*, denoting simple action in future time.   (The English *will* in the sense of exerting the power of choice translates the Greek verbs θέλω and βούλομαι.) The Greek future may also represent continuous action in future time. Thus λύσω may rightly be translated "I shall be loosing" as well as "I shall loose".

2. The personal endings are the same as those of the pres. indic. act.

3. There is no augment since the augment is the sign of an historic tense.

4. Between the verb stem and the ending σ is written. This may be regarded as a special sign of the future.[1]

When the stem ends in a consonant the σ of the future which follows causes certain changes.[2] At this stage the following should be noted:

(a) When the stem ends in a guttural (γ, κ, χ) the σ coalesces to make ξ (see p. 6).

ἄγω *I lead*; fut.: ἄξω *I shall lead* (γ + σ becomes ξ).

ἄρχω *I rule*; fut.: ἄρξω (not found in N.T.) (χ + σ becomes ξ).

(b) When the stem ends in a labial (β, π, φ) the σ coalesces to make ψ (see p. 7).

τρίβω *I rub*; fut.: τρίψω *I shall rub* (β + σ becomes ψ).

βλέπω *I look at, see*; fut.: βλέψω *I shall look at* (π + σ becomes ψ).

γράφω *I write*; fut. γράψω *I shall write* (φ + σ becomes ψ).

(c) When the stem ends in a dental (δ, τ, θ) the dental is dropped before σ (see p. 7).

πείθω *I persuade*; fut.: πείσω *I shall persuade* (θ before σ drops out).

(d) When the stem ends in a nasal (μ, ν) or liquid (λ, ρ) the following σ causes changes of a different kind which will be described on pp. 184 and 153 (see also pp. 8, 9).

5. The fut. indic. act. of verbs in -ω may easily be found by adding σ and the personal endings to the stem, care being taken to observe the rules in 4 above. Thus the fut. indic. act. of γράφω is:

γράψω	γράψομεν
γράψεις	γράψετε
γράψει	γράψουσι(ν)

Practise this with θύω *sacrifice*, κελεύω *order*, ἀνοίγω *open*, στρέφω *turn*, πέμπω *send*.

[1] It will be seen later, however, that the 1st, or Weak, Aorist Active and Middle Tenses also have the σ in the same position. The Future Indicative is closely dependent on the Aorist in the history of the development of the language.

[2] See p. 9.

6. Some verbs have seeming irregularities in the fut. indic. act. These will be mentioned as they occur, and can also be discovered in the lists of verbs. But notice immediately that the future of ἔχω *I have*, has a rough breathing: ἕξω.

THE FIRST DECLENSION—*continued*

The other forms of the first declension are as follows:

Nouns ending in α (*feminine*)

	α pure (i.e. following ε, ι or ρ)		α impure (i.e. following any other letter)	
	Singular	*Plural*	*Singular*	*Plural*
N.	ἡμέρα *day*	ἡμέραι	δόξα *glory*	δόξαι
V.	ἡμέρα	ἡμέραι	δόξα	δόξαι
A.	ἡμέραν	ἡμέρας	δόξαν	δόξας
G.	ἡμέρας	ἡμερῶν	δόξης	δοξῶν
D.	ἡμέρᾳ	ἡμέραις	δόξῃ	δόξαις

Nouns ending in -ης and -ας (*masculine*)

	Singular	*Plural*	*Singular*	*Plural*
N.	προφήτης *prophet*	προφῆται	νεανίας *youth*	νεανίαι
V.	προφῆτα	προφῆται	νεανία	νεανίαι
A.	προφήτην	προφήτας	νεανίαν	νεανίας
G.	προφήτου	προφητῶν	νεανίου	νεανιῶν
D.	προφήτῃ	προφήταις	νεανίᾳ	νεανίαις

NOTICE: 1. Where the α of the nom. sing. ending is *pure* (i.e. following ε, ι, or ρ) the α remains throughout the declension of the singular. Where α is *impure* (i.e. following any other letter than ε, ι, or ρ) -ης and -η are found in the gen. and dat. sing.[1]

2. The gen. sing. of masculine nouns of the 1st decl. ends in -ου.

[1] There are one or two exceptions to this, due to "levelling". Levelling is a process by which differing inflexions are brought into closer agreement, and is thus an attempt to make the language more regular. The most common exceptions are Σάπφειρα *Sapphira*; gen.: Σαπφείρης; πρῷρα *prow*; gen.: πρῴρης; σπεῖρα *cohort*; gen.: σπείρης; μάχαιρα *sword*; gen.: μαχαίρης. Levelling in the other direction appears in Λύδδα *Lydda*; gen.: Λύδδας; and Μάρθα *Martha*, gen.: Μάρθας.

3. The endings of the plural in all forms of the 1st decl. are -αι, -αι, -ας, -ῶν, -αις.

4. There are no other common nouns in the N.T. declined like νεανίας, and only a few proper nouns (see vocabulary before Exercise 7).

THE MASCULINE ARTICLE

The article used with the masculine noun is as follows:

	Singular	Plural
N.	ὁ	οἱ
A.	τόν	τούς
G.	τοῦ	τῶν
D.	τῷ	τοῖς

Thus: *the prophet* ὁ προφήτης; *of the prophet* τοῦ προφήτου; *the prophets* οἱ προφῆται; *to the prophets* τοῖς προφήταις.

VOCABULARY

Words which have been used as examples in the text will not always be given again in the vocabularies at the head of the exercises. The student should learn these words as he goes along.

Like ἀρχή:
ὀργή *anger*
φυλακή *guard, watch, prison*
Like ἡμέρα:
ἀλήθεια *truth*
βασιλεία *kingdom*
θύρα *door*
χρεία *need*
Like δόξα:
γλῶσσα *tongue*
τράπεζα *table*

Like προφήτης:
κλέπτης *thief*
μαθητής *disciple*
στρατιώτης *soldier*
'Ιορδάνης *Jordan*
'Ιωάνης *John*
τελώνης *publican*
Like νεανίας:
'Ανδρέας *Andrew*
Μεσσίας *Messiah*

κai (conjunction) *and*
ἀλείφω *anoint*
βουλεύω *take counsel*
δουλεύω (with dat.) *serve*
ἑρμηνεύω *interpret*

κηρύσσω *preach, announce*
λούω *wash*
νηστεύω *fast*
παίω *strike*

EXERCISE 7

Translate into English:

1. οἱ μαθηταὶ Ἰωάνου νηστεύουσιν. 2. ἐκήρυσσεν εἰρήνην τοῖς στρατίωταις. 3. ἡ θύρα χρείαν ἔχει φυλακῆς. 4. ὁ Ἀνδρέας τῷ Μεσσίᾳ δουλεύσει. 5. ἐβλέπομεν τὰς τραπέζας τῶν τελώνων. 6. οἱ προφῆται ἡρμήνευον τὰς ἐντολὰς τοῖς νεανίαις.

EXERCISE 8

Translate into Greek:

1. We were leading. 2. You (plur.) were strong. 3. I used to eat. 4. The young girl was naming the disciple. 5. You (sing.) used to lead the soldiers. 6. They will write the commandments. 7. The soldier was striking the thief's head. 8. The kingdom will have a guard. 9. We shall look at the table. 10. You (sing.) will hear the voice.

EXERCISE 9

Translate into Greek:

1. The bride was taking counsel. 2. The disciples used to interpret the scriptures for the soldiers. 3. We shall hear thunder. 4. The soldiers' spears were striking the door. 5. You (sing.) were insulting the bride. 6. We shall serve the kingdom. 7. You (plur.) tried to anoint the Messiah. 8. John was interpreting the scriptures for Andrew. 9. I shall wash the spear. 10. The voices of the thieves were arousing the land. 11. Anger will rule the tongue. 12. Truth is strong.

Chapter 5

PREPOSITIONS

A preposition is a word placed before a noun (or a pronoun) in order to make clear the relationship (usually a relationship of place or time) to some other word in a sentence. Thus in the sentence, "he put the book on the table", the preposition *on* relates *table* to *book*. In the sentence, "he walked towards the door", the preposition *towards* relates *door* to the action of walking. The use of prepositions in Greek is very similar to their use in English. Two differences, however, must be noticed:

1. Greek prepositions govern[1] different cases—accusative, genitive, or dative—and the student must carefully learn what case or cases each preposition takes.

2. Some English prepositions have no equivalent preposition in Greek. For example, as we have seen, the English preposition *of*, denoting possession, is rendered in Greek by the genitive of the noun without any preposition at all, and the English *to* or *for* of the indirect object are rendered by the dative case without a preposition.

Some Common Prepositions

εἰς	with the acc., *into, towards*: e.g., εἰς τὴν γῆν *into the land*.
ἀπό	with the gen., *from, away from*: e.g., ἀπὸ τῆς θαλάσσης *from the sea*.
ἐκ, ἐξ	with the gen., *from, out of*: e.g., ἐκ τῆς οἰκίας *out of the house*.
ἐν	with the dat., *in, on, among* (sometimes, *with*): e.g., ἐν τῇ καρδίᾳ *in the heart*.
σύν	with the dat., *with, together with*: e.g., σὺν ταῖς παιδίσκαις *with the young girls*.

NOTICE: 1. The difference between ἀπό and ἐκ, which may both correctly be translated *from*, is that ἀπό means *away from*, and ἐκ means *out*

---

1 Although it is a useful phrase, it is strictly incorrect to speak of a preposition *governing* a case. The *case* is the important factor in any phrase in which a preposition is employed, and in the early history of the language prepositions were not used at all. Different cases of the noun expressed different relations or positions. Then, in order to make the exact meaning clearer, adverbs were placed before the noun. Thus what we call prepositions are really adverbs employed in this particular way.

from or *out of.* Thus we should say that a swimmer comes ἐκ τῆς θαλάσσης *out of the sea.* But a man who has been standing by the sea shore walks ἀπὸ τῆς θαλάσσης *from,* or *away from, the sea.*

The difference can be expressed diagrammatically to aid memory:

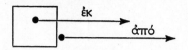

2. ἐκ is written before a consonant, ἐξ before a vowel or rough breathing. Thus ἐκ τῆς οἰκίας *out of the house,* but ἐξ οἰκίας *out of a house.*

3. σύν means *with* in the sense of accompaniment (Latin *cum*). *With* in the instrumental sense is translated by the dative case without a preposition, or by ἐν and the dative. Thus in the sentence, *he killed his enemy with a sword,* we might write μαχαίρᾳ or ἐν μαχαίρᾳ (μαχαίρῃ is sometimes found; see Note 1, p. 27).

THE NEGATIVE

The negative *not* when used in a statement is οὐ. Before a vowel οὐκ is used, and before a rough breathing οὐχ. It is placed before the word it governs:

οἱ μαθηταί οὐ γράφουσιν *the disciples do not write.*
οἱ μαθηταί οὐκ ἔγραφον *the disciples were not writing.*
οἱ μαθηταί οὐχ ἕξουσι τὴν δόξαν *the disciples will not have the glory.*

In some sentences οὐ is best translated by the English *no*: οὐ χρείαν ἔχομεν *we have no need.*

SECOND DECLENSION NOUNS

There are two classes of second declension nouns, each including a large number of words:

(*a*) Nouns ending in -ος which are masculine, apart from a few important exceptions which are feminine.
(*b*) Nouns ending -ον which are neuter.

	Singular	Plural	Singular	Plural
N.	λόγος *word*	λόγοι	ἔργον *work*	ἔργα
V.	λόγε	λόγοι	ἔργον	ἔργα
A.	λόγον	λόγους	ἔργον	ἔργα
G.	λόγου	λόγων	ἔργου	ἔργων
D.	λόγῳ	λόγοις	ἔργῳ	ἔργοις

NOTICE: 1. The endings of the gen. and of the dat., both sing. and plur., are the same for both -ος and -ον nouns.

2. The nom., voc., and acc. of the neuter noun are the same, both in the sing. and in the plur. This is true also of the neuter nouns of the 3rd decl.

3. It is a curious feature of Greek that a neuter plural subject has a singular verb. Thus τέκνα ἐσθίει *children eat*, not τέκνα ἐσθίουσιν.

The proper name ᾽Ιησοῦς *Jesus*, belongs to the second declension, although it is peculiar in having no iota in the dative: N. ᾽Ιησοῦς, V. ᾽Ιησοῦ,   A. ᾽Ιησοῦν,   G. ᾽Ιησοῦ,   D. ᾽Ιησοῦ.

THE NEUTER ARTICLE

The declension of the article with a neuter noun is as follows:

	Singular	Plural
N.	τό	τά
A.	τό	τά
G.	τοῦ	τῶν
D.	τῷ	τοῖς

Thus, τὸ βιβλίον *the book*; τὰ πλοῖα *the boats*.

THE ARTICLE IN FULL

We may now set out the definite article in full:

	Masculine		Feminine		Neuter	
	Singular	Plural	Singular	Plural	Singular	Plural
N.	ὁ	οἱ	ἡ	αἱ	τό	τά
A.	τόν	τούς	τήν	τάς	τό	τά
G.	τοῦ	τῶν	τῆς	τῶν	τοῦ	τῶν
D.	τῷ	τοῖς	τῇ	ταῖς	τῷ	τοῖς

Vocabulary

Masculine, declined like λόγος:

ἄγγελος *messenger, angel*
ἀγρός *field*; plur. *country places, farms*
ἀδελφός *brother*
ἄνθρωπος *man*
ἀπόστολος *apostle*
ἄρτος *bread, loaf*
δοῦλος *slave*
Θεός *God* (θεός *god*)
καιρός *time, season*

Κύριος *Lord* (κύριος *master*)
λαός *people*
λίθος *stone*
οἶκος *house*
ὄχλος *crowd*
τόπος *place*
υἱός *son*
φίλος *friend*
φόβος *fear*

Feminine, declined like λόγος:

ἄμπελος *vine*
βίβλος *book, roll*
ἔρημος *desert*
ὁδός *road, way*

τρίβος *path*
διάκονος, with fem. article *servant-girl*, but with masc. article, *servant*

Neuter, declined like ἔργον:

δαιμόνιον *evil spirit, demon, devil*
δεῖπνον *supper*
δένδρον *tree*
δίκτυον *net*
θηρίον *wild beast*
ἱερόν *temple*

παιδίον *child*
πλοῖον *boat*
πρόβατον *sheep*
σάββατον *sabbath*
τέκνον *child*

ἁμαρτία *sin*
καρδία *heart*
θάλασσα *sea*

ἐγγίζω *draw near*
κλαίω *weep*
κλείω *shut*

EXERCISE 10

Translate into English:

1. ὁ καιρὸς ἐγγίζει καὶ βλέψομεν τὴν βασιλείαν τοῦ Θεοῦ. 2. ὁ ἀπόστολος ἄξει τὸν λαὸν ἐν ταῖς τῆς ἀληθείας τρίβοις. 3. ἔπεμπε τὸν ὄχλον ἐκ τῆς ἐρήμου. 4. ὁ δοῦλος σὺν τοῖς διακόνοις δεῖπνον ἤσθιεν ἐν τῷ σαββάτῳ. 5. τὰ τέκνα λίθοις ἦγε τὰ πρόβατα εἰς τὸν ἀγρόν. 6. ὁ ἄνθρωπος ἔβαλλε τὰ βιβλία ἐν τῷ τοῦ υἱοῦ οἴκῳ.

EXERCISE 11

Translate into Greek:

1. The angel looses the apostle. 2. The apostles speak words of peace.
3. The slaves were eating bread. 4. The children are strong. 5. The Son of God brings righteousness and peace. 6. We were throwing the nets.
7. They will send a boat. 8. The slaves will drive the sheep. 9. Friends do not insult friends. 10. The sons of the kingdom will write words of truth.

EXERCISE 12

Translate into Greek:

1. We were driving the sheep into the place. 2. He throws the bread out of the house into the crowd. 3. They were eating supper on the sabbath. 4. The Lord will lead the people in the way of peace. 5. We shall shut the door with stones. 6. We were leading the children from the trees into the desert. 7. The disciples tried to loose the thieves from sin.
8. Truth will loose the man's tongue. 9. The brothers are weeping in the sister's house. 10. John and Andrew will send the Scriptures to Lydda.
11. The Lord brings fear into the hearts of men and casts the evil spirits into the sea. 12. The thieves were eating with the soldiers.

11 5 The Son
9 friends (acc.)

12

Chapter 6

THE VERB "TO BE"

We cannot get far in any language without a knowledge of the verb "to be". Its conjugation in Greek, as in other languages, shows irregularity. It is one of a class of verbs which will be dealt with fully later, whose 1st pers. sing. pres. indic. ends in -μι. But acquaintance with it cannot now be delayed. There are three tenses only, and we give them here in full.

PRESENT INDICATIVE

εἰμί	*I am*	ἐσμέν	
εἶ		ἐστέ	
ἐστί(ν)		εἰσί(ν)	

IMPERFECT

ἤμην	*I was*	ἦμεν or ἤμεθα	
ἦς or ἦσθα		ἦτε	
ἦν		ἦσαν	

FUTURE

ἔσομαι	*I shall be*	ἐσόμεθα	
ἔσῃ		ἔσεσθε	
ἔσται		ἔσονται	

NOTICE: 1. The stem is in reality ἐς-, but it is greatly modified in the pres. and imperf. The fut. shows it throughout.

2. The pres. tense of εἰμί, except for the 2nd pers. sing. εἶ, is enclitic, throwing its accent back on to the last syllable of the previous word if that syllable is able to receive it (see p. 273). Thus:

ἡ γλῶσσά ἐστι κακή *the tongue is evil.*
προφῆταί εἰσιν *they are prophets.*

But ἅγιος εἶ *thou art holy.*

The 3rd pers. sing. is, however, accented ἔστι(ν) when it stands first in its clause; when it means *exists*; and when it occurs after adverbs and conjunctions. Thus:

οὐκ ἔστι (John 10.12). The negative οὐ is an adverb.

It is also accented in the phrase τοῦτ᾽ ἔστι(ν) *that is* (1 Pet. 3.20).

ADJECTIVES

An adjective is a word which adds a description to a noun, as "the *pretty* girl", "*strong* horses". Adjectives may be used in two ways, as epithets to qualify a noun directly as in the preceding two examples, or predicatively,[1] as in "the girl is pretty", "horses are strong". Adjectives may also be used as equivalents of nouns as in "none but the *brave* deserves the *fair*"; but here they are properly adjectives with nouns understood: "brave *men*", "fair *ladies*".

In English adjectives are not inflected, i.e. there is no variation in them according as they qualify nouns of different number, gender, or case. In Greek, however, inflection does take place, and the ending of an adjective varies according to the number, gender and case of the noun with which it agrees.

ADJECTIVES OF THE SECOND AND FIRST DECLENSIONS

The most numerous class of adjectives is that which employs the same endings as for the 2nd decl. noun in the masculine and neuter, and the same ending as for the 1st decl. noun in -η for the feminine. Thus ἀγαθός, ἀγαθή, ἀγαθόν *good*.

		Masculine	Feminine	Neuter
Sing.	N.	ἀγαθός	ἀγαθή	ἀγαθόν
	V.	ἀγαθέ	ἀγαθή	ἀγαθόν
	A.	ἀγαθόν	ἀγαθήν	ἀγαθόν
	G.	ἀγαθοῦ	ἀγαθῆς	ἀγαθοῦ
	D.	ἀγαθῷ	ἀγαθῇ	ἀγαθῷ
Plur.	N., V.	ἀγαθοί	ἀγαθαί	ἀγαθά
	A.	ἀγαθούς	ἀγαθάς	ἀγαθά
	G.	ἀγαθῶν	ἀγαθῶν	ἀγαθῶν
	D.	ἀγαθοῖς	ἀγαθαῖς	ἀγαθοῖς

[1] The predicate is the part of a sentence which makes an assertion about the subject. In the sentence "the disciples *bring the colt to Jesus*" the words in italics are the predicate, since they tell us something about the disciples. In "horses are strong" the words *are strong* are the predicate, for they tell us something about the subject, horses. In this sentence the adjective *strong* is said to be used predicatively.

Practise writing out in full on the model of ἀγαθός some of the adjectives in the list which follows.

COMMON ADJECTIVES DECLINED LIKE ἀγαθός

ἀγαπητός *beloved*

ἀληθινός *true*

δυνατός *possible, powerful*

ἔσχατος *last*

καινός *new*

κακός *bad*

καλός *beautiful, good*

μόνος *alone, only*

ὀλίγος *small* (plur. *few*)

πιστός *faithful*

πρῶτος *first*

σοφός *wise*

τυφλός *blind*

φίλος *beloved, dear*

χαλεπός *hard, harsh*

Χριστός *anointed* (as noun, Χριστός *Christ*)

χωλός *lame*

ADJECTIVES WITH α PURE

When the feminine has α pure (i.e. α following ε, ι, or ρ: compare 1st decl. nouns like ἡμέρα, p. 27) the declension is as for ἅγιος *holy*:

		Masculine	Feminine	Neuter
Sing.	N.	ἅγιος	ἁγία	ἅγιον
	V.	ἅγιε	ἁγία	ἅγιον
	A.	ἅγιον	ἁγίαν	ἅγιον
	G.	ἁγίου	ἁγίας	ἁγίου
	D.	ἁγίῳ	ἁγίᾳ	ἁγίῳ
Plur.	N., V.	ἅγιοι	ἅγιαι	ἅγια

Other cases in the plural have the same endings as the plural of ἀγαθός.

COMMON ADJECTIVES DECLINED LIKE ἅγιος

ἄξιος *worthy*

δεύτερος *second*

δίκαιος *righteous*

ἐλεύθερος *free*

ἕτερος *different, other* (properly used of two)

ἰσχυρός *strong*

καθαρός *pure, clean*

μακάριος *blessed*

μικρός *small*

ὅμοιος *like*

παλαιός *old*

πονηρός *evil*

φανερός *manifest*

ADJECTIVES OF TWO TERMINATIONS

There is a smaller class of adjectives which have the -ος termination for both masc. and fem., and the neuter in -ον. Many of this class are compound adjectives, i.e. made up of two or more constituent parts: e.g. ἄπιστος *unbelieving*, is a compound of πιστός *faithful*, with α privative (i.e. with negative force, like the English prefix *un-*); ἀνθρωποκτόνος *manslaying*, is a compound of ἄνθρωπος *man* and κτονος which is connected with the root of κτείνω *I kill*. So ἀδύνατος *impossible*:

	Masc. and Fem.	Neut.
Sing. N.	ἀδύνατος	ἀδύνατον
V.	ἀδύνατε	ἀδύνατον
A.	ἀδύνατον	ἀδύνατον
G.	ἀδυνάτου	ἀδυνάτου
D.	ἀδυνάτῳ	ἀδυνάτῳ
Plur. N., V.	ἀδύνατοι	ἀδύνατα

Other cases in the plural have the same endings as the plural of ἀγαθός in the masc. and neut.

COMMON ADJECTIVES DECLINED LIKE ἀδύνατος

αἰώνιος *eternal*
ἀκάθαρτος *unclean*
ἁμαρτωλός *sinful*
ἀνθρωποκτόνος *manslaying*
ἄπιστος *faithless*

βάρβαρος *barbarous*
ἐπίγειος *earthly*
ἔρημος *desolate, desert*
ἕτοιμος *ready*
οὐράνιος *heavenly*

CONTRACTED ADJECTIVES

There are also a few adjectives of the first and second declensions which show contraction, an original -εος or -οος being contracted to -ους (see p. 10, and compare the contraction which takes place in some verbs; see pp. 109ff.).

Thus χρυσεος contracts to χρυσοῦς *golden*; διπλοος contracts to διπλοῦς *double*. Ἀργυρεος, contracting to ἀργυροῦς *silver*, shows α pure in the feminine, ἀργυρᾶ.

	Masculine	Feminine	Neuter	Masculine	Feminine	Neuter
Sing. N.	διπλοῦς	διπλῆ	διπλοῦν	ἀργυροῦς	ἀργυρᾶ	ἀργυροῦν
V.	διπλοῦς	διπλῆ	διπλοῦν	ἀργυροῦς	ἀργυρᾶ	ἀργυροῦν
A.	διπλοῦν	διπλῆν	διπλοῦν	ἀργυροῦν	ἀργυρᾶν	ἀργυροῦν
G.	διπλοῦ	διπλῆς	διπλοῦ	ἀργυροῦ	ἀργυρᾶς	ἀργυροῦ
D.	διπλῷ	διπλῇ	διπλῷ	ἀργυρῷ	ἀργυρᾷ	ἀργυρῷ
Plur. N., V.	διπλοῖ	διπλαῖ	διπλᾶ	ἀργυροῖ	ἀργυραῖ	ἀργυρᾶ
A.	διπλοῦς	διπλᾶς	διπλᾶ	ἀργυροῦς	ἀργυρᾶς	ἀργυρᾶ
G.	διπλῶν	διπλῶν	διπλῶν	ἀργυρῶν	ἀργυρῶν	ἀργυρῶν
D.	διπλοῖς	διπλαῖς	διπλοῖς	ἀργυροῖς	ἀργυραῖς	ἀργυροῖς

Other adjectives declined similarly are:

Like διπλοῦς, ἁπλοῦς, ῆ, οῦν *single*
τετραπλοῦς, ῆ, οῦν *fourfold*
χαλκοῦς, ῆ, οῦν *brazen*

Like ἀργυροῦς, πορφυροῦς, ᾶ, οῦν *purple*
σιδηροῦς, ᾶ, οῦν *iron*

Note that νέος *young, new*; στερεός *hard*; ὑπήκοος *obedient*; and ὄγδοος *eighth*, do not contract.

THE POSITION OF THE ARTICLE WITH ADJECTIVES

Greek usually follows the same order as English: article, adjective, noun: οἱ ἀγαθοὶ δοῦλοι *the good slaves*. The adjective, however, may follow the noun, but in this case it as well as the noun must have the article: οἱ δοῦλοι οἱ ἀγαθοί. If the adjective had no article, οἱ δοῦλοι ἀγαθοί, the meaning would be *the slaves are good*, ἀγαθοί being used predicatively (see p. 40) and εἰσίν (*are*) being understood. See also p. 36.

Where the indefinite article, *a* or *an*, would be used in English, no article is used in Greek with either noun or adjective. Thus οἰκία μικρά *a small house*.

ADJECTIVES AS NOUNS

As in English so in Greek an adjective with the article may serve as a noun. When necessary the words *man, men, woman, women, thing* or *things* may be supplied in translation. Thus ὁ ἀνθρωποκτόνος *the murderer* ("the manslaying man"); οἱ ἅγιοι *the saints* ("the holy men"); ἡ πιστή *the faithful woman*; αἱ δίκαιαι *the righteous women*; τὸ ὅλον *the whole*; τὰ ἔσχατα *the last things*.

4—N.T.G.

THE PREDICATIVE USE OF ADJECTIVES

The verb "to be" is a *copulative* verb[1]: that is to say, it joins together the two parts of a sentence in which there is expressed an identity between subject and predicate, or in which the predicate expresses some description of the subject. "God is a Spirit" expresses identity between *God* and *Spirit*. "I am the bread of life" expresses identity between *I* and *the bread of life*. "He seems unwell" expresses a description of the subject *he*. The verbs *is*, *am* and *seems* in these sentences are copulative verbs, linking (Latin: *copula*, "a link") the subject with the definition or description.

A *copulative* verb, therefore, has a very different function from a *transitive* verb[2] which expresses an action which takes place upon some object. In "they filled the waterpots" *filled* does not express any kind of identity between *they* and *the waterpots*. On the contrary it expresses something which *they* (the subject) *do* to the waterpots. Similarly in "the ruler of the feast calleth the bridegroom" the *ruler of the feast* is not the same person as *the bridegroom*. The transitive verb *calleth* expresses something which the ruler of the feast *does* in relation to the bridegroom. In these two sentences, in Greek *the waterpots* and *the bridegroom* would be put in the accusative case, clearly signifying the non-identity with the subjects of the sentences, which of course would be in the nominative case. But in such a sentence as "God is a Spirit", *Spirit* will be put in the nominative case, as is *God*, signifying the identity with the subject. In the same way when the predicate of a sentence contains an adjective describing the subject, that adjective will agree with the subject in case, number and gender. "The grass is green": it is the grass which is green, and the adjective *green* must therefore agree with *grass*.

All this may be summed up in two simple but very important rules:

1. The verb "to be" and other copulative verbs[3] take the same case after them as before.

---

[1] It is also in Greek as in English used absolutely in the sense of "to exist": e.g. *God is*.

[2] A transitive verb is one which denotes an action which "passes over" (Latin: *transeo*, supine *transitum*) from the subject to an object. The verb "to love" is transitive because it requires a direct object, either expressed or implied, to complete its meaning. On the other hand a verb like "to go" is intransitive, since it cannot have a direct object.

[3] Other common English copulative verbs are "to seem", and the passives of verbs of thinking and naming. Thus: "he *seems* unwell"; "he *is thought* intelligent"; "he *was called* king". In translating these sentences into Greek, *unwell*, *intelligent* and *king* would have to be put in the nominative case.

2. An adjective used predicatively, like an adjective used as an epithet, must agree with its noun in case, gender and number.

σὺ οὐκ εἶ ὁ Χριστός *thou art not the Christ.*

σκληρός ἐστιν ὁ λόγος *the saying is hard.* Notice that the subject need not be put first: so in English we might say "hard is the saying".

ἡ διδαχὴ οὐκ ἔστιν ἐμή *the teaching is not mine* (ἐμή, fem. of possessive adj. ἐμός, ή, όν).

μακάριοί εἰσιν οἱ καθαροί *blessed are the pure.* Notice that where two adjectives are joined by a copulative verb that which has the article is the subject.

ADVERBS

An adverb is a word which adds a qualification to a verb (hence its name), an adjective or another adverb. Examples: "He *quickly* returned to his place." (*Quickly* here qualifies the verb *returned*.) "It is a *very* beautiful picture." (*Very* is an adverb qualifying the adjective *beautiful*.) "He walked terribly slowly." (In this colloquial sentence the adverb *terribly* qualifies the second adverb *slowly*.)

The Greek adverb is not inflected: that is to say, its spelling remains the same however it is used. Its use to qualify verbs is much more frequent than its use with adjectives or adverbs. The position of the adverb in the sentence is often close to the verb: καὶ εὐθὺς ἐκάλεσεν αὐτούς *and immediately he called them* (Mark 1.20); but not necessarily so: καὶ ἤρχοντο πρὸς αὐτὸν πάντοθεν *and they came to him from every quarter* (Mark 1.45).

The following is a list of some of the adverbs which appear in the early verses of St Mark:

εὐθύς *immediately, straightway*   ἔξω *outside*
ἐκεῖ *there*   πάντοθεν *from all sides*
πρωΐ *early, in the morning*   πάλιν *again*

CONJUNCTIONS

A conjunction is a word whose function is to join (Latin *conjungo*, "I join"). It may join together two words of like nature, e.g. two nouns as in "the boy *and* the girl"; two adjectives as in "plain *or* coloured"; two adverbs as in "quickly *but* thoroughly". Or it may join together two clauses or two sentences: e.g. "Some fell by the way side, *and* the birds came *and* devoured it."

We have already had occasion to use the frequent conjunction καί *and*. Another common conjunction is ἀλλά *but*: τό παιδίον οὐκ ἀπέθανεν ἀλλά καθεύδει *the child is not dead but sleepeth* (Mark 5.39).

For other conjunctions, see pp. 55ff.

Vocabulary

κόσμος (2) *world*
μισθός (2) *hire, wages, reward*
Χριστός, ὁ (2) *Christ*
ὅλος, η, ον *whole*
ὅμοιος, α, ον *like* (with dat., e.g.
  ὅμοιος ἀγγέλῳ like an angel)

σκληρός, ά, όν *hard*
ἐγώ (pronoun) *I* (used only where
  there is emphasis)
σύ (pronoun) *you* (sing.: used only
  where there is emphasis)
νῦν (adverb) *now*

EXERCISE 13

Translate into English: ·

1. ὁ Ἰησοῦς ἦν ἐν ἐρήμῳ τόπῳ σὺν τοῖς θηρίοις. 2. ἡ διδαχὴ τοῦ προφήτου καινή ἐστιν. 3. ἐγώ εἰμι ἡ ἄμπελος ἡ ἀληθινή. 4. ὦ ἀγαπητοί, νῦν τέκνα θεοῦ ἐσμεν καὶ ὅμοιοι τῷ θεῷ ἐσόμεθα. 5. κύριός ἐστιν ὁ Υἱὸς τοῦ ἀνθρώπου τοῦ σαββάτου. 6. ἅγιοί ἐστε ἐν πονηρῷ κόσμῳ.

EXERCISE 14

Translate into Greek:

1. The way will be hard. 2. The soldiers were lame. 3. Sheep are not unclean. 4. Ye are murderers. 5. Few boats are in the sea. 6. The holy book is in the earthly temple. 7. He was a slave, but now he is free. 8. The righteous man does not strike a slave. 9. I shall be first, you will be second, and the young girl will be last. 10. The words of the prophet will be manifest to the whole world.

EXERCISE 15

Translate into Greek:

1. The blessed apostle was eating small loaves in the house. 2. We shall immediately behold the eternal kingdom of the eternal God. 3. You speak hard words, but the whole world hears. 4. The blind now see and the lame are strong. 5. God will again send an apostle like unto John. 6. The brother was striking the wicked thief's head with a stone. 7. The apostles are few, but the reward is eternal. 8. True was the word of the prophet. 9. We tell the wise disciple the truth. 10. Christ looses the sinful man from (ἐκ) sin.

Chapter 7

AORIST INDICATIVE ACTIVE TENSES

The word *aorist* is from the Greek ἀόριστος *indefinite*, and signifies that this tense presents us with the idea of the verb, regarded usually in past time, but not defined in any other way: it does not tell us whether the action was momentary or prolonged. Usually the aorist indicative active can be translated by the English past simple tense. Other uses of the aorist indicative are noted on pp. 251ff.

There are two forms of the Greek aorist, the weak or first aorist, and the strong or second aorist. Very few verbs have both. There is no distinction of meaning between weak and strong aorists except in those verbs which have both (see p. 222). Thus ἔλυσα (a weak aorist) means "I loosed", and ἔβαλον (a strong aorist), means "I threw". With the negative the aorist indicative is often best translated with the English "did": οὐκ ἔλυσαν *they did not loose*.

THE WEAK (1st) AORIST INDICATIVE ACTIVE OF λύω

ἔ-λυ-σ-α *I loosed*	ἐ-λύ-σ-αμεν
ἔ-λυ-σ-ας	ἐ-λύ-σ-ατε
ἔ-λυ-σ-ε(ν)	ἔ-λυ-σ-αν

NOTICE: 1. The augment is prefixed, indicating past time.

2. The σ following the stem λυ- is characteristic of the weak aorist, as it is also of the future (see p. 26).

When the stem ends in a consonant this σ causes changes similar to those in the future (p. 26).

In some verbs, e.g. those with stems ending in nasal and liquid consonants, the σ disappears, but its influence is felt, and is shown in a compensating lengthening of the stem. The aorists of such verbs will be dealt with fully on p. 153 (liquids) and p. 184 (nasals).

Examples

διώκω *pursue* has aor. indic. act.	ἐδίωξα	for	ἐ-δίωκ-σ-α
τρίβω *rub*	,,	,,	,,	,,	ἔτριψα	,,	ἔ-τριβ-σ-α
πείθω *persuade*	,,	,,	,,	,,	ἔπεισα	,,	ἔ-πειθ-σ-α
μένω *remain*	,,	,,	,,	,,	ἔμεινα	,,	ἔ-μεν-σ-α

The Strong Aorist Indicative Active of βάλλω

ἔ-βαλ-ον *I threw, put*	ἐ-βάλ-ομεν
ἔ-βαλ-ες	ἐ-βάλ-ετε
ἔ-βαλ-ε(ν)	ἔ-βαλ-ον

This aorist is formed from the verb stem (see below), prefixed by the augment, and with the same endings in the indic. as are used in the imperf. indic.

The Verb Stem. The verb stem is that part of the verb to which additions are made as the verb inflects. Thus the verb stem λυ- (*loose*) may receive additions by way of prefix (e.g. the augment ἐ-) or suffix (e.g. the personal endings). The stem used in the present tense of some verbs, however, is a lengthened form of the verb stem. (From the *present stem* are formed only the present tense, all voices and moods, and the imperfect, all voices, but with no mood but the indicative. All other parts of the verb are formed from the *verb stem.*) For example the verb stem of βάλλω is not βαλλ-, but βαλ-. In εὑρίσκω *find*, the verb stem εὑρ- is lengthened by the addition of -ισκ-. It is convenient, therefore, in the case of these verbs to speak of a present stem (which is also used for the imperfect) and a verb stem. (It is not always easy to distinguish the present stem from the verb stem: indeed there may be, especially with verbs whose stems end in a vowel, no difference between them; but, except for the present and imperfect, *all* forms of a verb are based on the verb stem however it may be modified. The simplest way of finding it is from the 2nd aorist (if the verb has one), by taking off the augment and termination. E.g. from λαμβάνω, ἐ-λάβ-ον. See also what is said on this subject above, p. 16.) Other ways in which a verb stem may be lengthened into the present stem are noted on pp. 190ff.

Practise writing out the aor. indic act. of the following verbs (all strong aor.):

ἁμαρτάνω *sin.* Verb stem ἁμαρτ- (when augmented, ἡμαρτ-)

εὑρίσκω *find.* Verb stem εὑρ- (when augmented, both ηὑρ- and εὑρ- are used)

-θνήσκω *die.* Verb stem -θαν-. Only found as a compound, usually ἀποθνήσκω *I die* or *I am dying.* The (2nd) aorist, ἀπέθανον, means *I died,* and consequently also "*I am dead*"

λαμβάνω *take.* Verb stem λαβ-

λείπω *leave.* Verb stem λιπ-

μανθάνω *learn.* Verb stem μαθ-

πίνω *drink.*   Verb stem πι-
-τέμνω *cut.*   Verb stem -τεμ-
τίκτω *bring forth, bear.*   Verb stem τεκ-
φεύγω *flee.*   Verb stem φυγ-

(A hyphen in front of a verb indicates that it is only found in the New Testament compounded with a preposition. Thus θνῄσκω and τέμνω are not found, but ἀποθνῄσκω (*die*) and περιτέμνω (*circumcise*) appear.)

PRONOUNS

A pronoun is a word which stands in place of a noun, as the English *he, it, they.* They are useful words, for they enable us to avoid continual repetition of proper and common nouns. "Mr Smith took the flag in his strong hands, and marched down the street in which he lived, gaily waving it in the faces of all who met him" sounds better than "Mr Smith took the flag in Mr Smith's strong hands, and marched down the street in which Mr Smith lived, gaily waving the flag in the faces of all who met Mr Smith". Every language must have its pronouns.

PERSONAL PRONOUNS, i.e. pronouns which stand in place of persons.

1st person

The pronoun for the 1st pers., *I*, is ἐγώ, plur. ἡμεῖς *we.* There is no distinction of gender.

	Sing.	*Plur.*
N.	ἐγώ	ἡμεῖς
A.	ἐμέ or με	ἡμᾶς
G.	ἐμοῦ or μου	ἡμῶν
D.	ἐμοί or μοι	ἡμῖν

NOTICE: 1. The nominative is only used for emphasis. λέγω means "I say", the personal ending -ω sufficiently indicating the subject. If ἐγὼ λέγω is written there is special emphasis, as in Matt. 5.22, "*I* say unto you" (where "I" is contrasted with "them of old time").

2. There are no vocatives.

3. The forms με, μου, μοι are enclitics (see p. 273 and cf. the enclitic parts of εἰμί, p. 35). They are shown without accent since they "lean on" (Gk. ἐγκλίνω) a preceding word by throwing their accent on it. The

last syllable of the preceding word shows the accent unless the word already possesses some accent which will not allow it to be shown (see pp. 273f.). They are used when there is no particular emphasis on the pronoun, e.g. δείξατέ μοι δηνάριον *show me a penny*.[1] The enclitic is used, since there is little emphasis on the pronoun. On the other hand in ὁ ἑωρακὼς ἐμὲ ἑώρακεν τὸν Πατέρα *he that hath seen me hath seen the Father*, the accented pronoun is used since there is emphasis upon it.

2nd Person

The pronoun for the 2nd pers., *thou*, is σύ, plur. ὑμεῖς *you*. There is no distinction of gender.

	Sing.	*Plur.*
N.	σύ	ὑμεῖς
V.	σύ	ὑμεῖς
A.	σέ or σε	ὑμᾶς
G.	σοῦ or σου	ὑμῶν
D.	σοί or σοι	ὑμῖν

NOTICE: 1. The nominatives σύ and ὑμεῖς are only used where there is some emphasis.

2. The vocative is the same as the nominative.

3. The enclitic forms, σε, σου, σοι, like those of the 1st pers., are less emphatic than the accented forms. But whether emphatic or not, the accented forms are used after prepositions, ἀπὸ σοῦ *from thee*; ἐν σοί *within thee*. But with πρός *to, towards*, both forms are found.

3rd Person

Greek, like Latin but unlike English, has no 3rd pers. pronoun meaning *he, she, it*, or *they*. As Latin had to borrow a demonstrative adjective (Latin *demonstro*, point out) and use it as a pronoun (e.g. *is, ea, id; hic, haec, hoc; ille, illa, illud*) so Greek uses adjectives as pronouns of the 3rd person as follows:

αὐτός, αὐτή, αὐτό (the equivalent of the Latin *is, ea, id*), *he, she, it*, and in the plur., *they*.

---

[1] In some of the illustrations forms are used which will be explained later, e.g. the 1st aor. imperative δείξατε here.

οὗτος, αὕτη, τοῦτο (Latin *hic, haec, hoc*), *this man* (*woman, thing*), *he, she, it*, and in the plur., *these*.

ἐκεῖνος, ἐκείνη, ἐκεῖνο (Latin *ille, illa, illud*), *that man* (*woman, thing*), *he, she, it*, and in the plur., *those*.

	Masculine	*Feminine*	*Neuter*
Sing. N.	αὐτός	αὐτή	αὐτό
A.	αὐτόν	αὐτήν	αὐτό
G.	αὐτοῦ	αὐτῆς	αὐτοῦ
D.	αὐτῷ	αὐτῇ	αὐτῷ
Plur. N.	αὐτοί	αὐταί	αὐτά
A.	αὐτούς	αὐτάς	αὐτά
G.	αὐτῶν	αὐτῶν	αὐτῶν
D.	αὐτοῖς	αὐταῖς	αὐτοῖς

NOTICE: 1. There are no vocatives.

2. The nom. and acc. neut. sing. have no final ν: αὐτό, not αὐτόν.

3. Again the nominatives are only used where there is emphasis.

4. The genitive, representing the English *his, her, its,* or *their*, may precede or follow the noun of which it denotes possession, and this noun usually has the article: οἱ μαθηταὶ αὐτοῦ *his disciples*; αὐτοῦ ἡ μαρτυρία *his testimony*. So also with the genitives of the 1st and 2nd pers. pronouns: ὁ ἄρτος ἡμῶν *our bread*; ὑμῶν ἡ ὥρα *your hour*.

5. Αὐτός, ή, ό takes the gender of the noun it represents. Αὐτός can represent any masculine noun whether or not it is a male proper name or the name of a male animal. Αὐτός, αὐτόν, etc., are not, therefore, automatically to be translated *he, him,* etc. Nor are αὐτή, αὐτήν, etc., always to be rendered by *she, her,* etc. They must sometimes be translated by *it*. Thus ὁ κλάδος αὐτῆς (Mark 13.28) *its branch*, since αὐτῆς represents the feminine noun συκῆ *fig-tree*.

Other Meanings of αὐτός

1. As an adjective αὐτός means *self* (i.e. *himself, herself, itself, themselves*, as the case may be):

ὑμεῖς αὐτοί or αὐτοὶ ὑμεῖς *you yourselves*
᾿Ιησοῦς αὐτός *Jesus himself*
αὐτὸς ὁ κόσμος *the world itself*

2. When αὐτός follows the article it frequently means *same*: ἡ αὐτὴ ἐπαγγελία, or ἡ ἐπαγγελία ἡ αὐτή *the same promise*. This use, however, is not frequent in the New Testament. Αὐτὴ ἡ ἐπαγγελία or ἡ ἐπαγγελία αὐτή would mean "the promise itself".

But τὸ αὐτό *the same*, or *the same thing*, is common.

Examples of the use of 1st, 2nd, *and* 3rd *pers. pronouns*
  ἐν αὐτῷ *in him*
  ἔλαβον αὐτόν *they received him*; ὁ κόσμος ὑμᾶς μισεῖ *the world hateth you*
  ἐγὼ βαπτίζω *I baptize* (see John 1.26: ἐγώ emphasizes the contrast in the sentence)
  λέγει αὐτοῖς *he saith unto them*

Οὗτος AND ἐκεῖνος AS PRONOUNS

When something more definite than αὐτός is needed for the 3rd pers. pronoun, Greek uses the demonstrative adjectives οὗτος or (though rarely in the N.T.) ὅδε *this* (Latin *hic*); or ἐκεῖνος *that* (Latin *ille*).

	Masculine	*Feminine*	*Neuter*
Sing. N.	οὗτος	αὕτη	τοῦτο
A.	τοῦτον	ταύτην	τοῦτο
G.	τούτου	ταύτης	τούτου
D.	τούτῳ	ταύτῃ	τούτῳ
Plur. N.	οὗτοι	αὗται	ταῦτα
A.	τούτους	ταύτας	ταῦτα
G.	τούτων	τούτων	τούτων
D.	τούτοις	ταύταις	τούτοις

	Masculine	*Feminine*	*Neuter*
Sing. N.	ἐκεῖνος	ἐκείνη	ἐκεῖνο
A.	ἐκεῖνον	ἐκείνην	ἐκεῖνο
G.	ἐκείνου	ἐκείνης	ἐκείνου
D.	ἐκείνῳ	ἐκείνῃ	ἐκείνῳ
Plur. N.	ἐκεῖνοι	ἐκεῖναι	ἐκεῖνα
A.	ἐκείνους	ἐκείνας	ἐκεῖνα
G.	ἐκείνων	ἐκείνων	ἐκείνων
D.	ἐκείνοις	ἐκείναις	ἐκείνοις

NOTICE: 1. The gen. plur. of οὗτος is τούτων in all genders. The fem. is not ταύτων.

2. The nom. and acc. neut. sing. have no final ν: τοῦτο and ἐκεῖνο, not τοῦτον, ἐκεῖνον.

3. The oblique cases of οὗτος (i.e. cases other than the nom.) have an initial τ, as do the nom. neut. sing. and plur.

4. Οὗτος and ἐκεῖνος (like αὐτός, see above) take the gender of the noun they represent. Οὗτος is not, therefore, automatically to be translated *this man*, or *he*. It may have to be rendered *this thing*, or *it*.

Both οὗτος and ἐκεῖνος are originally demonstrative *adjectives*, as is ὅδε (see below), and can be so used (cf. the remarks on αὐτός as an adjective, p. 46). When they are thus used to qualify a noun, the noun always has the article. The adjective may either precede or follow the noun. Thus οὗτοι οἱ ἄνθρωποι or οἱ ἄνθρωποι οὗτοι *these men*; ἐν τῇ ἡμέρᾳ ἐκείνῃ or ἐν ἐκείνῃ τῇ ἡμέρᾳ *in that day*.

Notice carefully the difference between this usage and that with ordinary adjectives. Briefly stated it is that with the ordinary adjective the order is article, adjective, noun (the adjective may follow the noun, but if so the article must be repeated: article, noun, article, adjective). But οὗτος and ἐκεῖνος are written *outside* the article and the noun.

The following are examples of οὗτος and ἐκεῖνος used as pure pronouns:

οὗτός ἐστιν ὁ Υἱός μου ὁ ἀγαπητός (Matt. 3.17) *this is my beloved Son*.

μετὰ τοῦτο λέγει τοῖς μαθηταῖς (John 11.7) *after this he saith to the disciples*.

σὺ μαθητὴς εἶ ἐκείνου (John 9.28) *thou art his disciple*, or *thou art that man's disciple*.

ἐκεῖνος κλέπτης ἐστίν (John 10.1) *he (or that man) is a thief*. But ἐκεῖνος ὁ κλέπτης would mean "that thief".

Ὅδε, *this*

The common classical Greek demonstrative adjective ὅδε, ἥδε, τόδε *this* (which is the definite article with enclitic δε added) is not common in the N.T. It is found in Luke 10.39: καὶ τῇδε ἦν ἀδελφή *and she had a sister* (literally, "to this woman was a sister"). The neut. plur. τάδε is, however, frequent in the Revelation, meaning *these things*; e.g. τάδε λέγει ὁ Υἱὸς τοῦ Θεοῦ (Rev. 2.18) *these things saith the Son of God.*

For the full declension, see p. 289.

PARSING

Translation into English of Greek of which the student, through his acquaintance with the New Testament, finds it comparatively easy to guess the meaning is no proof of proficiency. As we have said in the Introduction, ability to translate English into Greek is a surer sign of progress. So also is the ability to parse. To parse is to state precisely what part of speech a word is. If the word is a verb, its person and number, and the tense, mood and voice must be indicated, together with the 1st pers. sing. of the present indicative active.

If the word is a noun, the case and number must be given, together with the nominative singular (and, for good measure, the genitive singular, showing that the student knows his declensions).

If the word is an adjective or pronoun, the gender, case and number should be given, together with the nominative singular forms.

The following are examples of parsing of some of the words in Exercise 16 which follows:

εὗρεν: 3rd pers. sing. of the 2nd (or strong) aor. ind. act. of εὑρίσκω *I find.*

οἴκῳ: dat. sing. of οἶκος, -ου, (ὁ) *house.*

ἐκείνης, gen. sing. fem. of the demonstrative adjective or pronoun ἐκεῖνος, -η, -ο.

The student is advised to practise by parsing a few of the words in each of the Greek exercises in the book. When he comes to the translation of passages from the New Testament he should never leave a word until he is sure that he can parse it.

Vocabulary

μαρτυρία, ἡ *testimony*	μετά (preposition with acc.) *after*
οἶνος, ὁ *wine*	πρός (preposition with acc.) *to, to-*
Πέτρος, ὁ *Peter*	*wards*
ποταμός, ὁ *river*	εἰ (conjunction) *if*
καί *and* (may be translated *also*)	

εἶπον *I said, told* (strong aor. ind. No present of this stem is found, and
λέγω takes its place.)

ἦλθον *I came* (str. aor. indic. No present of this stem is found, and
ἔρχομαι (deponent, see p. 85) takes its place.)

ἦν (imperf. indic. of εἰμί *I am*) *he, she,* or *it was.*

ἀκούω *I hear, listen to* (takes the acc. of the thing heard and gen. of the
person).

λέγω *I say, tell* (takes the acc. of the thing said, and dat. of the person to
whom said: ταῦτά σοι λέγω *these things I say to you, I tell you these
things*).

λείπω *I leave, leave behind* (used of leaving things, not places).

πείθω *I persuade* (weak aor.: ἔπεισα; takes acc.).

πιστεύω *I believe* (takes the dat. of the object of belief though the acc.
is also found. But εἰς with acc. and ἐν with dat. are found, especially
when personal trust is implied. In this case "believe in" is the best
translation: πιστεύω αὐτῷ *I believe him*; πιστεύω εἰς αὐτόν or ἐν
αὐτῷ *I believe in him*.)

EXERCISE 16

Translate into English:

1. ὁ Πέτρος εἶπεν αὐτῷ, σὺ εἶ ὁ Χριστός. 2. ὁ δοῦλός μου εὗρεν
αὐτοὺς ἐν τῷ οἴκῳ ἐκείνῳ. 3. ταῦτα εἶπεν, καὶ μετὰ τοῦτο λέγει
αὐτοῖς, ὁ φίλος ἡμῶν ἀπέθανεν. 4. ἀπ᾽ ἐκείνης τῆς ἡμέρας οὐκ ἦλθον
οἱ μαθηταὶ αὐτοῦ εἰς τοῦτον τὸν τόπον. 5. εἰ ἐμὲ ἐδίωξαν, καὶ ὑμᾶς
διώξουσιν. 6. καὶ πάλιν τοὺς αὐτοὺς λόγους εἶπεν.

EXERCISE 17

Translate into Greek:

1. This man came for (use εἰς with acc.) witness. 2. We received him.
3. You heard his voice. 4. The disciples heard him. 5. He brought
them to Jesus. 6. His disciples believed on him. 7. He found sheep in
the temple. 8. Ye receive not our witness. 9. It was the sabbath on that
day. 10. We pursued the thieves into the desert.

EXERCISE 18

Translate into Greek:

1. In those days John was preaching in the desert. 2. From that place
you led us to the sea. 3. They persuaded us and we remained in the same

place. 4. This woman brought forth a son. 5. You yourselves learned the Scriptures in this house. 6. We cut our loaves and you drank your wine. 7. You left your net, but I found it. 8. The apostles told you the same things. 9. I am throwing the same stones into the river. 10. The Lord himself will loose us from those sins.

Chapter 8

THE PERFECT INDICATIVE ACTIVE

The Perfect Tense denotes an action which is perfected or completed in the past, but the effects of which are regarded as continuing into the present. "I have seen" is an English perfect. When we say "I have seen this play" we are implying that, whilst it was in the past that the act of seeing took place, we still have it in mind. But if we use the English past simple tense "I saw", we are referring solely to the act of seeing in the past without any reference to the present, and we often add some words which define the time as past, as "I saw this play last year" or "I saw this play when I was on holiday". Here the Greek aorist would rightly be used. See also pp. 251ff.

There are two forms of the Greek perfect: the First, or Weak, Perfect; and the Second, or Strong, Perfect. As in the case of the aorist few verbs have both.

PERFECT INDICATIVE ACTIVE OF λύω (WEAK)

λέ-λυ-κα *I have loosed*	λε-λύ-καμεν
λέ-λυ-κας	λε-λύ-κατε
λέ-λυ-κε(ν)	λε-λύ-κασι(ν)

NOTICE: 1. Reduplication, which is the mark of a completed action, occurs in the perfect and pluperfect (see p. 61) tenses. It is usually a doubling of the initial consonant of the stem with the help of the vowel ε, but some very common verbs form the reduplication differently (see p. 54).

2. The personal endings of the 1st perf. (which belongs especially to verbs with vowel stems), are added to the verb stem after the letter κ.

3. The personal endings (for both perfects) are the same as those for the weak aor. indic. act. with the exception of the 3rd pers. plur.

PERFECT INDICATIVE ACTIVE OF λαμβάνω (STRONG)

εἴληφα *I have taken*	εἰλήφαμεν
εἴληφας	εἰλήφατε
εἴληφε(ν)	εἰλήφασι(ν)

NOTICE: 1. The personal endings are the same as for the wk. perf., but the κ does not appear. The prefixed εἰ- instead of reduplication in this perfect is an irregularity.

2. In εἴληφα (a) a considerable modification of the verb stem (λαβ-) occurs. This is true of many str. perfs. (b) The final consonant of the stem is aspirated before the personal ending, β becoming φ. This also is true of many str. perfs. Study the list below for both these points.

Pres. Indic.	English	Aor. Act.	Stem	Perf.
γράφω	write	ἔγραψα	γραφ-	γέγραφα
κράζω	cry	ἔκραξα	κραγ-	κέκραγα
πάσχω	suffer	ἔπαθον	παθ-	πέπονθα
πείθω	persuade	ἔπεισα	πιθ-	πέποιθα[1]
τάσσω	arrange, draw up	ἔταξα	ταγ-	τέταχα

REDUPLICATION

Reduplication may be formed in other ways than that mentioned on p. 53:

(a) If the initial consonant of a stem is an aspirate, the corresponding unaspirated consonant is used in the reduplication. Thus θύω sacrifice, has perf. τέ-θυ-κα; φυτεύω plant, has perf. πε-φύτευ-κα.

(b) Verbs whose stems begin with a vowel lengthen the vowel instead of reduplicating. Thus ὁμολογέω agree, confess, has perf. ὡμολόγηκα.

(c) Verbs whose stems begin with two consonants, the second being a mute (i.e. a guttural, κ, γ, χ, dental, τ, δ, θ, or labial, π, β, φ) prefix ἐ instead of reduplicating. The same is true of stems which begin with a double consonant, ζ, ξ, ψ, or with ρ. ρ is usually doubled. The following examples show (as did the examples of the str. perf. given above) how frequently the verb stem is modified in the perfect. With a little attention to the lists of verbs given on pp. 318ff. and by reading in the Greek Testament later on, the student will gradually master these apparent irregularities.

Thus: στέλλω send, has perf. ἔσταλκα.
φθάνω anticipate, come suddenly, has perf. ἔφθακα.
ζητέω seek, has perf. ἐζήτηκα.

---

1 This perfect has the present sense, I trust. It takes dat. or ἐν with dat.

(d) Verbs whose stems begin with two consonants, the second being a liquid or a nasal (λ, ρ, μ, ν), repeat the first consonant in reduplication.

Thus, -θνήσκω *die*, has perf. τέθνηκα. The unaspirated consonant τ is used here in the reduplicating prefix instead of θ, since Greek does not tolerate a repetition of an aspirate. See under (a) above. See also p. 9. This perfect has the meaning "I have died", and so "I am dead". It well illustrates the exact meaning of the perfect tense with a reference to the past (the death being in the past) and to the present effect (the state of being dead).

> Thus: γράφω *write*, has perf. γέγραφα.
> κλίνω *incline*, has perf. κέκλικα.

(e) The so-called Attic reduplication. This is the prefixing of the first *syllable* of the stem, and the lengthening of the first vowel of the stem itself. Thus the perfect of ἐλαύνω (stem ἐλα-) *I drive*, is ἐλ-ήλα-κα. Another instance of it is ἀκήκοα, the perfect of ἀκούω.

MORE CONJUNCTIONS

A conjunction, as we have seen (p. 41), is a part of speech whose function is to join together words, clauses or sentences. It may also be used to point to a connection between *separate* sentences; e.g. "They worked for hours. Therefore they were ready for a rest." Here *therefore* gives the connection with the previous sentence.

In Greek, as in English, conjunctions are many and varied. Some of them involve special constructions in clauses which they introduce. These we shall leave until later, and notice here only a few of the conjunctions which simply join words or phrases. Of these καί and ἀλλά we have already had occasion to use.

καί *and*:

> ἡ ἀλήθεια καὶ ἡ δικαιοσύνη *truth and righteousness.*
> βλέπει τὸν ᾿Ιησοῦν καὶ λέγει . . . *he sees Jesus and says . . . .*

In a few instances, but of common occurrence, where καί precedes a word beginning with a vowel, a process called *crasis* (mixing) takes place. The ι of καί is dropped, and the vowels thus brought together contract, and are marked with '. In the following N.T. examples the dropping of

the ι brings the vowels α and ε together, and the regular contraction into
α takes place (see p. 109).

καὶ ἐγώ	and I,	becomes	κἀγώ
καὶ ἐμοί	and to me,	,,	κἀμοί
καὶ ἐκεῖ	and there,	,,	κἀκεῖ
καὶ ἐκεῖθεν	and thence,	,,	κἀκεῖθεν
καὶ ἐκεῖνος	and he,	,,	κἀκεῖνος
καὶ ἐάν or ἄν	and if,	,,	κἄν

τε ... καί *both ... and*: sometimes καί is preceded by the enclitic τε.
This denotes a somewhat closer connection than καί by itself: "both
... and" or "as well ... as" are possible translations.

ἐσθίει τε καὶ πίνει *he both eats and drinks*.
ποιεῖν τε καὶ διδάσκειν *as well to do as to teach*.

The τε may be separated from the καί, as in Acts 21.30: ἐκινήθη τε ἡ
πόλις ὅλη καὶ ἐγένετο συνδρομὴ τοῦ λαοῦ *both the whole city was moved
and the people ran together* (literally: "there was a running together of the
people").

ἔβαλόν τε οἱ λῃσταὶ λίθους καὶ ἐβόησαν *the robbers both threw stones
and shouted*.
ὅ τε ἁγιάζων καὶ οἱ ἁγιαζόμενοι *both he that sanctifieth and they that
are sanctified* (Heb. 2.11).

Notice how wherever possible τε, being enclitic, throws an accent back
on the last syllable of the preceding word (see p. 45).

ἀλλά *but*:

τὸ παιδίον οὐκ ἀπέθανεν ἀλλὰ καθεύδει, *the child is not dead but
sleepeth* (Mark 5.39).

A common phrase is οὐ μόνον ... ἀλλὰ καί *not only ... but also*.

οὐ μόνον ἔλυεν τὸ σάββατον, ἀλλὰ καὶ ἔλεγεν ... *he not only brake
the sabbath but (also) said* ... (John 5.18).

δέ also means "but". δέ is weaker than ἀλλά, often having no stronger
meaning than *and*, being used simply to provide a link with what has
gone before. δέ cannot stand first in a clause or sentence, but is
usually the second word.

οὐκ ἐπιστεύσατε αὐτῷ· οἱ δὲ τελῶναι καὶ πόρναι ἐπίστευσαν αὐτῷ *ye believed him not: but the publicans and harlots believed him* (Matt. 21.32).

ὁ δὲ Πέτρος ἠκολούθει αὐτῷ *but Peter followed him* (Matt. 26.58).

Notice how often δέ is used in a typical piece of N.T. narrative, e.g. Mark 15.1–15. In many places it can be translated equally well as *and* or *but.*

μέν . . . δέ: the particles[1] μέν and δέ may be used to call attention to two parallel or contrasting words or ideas which in English the tone of voice is often sufficient to indicate. "On the one hand . . . on the other hand" is a clear translation, but often this would be cumbrous, and no more is necessary than to translate the δέ by *but,* and to ignore the μέν in translation: e.g.

ὁ μὲν θερισμὸς πολύς, οἱ δὲ ἐργάται ὀλίγοι *the harvest* (is) *plenteous, but the labourers* (are) *few* (Matt. 9.37).

In English a slight emphasis on the words *harvest* and *labourers* brings out the contrast. But a word like *indeed* or *truly,* as in the A.V. and R.V., may be used: "the harvest indeed . . .".

τὸ μὲν πνεῦμα πρόθυμον, ἡ δὲ σὰρξ ἀσθενής *the spirit indeed* (is) *willing, but the flesh* (is) *weak* (Matt. 26.41).

οἱ μέν followed by ἄλλοι δέ or οἱ δέ, means "some . . . others":

οἱ μὲν ἔλεγον ὅτι Ἀγαθός ἐστι· ἄλλοι δὲ ἔλεγον Οὔ *some said* (that) *He is* (a) *good* (man); *others said, Not so* (John 7.12).
οἱ μὲν ἦσαν σὺν τοῖς Ἰουδαίοις, οἱ δὲ σὺν τοῖς ἀποστόλοις *part* (or *some*) *held* (were) *with the Jews, and part with the apostles* (Acts 14.4).

ὁ δέ *and he, but he.* δέ with the article having no noun agreeing with it is translated *and he* or *but he:* οἱ δέ, *and they, but they.*

ὁ δὲ λέγει αὐταῖς *and he saith to them* (Mark 16.6).

γάρ *for:* γάρ usually stands after the first word in a clause which gives a reason for something previously stated.

[1] Particle in grammar is a general term for any small word which is indeclinable. Thus the negatives οὐ and μή, prepositions, and conjunctions are particles. More specifically those words are called particles which are used to give some emphasis to a particular word or expression, like μέν and δέ, γέ (enclitic), *at least,* and ἄν (see p. 230).

αὐτὸς γὰρ σώσει τὸν λαὸν αὐτοῦ *for he shall save his people* (Matt. 1.21).

Δαυεὶδ γὰρ λέγει *for David says* . . . (Acts 2.25).

οὖν, *therefore, then*: οὖν expresses consequence or sequence. It does not stand first in a clause.

λέγει οὖν αὐτοῖς ὁ 'Ιησοῦς *Jesus therefore saith unto them* . . . (John 7.6).
ἐθαύμαζον οὖν οἱ 'Ιουδαῖοι *the Jews therefore marvelled* (John 7.15).
εἶπεν οὖν πάλιν ὁ 'Ιησοῦς *Jesus then said again* . . . (John 10.7).

οὐδέ, *and not, neither*. οὐδέ as a conjunction is the equivalent of καὶ οὐ.

οὐδὲ καίουσιν λύχνον *and (men) do not light a candle* . . . (Matt. 5.15).
οὐδέ is also used as an adverb, *not even*:
καὶ οὐδὲ φαγεῖν εὐκαίρουν *and they had opportunity not even to eat* (Mark 6.31).

ὅτι, *because*: ὅτι is a causal conjunction and introduces a clause which expresses the reason for the action of the main clause.

μεταβεβήκαμεν (perf. of μεταβαίνω) ἐκ τοῦ θανάτου εἰς τὴν ζωήν, ὅτι ἀγαπῶμεν τοὺς ἀδελφούς, *we have passed from death unto life because we love the brethren* (1 John 3.14).

Here the clause introduced by ὅτι is a causal clause, giving the reason for the statement in the main clause "we have passed . . .".

Vocabulary

A number in brackets following a noun denotes its declension.

κωμή (1) *village*
ἀδικία (1) *unrighteousness, wickedness*
ἄνεμος (2) *wind*
δεσμός (2) *bond, chain*
θάνατος (2) *death*

καρπός (2) *fruit*
Λάζαρος (2) *Lazarus*
Φαρισαῖος (2) *Pharisee*
Φίλιππος (2) *Philip*
βιβλίον (2) *book*

Ναθαναήλ *Nathanael* (indeclinable noun; i.e. the ending is the same in whatever case it is used).
ἐπί (preposition) *upon*; followed by acc. when it conveys any idea of movement.

	Aorist	Perfect
ἁγιάζω I sanctify	ἡγίασα	ἡγίακα
ἁμαρτάνω I sin	ἥμαρτον	ἡμάρτηκα
	(Weak aor. ἡμάρτησα	
	is also found)	
βάλλω I throw, put	ἔβαλον	βέβληκα
ἐλαύνω I drive	-ήλασα	ἐλήλακα
εὑρίσκω I find	εὗρον	εὕρηκα
θάπτω I bury	ἔθαψα	
ὑβρίζω I insult	ὕβρισα	

EXERCISE 19

Translate into English:

1. ἀκηκόαμέν τε τὸν λόγον τοῦ Θεοῦ, καὶ ὑμῖν λέγομεν. 2. τέθνηκεν ὁ Λάζαρος, ἀλλὰ λύσει αὐτὸν ὁ Ἰησοῦς ἐκ τοῦ θανάτου. 3. εἶπον οὖν αὐτῷ οἱ Φαρισαίοι Πῶς (how) οὖν τεθεράπευκέ σε; 4. ἀπόστολοί ἐσμεν, ὁ γὰρ Κύριος ἡμᾶς ἔσταλκεν εἰς τὰς κωμάς. 5. ὁ δὲ εἶπεν, Ταῦτα πέπονθα ἐν τῇ καρδίᾳ μου ὅτι ἡμάρτηκας εἰς ἐμέ. 6. εὑρίσκει Φίλιππος τὸν Ναθαναὴλ καὶ λέγει αὐτῷ, Τὸν Χριστὸν εὑρήκαμεν.

EXERCISE 20

Translate into Greek:

1. We have sacrificed our sheep. 2. You (plur.) have written these things in a book. 3. I have loosed the slaves from their bonds. 4. They have sinned and suffered. 5. He has found the child in the desert. 6. You (sing.) have planted, but I have taken the fruit. 7. They have confessed their sins in the temple. 8. They have thrown the bread into the river. 9. We have both heard and believed. 10. The kingdom of God has come suddenly upon you.

EXERCISE 21

Translate into Greek:

1. We told you the truth, but you did not believe. 2. Some are bringing loaves, but others are taking them (them may be omitted here in translation). 3. And he said to the disciples, I send you into the land.

4. We are not eating, for we have no bread.　5. He not only insulted me, but he also struck you.　6. You, therefore, speak wickedness, but this man speaketh truth.　7. You have thrown stones into the house; therefore the soldiers pursued you.　8. The Jews do not break the Sabbath, for on that day they sanctify their hearts.　9. They are dead, and that man buried them.　10. The wind has driven the boats away from the land.

Chapter 9

THE PLUPERFECT INDICATIVE ACTIVE

The pluperfect indicative tense is much less frequent in the N.T. than any other tense, but it is frequent enough to make acquaintance with it on the part of the student necessary.   It is the past or *historic* tense of the perfect which is a primary tense (see p. 53).   The perfect tense looks back to the past from the standpoint of the present: "*I have read* this book" (which you now show me).   The pluperfect tense looks back to the past from a point in the past: "*I had read* this book" (before it appeared in a cheap edition).   The pluperfect tense, like the perfect, is a tense of completion, denoting the completion or perfecting of an action.   The perfect tense denotes the action as perfected in relation to the present: *he has finished*, i.e. as I now speak, or as I contemplate the work.   The pluperfect describes the action as perfected in relation to some point in past time: *he had finished*, e.g. by the time I arrived.

The pluperfect is formed by prefixing the augment to the reduplication.   The terminations are given below.   The pluperfect, like the perfect, may be weak or strong in form.

WEAK PLUPERFECT INDICATIVE ACTIVE OF λύω

(ἐ)λελύκειν *I had loosed*	(ἐ)λελύκειμεν
(ἐ)λελύκεις	(ἐ)λελύκειτε
(ἐ)λελύκει	(ἐ)λελύκεισαν

NOTICE: 1. Reduplication signifies a tense of completion.

2. The augment signifies historic time.   The augment of the pluperfect may, however, be omitted in N.T. Greek.

3. If the perfect indic. act. of a verb is known, the pluperfect can be found by substituting the endings -ειν, -εις, etc. for -α, -ας, etc., and prefixing (or not) the augment.

THE STRONG PLUPERFECT INDICATIVE ACTIVE

A strong pluperfect may be formed from a strong perfect in the same way as a weak pluperfect is formed from a weak perfect.

(ἐ)γεγράφειν *I had written.*   (ἐ)πεπόνθειν *I had suffered.*
(ἐ)τετάχειν *I had arranged.*

FURTHER USES OF THE CASES

THE VOCATIVE

We have given the vocative case in the declension of nouns, but so far we have not had occasion to use it. It is used when a person (or sometimes a thing) is addressed. "These, *gentlemen*, are my reasons"; "Please pass the mustard, *John*"; "*Father*, I cannot tell a lie". In these sentences the words in italics would be in the vocative case in Greek. Frequently ὦ (compare the English *o*) precedes the vocative. Occasionally a vocative adjective is found, as in John 17.25: Πατὴρ δίκαιε *righteous Father*, and Luke 4.34 Ἰησοῦ Ναζαρηνέ.

Examples of the use of the vocative may be studied in Mark 2.5: τέκνον, and 7.28: κύριε (and frequently).

THE DATIVE (Possessive Use)

We have noted the use of the dative for the indirect object (p. 21) and the instrumental dative (p. 31). The dative may also be used to express possession, especially with εἰμί *I am*, or its equivalents:

τὸ ὄνομα ἦν αὐτῷ Ἰωάνης *his name was John* (literally, "the name to him was John"). The dative is here used instead of the more normal genitive, τὸ ὄνομα αὐτοῦ ἦν Ἰωάνης.

τούτῳ τῷ ἀνθρώπῳ ἑκατὸν πρόβατά ἐστιν *this man has a hundred sheep* (literally, "to this man is a hundred sheep"—note that in Greek a neuter plural subject has its verb in the singular). The usual way of expressing this would be οὗτος ὁ ἄνθρωπος ἔχει ἑκατὸν πρόβατα.

PREPOSITIONS—*continued*

The prepositions which we dealt with on p. 30 could be followed by one and only one case of the noun or pronoun. Other prepositions can be followed by one of two, or even three, different cases, the meaning varying accordingly. The following table presents some prepositions which may be found with either the acc. or gen. cases, and gives their most frequent meanings:

Preposition	Before vowel	Before aspirate	Meanings with Acc.	Meanings with Gen.
διά	δι'	δι'	on account of, for the sake of	through (of place and time), by means of
κατά	κατ'	καθ'	down to, towards, through, throughout, during, about (time), in relation to, according to, after	down from, down, against
μετά	μετ'	μεθ'	after (place and time)	among, with (accompaniment)
περί	περί	περί	about (place and time), around (place and time)	about, concerning, on account of
ὑπέρ	ὑπέρ	ὑπέρ	over, beyond, more than	for, on behalf of, concerning
ὑπό	ὑπ'	ὑφ'	under	under, by (of agent)

Examples

διά *with acc.*:

τὸ σάββατον διὰ τὸν ἄνθρωπον ἐγένετο (Mark 2.27) *the Sabbath was made for the sake of man.*

διὰ τοῦτο *therefore* (lit. "on account of this").

Note the many equivalents in English of the phrase "on account of"; e.g. "for the sake of", "by reason of", "because of", "owing to".

διά *with gen.*:

διὰ τῆς Σαμαρίας (John 4.4) *through Samaria.*

δι' ὅλης νυκτός (Luke 5.5) *through the whole night.*

κρίνει ὁ Θεός ... διὰ Χριστοῦ Ἰησοῦ (Rom. 2.16) *God judges ... by Jesus Christ.*

ἐλάλησεν διὰ στόματος τῶν ἁγίων προφητῶν (Luke 1.70) *he spake by the mouth of his holy prophets.*

διὰ παντός (sometimes written as one word διαπαντός) occurs frequently, meaning *always, continually.* Χρόνου is to be understood with παντός, "throughout all time".

κατά *with acc.*:

καὶ Λευείτης κατὰ τὸν τόπον ἐλθών ... (Luke 10.32) *and a Levite having come to* (or *down to*) *the place* ...

καὶ ἀπῆλθεν καθ' ὅλην τὴν πόλιν κηρύσσων ... (Luke 8.39) *and he went his way, publishing throughout the whole city* ...

κατὰ σκοπὸν διώκω (Phil. 3.14) *I press on* (pursue) *toward the goal.*

ἐγένετο δὲ κατὰ τὸν καιρὸν ἐκεῖνον τάραχος οὐκ ὀλίγος περὶ τῆς ὁδοῦ (Acts 19.23) *and about that time there arose no small stir concerning the way.*

οὐ πολλοὶ σοφοὶ κατὰ σάρκα (1 Cor. 1.26) *not many wise (men) after (in relation to) the flesh.*

κατὰ τὰ στοιχεῖα τοῦ κόσμου καὶ οὐ κατὰ Χριστόν (Col. 2.8) *after (according to) the rudiments of the world, and not after Christ.*

κατὰ τὴν παράδοσιν τῶν πρεσβυτέρων (Mark 7.5) *according to the tradition of the elders.*

Κατά is also used with the acc. in a number of adverbial phrases (i.e. phrases which are the equivalent of adverbs): κατ' οἶκον *at home*; κατ' ἰδίαν (acc. fem. sing. of ἴδιος, *private*), *by oneself, privately*. It may also have the effect of "distributing" the word which follows: κατὰ πόλιν *from city to city*; κατ' ἔτος *year by year*; κατὰ τόπους *from place to place*; καθ' ἕνα *one by one*; κατ' ὄνομα *name by name*. This distributive use of κατά is idiomatic, and accordingly the English translation varies.

κατά *with gen.*:

καὶ ὥρμησεν ἡ ἀγέλη κατὰ τοῦ κρημνοῦ εἰς τὴν θάλασσαν (Mark 5.13) *and the herd rushed down the steep into the sea.*

ὃς γὰρ οὐκ ἔστιν καθ' ἡμῶν, ὑπὲρ ἡμῶν ἐστιν (Mark 9.40) *for he who is not against us is for us.*

μετά *with acc.*:

μετὰ δὲ τὸ δεύτερον καταπέτασμα σκηνή (Heb. 9.3) *and after the second veil, (the) tabernacle.*

ἦν δὲ τὸ πάσχα... μετὰ δύο ἡμέρας (Mark 14.1) *now after two days was the passover.*

μετὰ τοῦτο and μετὰ ταῦτα occur frequently for "after this", "after these things".

μετά *with gen.*:

τί ζητεῖτε τὸν ζῶντα μετὰ τῶν νεκρῶν; (Luke 24.5) *why seek ye the living among the dead?*

τότε προσῆλθεν αὐτῷ ἡ μήτηρ τῶν υἱῶν Ζεβεδαίου μετὰ τῶν υἱῶν αὐτῆς (Matt. 20.20) *then came to him the mother of the sons of Zebedee with her sons.*

περί *with acc.*:

ὁ 'Ιωάνης εἶχεν... ζώνην δερματίνην περὶ τὴν ὀσφὺν αὐτοῦ (Matt. 3.4) *John had ... a leathern girdle about his loins.*

περὶ τρίτην ὥραν (Matt. 20.3) *about the third hour.*

περί *with gen.*:

ἠρώτησαν αὐτὸν περὶ αὐτῆς (Luke 4.38) *they besought him for (con-cerning) her.*

ἐξετάσατε ἀκριβῶς περὶ τοῦ παιδίου (Matt. 2.8) *search out carefully concerning the young child.*

ὑπέρ *with acc.*:

οὐκ ἔστιν μαθητὴς ὑπὲρ τὸν διδάσκαλον (Matt. 10.24) *a disciple is not above his master.*

κεφαλὴν ὑπὲρ πάντα τῇ ἐκκλησίᾳ (Eph. 1.22) . . . *head over all things to the church.*

ὑπέρ *with gen.*:

προσεύχεσθε ὑπὲρ ἀλλήλων (James 5.16) *pray one for another.*

ὑπὲρ αὐτοῦ πάσχειν (Phil. 1. 29) *to suffer for his sake (on behalf of him).*

ὑπὲρ τούτου τρὶς τὸν Κύριον παρεκάλεσα (2 Cor. 12.8) *concerning this thing I besought the Lord thrice.* (ὑπέρ here is the equivalent of περί.)

ὑπό *with acc.*:

ὑπὸ τὴν κλίνην (Mark 4.21) *under the bed.*

ἄνθρωπός εἰμι ὑπὸ ἐξουσίαν (Matt. 8.9) *I am a man under authority.*

ὑπό *with gen.*:

ἀγαπηθήσεται ὑπὸ τοῦ Πατρός μου (John 14.21) *he will be loved by my Father.*

καὶ ἐβαπτίζοντο ὑπ’ αὐτοῦ ἐν τῷ ’Ιορδάνῃ ποταμῷ (Mark 1.5) *and they were baptized by him in the river Jordan.*

ὑπό is frequently used with the agent with verbs in the passive (see p. 72).

COMPOUND VERBS

Different shades of meaning may be attached to many Greek verbs by prefixing different prepositions. This practice is common to many languages, and is found in English; e.g. go, undergo; pose, propose, impose; fuse, confuse, infuse. Most English examples of this, like the last two, are in words of Latin origin.

The great use to which Greek puts this useful device is shown by the compounds of ἄγω and βάλλω.

Compounds of ἄγω *I drive, I lead*: ἀπάγω *drive away*
                                        συνάγω *drive together*
                                        ὑπάγω *depart*
                                        εἰσάγω *bring in*
                                        ἐξάγω *lead out*
                                        παράγω *pass by*

Compounds of βάλλω *I cast, I put*: ἀμφιβάλλω *cast round*
                                        ἐκβάλλω *cast out*
                                        ἐπιβάλλω *put upon*
                                        περιβάλλω *put round*
                                        προβάλλω *put forward*
                                        συνβάλλω *put together, confer, consider*

The shade of meaning thus given to the verb is frequently made clear by the particular meaning of the preposition which is prefixed. This, however, is not always so; e.g. the force of the παρά in παραγγέλλω *command* (compound of παρά and ἀγγέλλω *announce*) is not at once obvious.[1]

A few compound verbs have no *simplex* (i.e. uncompounded form); e.g. the common verb ἀποκτείνω *kill*: κτείνω is not found in the N.T.

When the verb stem begins with a vowel, the final vowel of the prefix is omitted; thus, ἀπάγω not ἀποάγω. But περί, πρό, and ἀμφί are exceptions; thus περιάγω *I lead about*; περιέχω *I encompass*; προάγω *I lead forward*; προέχω *I hold before*; ἀμφιέννυμι *I clothe*.

ἐκ before a vowel becomes ἐξ; thus, ἐκβάλλω: but ἐξάγω.

συν before a labial is in some texts changed to συμ; συμβάλλω, and before λ to συλ; συλλαμβάνω.

THE AUGMENT IN COMPOUND VERBS

The augment comes after the prefix, immediately before the stem. Thus, the imperf. indic. of συνβάλλω is συνέβαλλον not ἐσυνβαλλον. The final vowel of a preposition is omitted, except in the case of περί and πρό.

ἐπιβάλλω: imperf. indic., ἐπέβαλλον.
περιβάλλω: ,,   ,,   περιέβαλλον.
προβάλλω: ,,   ,,   προέβαλλον.

[1] The more advanced student will find the tracing of the root meanings of prepositions in compound words a fascinating study. Part III, "Word Formation", of Moulton's *Grammar of New Testament Greek*, Vol. II, is of great help. The subject is treated more fully, pp. 261ff. below.

COMPOUND VERBS   67

Some other common Compound Verbs

ἀναγινώσκω read:	Simplex	γινώσκω perceive, know.
ἀνοίγω open:	,,	No simplex found in N.T.
ἀπαγγέλλω announce:	,,	ἀγγέλλω announce.
ἀποκαλύπτω reveal:	,,	καλύπτω cover.
ἀπολύω release:	,,	λύω loose.
ἀποστέλλω send:	,,	στέλλω restrain (middle voice only in N.T.).
ἐκκόπτω cut down:	,,	κόπτω cut.
ἐνδύω put on:	,,	δύω sink, set (of the sun).
κατακρίνω condemn:	,,	κρίνω judge.
καταλείπω leave behind:	,,	λείπω leave.
κατεσθίω devour:	,,	ἐσθίω eat.
ὑπακούω obey (with dat.):	,,	ἀκούω hear.

Vocabulary

ἀγέλη (1) herd
ἀστραπή (1) lightning
κλίνη (1) bed
σκηνή (1) tent, tabernacle
ἀπιστία (1) unbelief
ἐκκλησία (1) assembly, church
ἐξουσία (1) authority
Σαμαρία (1) Samaria
ἑκατοντάρχης (1) centurion
Λευείτης (1) Levite
διδάσκαλος (2) teacher
Ζεβεδαῖος (2) Zebedee
κρημνός (2) steep bank, cliff
σκοπός (2) mark, goal
τάραχος (2) disturbance

στοιχεῖον (2) element, rudiment
ἁμαρτωλός, όν sinful. Masc. as noun: sinner
ἐχθρός, ά, όν hated, hostile. Masc. as noun: enemy
νεκρός, ά, όν dead
πρεσβύτερος, α, ον elder. Masc. as noun: an elder
τρίτος, η, ον third
ἀκριβῶς (adverb) carefully
ἐκεῖθεν (adverb) thence
ἤδη (adverb) already
τότε (adverb) then
τρίς (adverb) thrice
κατὰ καιρόν in due time

πολλά (neut. plur. of πολύς much; see p. 151) many things.
διδάσκω teach (aor. ἐδίδαξα: takes acc. of person and of thing taught).
ἐξῆλθον I came out (strong aor. of ἐξέρχομαι, compound of ἔρχομαι; see p. 86).

EXERCISE 22

Translate into English:

1. ὁ Ἰησοῦς ἐσθίει μετὰ τῶν ἁμαρτωλῶν καὶ τελωνῶν. 2. καὶ ἐξῆλθεν ἐκεῖθεν διὰ τὴν ἀπιστίαν αὐτῶν. 3. περὶ τὴν τρίτην ἡμέραν

εἰλήφεισαν τὰ βιβλία ἐκ τῆς συναγωγῆς.   4. ἐτετάχει τοὺς δούλους
ὑπὸ τὴν ἐξουσίαν κυρίου ἀγαθοῦ.   5. ταῦτα μὲν ἐμοί ἐστιν, ἐκεῖνα δὲ
σοί.   6. οἱ στρατιῶται ἐξέκοπτον τὰ δένδρα κατὰ τὰς τοῦ ἑκατοντάρχου ἐντολάς.

EXERCISE 23

Translate into Greek:

1. The assembly cast out the sinners.   2. The Levite drew near the dead
man and passed by.   3. The people obeyed the authority of the elders.
4. After the lightning the wind knocked down (use ἐκκόπτω) the tents.
5. God does not condemn you for this.   6. The Lord judges sinners
according to the scriptures.   7. Jesus suffered for his sheep.   8. Lord, I
will open the door of my heart.   9. The elder departed through Samaria
to the Jordan.   10. He had loosened his belt, and had put it in the tent.

EXERCISE 24

Translate into Greek:

1. We had already written to (use dat.) the church.   2. Because of the
disturbance we drove the herd together under the cliff.   3. The teachers
carefully taught the children the elements of truthfulness.   4. Zebedee
with his sons then brought the boat in towards the land.   5. I have already
written concerning the second book, but of the third I will write in due
time.   6. Jesus Christ suffered many things at the hands of (use ὑπό, by)
his enemies.   7. Three times you have thrown the spear beyond the mark.
8. About the second day he had written the whole book.   9. This man's
name (τὸ ὄνομα) was Andrew, but that man's was John.

Chapter 10

THE PASSIVE VOICE

Hitherto we have only been concerned with the active voice of the verb, which expresses the idea that the subject of the verb is the doer of the action described. The passive voice expresses the idea that the subject is the recipient of the action, and is therefore passive in relation to the action of the verb. Examples in English are: "the cow is being milked" (present); "he will be killed" (future); "it was sold" (past).

THE PASSIVE INDICATIVE TENSES OF λύω[1]

	Present indicative passive	Imperfect indicative passive	Future indicative passive
Sing. 1	λύομαι *I am being loosed*	ἐλυόμην *I was being loosed*	λυθήσομαι *I shall be loosed*
2	λύῃ	ἐλύου	λυθήσῃ
3	λύεται	ἐλύετο	λυθήσεται
Plur. 1	λυόμεθα	ἐλυόμεθα	λυθησόμεθα
2	λύεσθε	ἐλύεσθε	λυθήσεσθε
3	λύονται	ἐλύοντο	λυθήσονται

	Weak aorist indicative passive	Perfect indicative passive	Pluperfect indicative passive
Sing. 1	ἐλύθην *I was loosed*	λέλυμαι *I have been loosed*	(ἐ)λελύμην *I had been loosed*
2	ἐλύθης	λέλυσαι	(ἐ)λέλυσο
3	ἐλύθη	λέλυται	(ἐ)λέλυτο
Plur. 1	ἐλύθημεν	λελύμεθα	(ἐ)λελύμεθα
2	ἐλύθητε	λέλυσθε	(ἐ)λέλυσθε
3	ἐλύθησαν	λέλυνται	(ἐ)λέλυντο

NOTICE: 1. The historic tenses, imperfect, aorist, and pluperfect (in its complete form), have the augment. The perfect and pluperfect reduplicate the initial consonant of the stem.

[1] The complete paradigm of λύω on pp. 294–8 should be studied, so that the relation of the passive to the active and middle may be seen.

2. In the weak aor. pass. -θη- is added to the stem. This tense is unthematic: i.e. the person-endings are added directly, without the thematic vowels o and ε which appear, for instance in the thematic pres. indic. act. and pass. (see p. 17, note).

3. The -θη- is also added to the stem in the fut. pass. indic. The future tense is historically derived from the aorist, and normally shows the same stem: cf. aorist active ἔ-λυ-σ-α and future λύ-σ-ω.

4. The endings of the present and future passive tenses are the same.

5. The perfect tense, it is to be remembered, is not a past tense. It is in primary, not historic, time, denoting as it does an action which, while performed in the past, continues in its effects into the present: in λέλυμαι *I have been loosed*, the emphasis is not on the past act of deliverance, but on the present state of freedom.

The pluperfect tense (which is not frequent in the N.T.) is a past tense, and properly has the augment. It is, however, sometimes written without it.

6. In translating passives care must be taken not to fall into one or two common errors. For example, "I am loosing", which is active, must not be confused with "I am being loosed". Care must also be taken in translating into Greek such a sentence as "they are loosed". It is possible that this may refer to something which is actually being done at the moment of speaking, and that it is the equivalent of "they are being loosed", which must be translated λύονται; but it is more likely that it refers to something done in the past, the effects of which continue in the present, "they were loosed, and are now free". For this the perfect is the correct tense: λέλυνται.

THE STRONG AORIST PASSIVE

A strong instead of a weak aorist passive is found in some verbs (not necessarily those which have a strong aorist active). In the wk. aor. pass. -θη- is added to the verb stem before the ending, ἐ-λύ-θη-ν. In the str. aor. pass. -η- replaces -θη-. The endings are the same. Strong aorist passives in the following list are: ἐ-γράφ-η-ν, -ἐ-στάλ-η-ν, ἐ-στράφ-η-ν, -ἐ-τρίβ-η-ν.

Below are given the present, aorist, and perfect indicative passive tenses of some of the verbs already listed. The imperfect passive may easily be obtained by prefixing the augment, and substituting -όμην for the -ομαι of the present. To obtain the future passive omit the augment and substitute the ending -σομαι for the -ν of the aorist passive. Thus: ἠγέρθην, aor. indic. pass. of ἐγείρω: ἐγερθήσομαι, fut. indic. pass. -ἐτρίβην, aor.

(str.) indic. pass. of -τρίβω: -τριβήσομαι, fut. indic. pass. The pluperfect passive has the ending -μην for the -μαι of the perfect, and the augment may be prefixed to the reduplicated stem.

In the following list some irregularities will be noticed, especially variations of the stem in the aor. and perf. pass. of some verbs. Too much attention need not be paid to these variations at this stage. They will become clearer later. They are included here in order that the student may begin to become familiar with the way in which the passive tenses are formed in some of the verbs in common use, and that he may be able to gain some practice in the formation of the passive in the exercises which follow.

A space left blank indicates that the part is not found in the N.T. A word in brackets also signifies that the part is not found in the N.T.; but it is given in order to help the student to form other tenses: e.g. ἐδιώχθην, the aor. pass. of διώκω, is not actually found in the N.T. but the form is needed to arrive at the fut. pass. (omit the augment, and substitute -σομαι for -ν: διωχθήσομαι *I shall be pursued*).

A hyphen in front of a word indicates that it is only found in the N.T.

	Pres. indic. pass.	Aor. indic. pass.	Perf. indic. pass.
ἁγιάζω *sanctify*	ἁγιάζομαι	ἡγιάσθην	ἡγίασμαι
ἄγω *lead*	ἄγομαι	ἤχθην	ἦγμαι
ἀκούω *hear*	ἀκούομαι	ἠκούσθην	
ἀλείφω *anoint*		-ἠλείφθην	
βάλλω *throw, put*	βάλλομαι	ἐβλήθην	βέβλημαι
βαπτίζω *baptize*	βαπτίζομαι	ἐβαπτίσθην	βεβάπτισμαι
βλέπω *look*	βλέπομαι		
γράφω *write*	γράφομαι	ἐγράφην	γέγραμμαι
διδάσκω *teach*		ἐδιδάχθην	
διώκω *pursue*	διώκομαι	(ἐδιώχθην)	δεδίωγμαι
ἐγείρω *rouse*	ἐγείρομαι	ἠγέρθην	ἐγήγερμαι
ἑρμηνεύω *interpret*	ἑρμηνεύομαι		
εὑρίσκω *find*	εὑρίσκομαι	εὑρέθην	
θύω *sacrifice*	θύομαι	ἐτύθην	τέθυμαι
κελεύω *bid*	κελεύομαι	ἐκελεύθην	κεκέλευμαι
κλείω *shut*		-ἐκλείσθην	κέκλεισμαι
λούω *wash*	λούομαι		λέλουμαι
πείθω *persuade*	πείθομαι	ἐπείσθην	πέπεισμαι
πέμπω *send*	πέμπομαι	-ἐπέμφθην	
πιστεύω *believe* (pass. be entrusted with)	πιστεύομαι	ἐπιστεύθην	πεπίστευμαι
-στέλλω *send*	στέλλομαι	-ἐστάλην	-ἔσταλμαι
στρέφω *turn*	στρέφομαι	ἐστράφην	ἔστραμμαι
τάσσω *arrange*	τάσσομαι	ἐτάχθην	τέταγμαι
-τρίβω *rub*		(-ἐτρίβην)	-τέτριμμαι

as a compound: e.g. -τρίβω. The simplex τρίβω is not in the N.T., but διατρίβω *rub hard, spend time,* and συντρίβω *break in pieces* are found.

The conjugation of the perf. and pluperf. indic. pass. of verbs whose stems end in a labial (e.g. γέγραμμαι, stem γραφ-), a guttural (e.g. δεδίωγμαι, stem διωκ-), or a dental (e.g. πέπεισμαι, stem πειθ-) shows peculiarities which the student will not need to know at present. They will be explained on pp. 126, 134, 145.

AGENT AND INSTRUMENT

"The soldier mounts the horse": the meaning of this sentence can be exactly expressed by turning the sentence round, making the object into the subject and the verb passive: "the horse is mounted by the soldier". "A stone kills the dog" can be similarly treated: "the dog is killed by a stone".

With a passive verb we frequently have occasion to express agent or instrument, as, for example, "by the soldier" and "by a stone" in the sentences above. The distinction between *agent* and *instrument* is that the former refers to living beings, while *instrument* refers to inanimate things.

To express agent Greek uses the preposition ὑπό followed by the genitive case of noun or pronoun: ὑπὸ τοῦ στρατιώτου *by the soldier;* ὑπὸ σοῦ *by you.* The final o of ὑπό is elided before a vowel: ὑπ’ αὐτοῦ *by him.* Before a rough breathing the o is elided and the π is aspirated: ὑφ’ ἡμῶν *by us.* ἀπό and διά, both with the genitive, may also be used to express agent: ἀπὸ τῶν ἀποστόλων *by the apostles* (Acts 4.36); διὰ τῶν μαθητῶν αὐτοῦ *by his disciples* (Matt. 11.2). But ὑπό is the usual word to express agency with a passive verb.

To express instrument the dative case is used without a preposition: λίθῳ *by a stone,* or, *with a stone;* λόγχαις *with spears.* In the New Testament, but not in classical Greek, the preposition ἐν may also be used with the dative case to express instrument, ἐν μαχαίρᾳ *with a sword.*

Vocabulary

ὥρα (1) *hour*
ληστής (1) *robber*
αἰγιαλός (2) *sea-shore*
ἀμνός (2) *lamb*
διάβολος (2) *accuser, Satan, the Devil*
κῆπος (2) *garden*

κωφός, ή, όν *dumb, deaf* (according to context)
λεπρός, ά, όν *leprous.* Masc. as noun: *leper*
πτωχός, ή, όν *poor.* Masc. as noun: *poor man*

ἐγείρω *arouse* (pass., *be aroused, arise*).

εὐαγγελίζω *bring good news, preach the gospel.* (But the verb is deponent
in N.T.: see p. 86. Pass.: *have the gospel preached to one.*)

καθαρίζω *cleanse.*

πειράζω *prove, test, tempt*

πιστεύω *believe* (pass.: *be entrusted with*).

EXERCISE 25

Translate into English:

1. καὶ ἐβαπτίζοντο ὑπ᾽ αὐτοῦ ἐν τῷ Ἰορδάνῃ ποταμῷ. 2. ὁ
Ἰησοῦς ἐγήγερται ἐκ νεκρῶν. 3. ὁ ἀνθρωποκτόνος εὐθὺς εἰς φυλακὴν
βληθήσεται. 4. τότε ὁ Ἰησοῦς ἀνήχθη εἰς τὴν ἔρημον ὑπὸ τοῦ
Πνεύματος (genitive of πνεῦμα (3) *spirit*) καὶ ἐπειράσθη ὑπὸ τοῦ
διαβόλου. 5. λεπροὶ καθαρίζονται καὶ κωφοὶ ἀκούουσιν, καὶ νεκροὶ
ἐγείρονται καὶ πτωχοὶ εὐαγγελίζονται. 6. οἱ δὲ μαθῆται ἐν ἐκείνῃ τῇ
ὥρᾳ ἀπεστάλησαν ἐκ τοῦ οἴκου.

EXERCISE 26

Translate into Greek:

1. They were being sanctified. 2. You (sing.) will be led. 3. It was
heard. 4. You (plur.) have been put. 5. We were baptized. 6. They
are being looked at. 7. It was written. 8. You (sing.) were taught.
9. I was being pursued. 10. They were aroused. 11. We shall be found.
12. It has been sacrificed. 13. You (sing.) were bidden. 14. It had been
shut. 15. They are being washed. 16. We were persuaded. 17. You
(plur.) are being sent. 18. He has been entrusted. 19. We were turned.
20. They will be broken in pieces.

EXERCISE 27

Translate into Greek:

1. Jesus was baptized by John in the Jordan. 2. They were gathered
together in the house. 3. And immediately he arose from his bed.
4. Thou wilt be cast into prison. 5. The good tree is not cut down.
6. These stones will not be broken. 7. I have been pursued by robbers.
8. You, O disciples, were being sent to the villages. 9. The lamb was
sacrificed in the temple. 10. The boats are being arranged on the sea-
shore. 11. Thy brother was found. 12. Jesus was led from the garden.

Chapter 11

THIRD DECLENSION NOUNS

The student will find the third declension nouns more difficult to master than either the first or second declensions. Their variety seems bewildering. This is due to the great variety of the stems. There are, however, constant features in the endings. The gen. sing. always ends in ς (and in ος most frequently); the dat. sing. in ι; the nom., voc., and acc. plur. of masculine and feminine nouns end in ς (and in ες, ες, ας most frequently); and the dat. plur. in σι(ν). Neuter nouns of the 3rd decl., as of the 2nd, have voc. and acc. the same as the nom. both in sing. and plur.

Masc. and Fem. Third Declension Nouns with stems ending in Gutturals, Labials and Dentals

(The stem is to be found by striking off the -ος ending of the genitive singular.)

	Stem ending in Guttural κ	Stem ending in Guttural γ	Stem ending in Labial β	Stem ending in Dental δ
Stem	ὁ φύλαξ guard φυλακ-	ἡ σάλπιγξ trumpet σαλπιγγ-	ἡ φλέψ vein φλεβ-	ὁ, ἡ παῖς boy, girl παιδ-
Sing. N. V. A. G. D.	φύλαξ φύλακα φύλακος φύλακι	σάλπιγξ σάλπιγγα σάλπιγγος σάλπιγγι	φλέψ φλέβα φλεβός φλεβί	παῖς παῖδα παιδός παιδί
Plur. N. V. A. G. D.	φύλακες φύλακας φυλάκων φύλαξι(ν)	σάλπιγγες σάλπιγγας σαλπίγγων σάλπιγξι(ν)	φλέβες φλέβας φλεβῶν φλεψί(ν)	παῖδες παῖδας παίδων παισί(ν)

	Stem ending in Dental τ (1)	Stem ending in Dental τ (2)	Stem ending in Dental τ (3)
Stem	ὁ ἄρχων ruler ἀρχοντ-	ὁ ὀδούς tooth ὀδοντ-	ὁ ἱμάς strap ἱμαντ-
Sing. N. V. A. G. D.	ἄρχων ἄρχοντα ἄρχοντος ἄρχοντι	ὀδούς ὀδόντα ὀδόντος ὀδόντι	ἱμάς ἱμάντα ἱμάντος ἱμάντι
Plur. N. V. A. G. D.	ἄρχοντες ἄρχοντας ἀρχόντων ἄρχουσι(ν)	ὀδόντες ὀδόντας ὀδόντων ὀδοῦσι(ν)	ἱμάντες ἱμάντας ἱμάντων ἱμᾶσι(ν)

NOTICE: 1. In all the above nouns the vocative, sing. and plur., is the same as the nominative.

2. The dat. plur. is obtained by adding -σι (-σιν before vowels) to the stem. Notice the effect when -σι is added to a guttural (φύλακ-σι becomes φύλαξι: σάλπιγγ-σι becomes σάλπιγξι); and when added to a labial (φλεβ-σί becomes φλεψί). When -σι is added to a dental, the dental is dropped. Thus παιδ-σί becomes παισί. When it is added to a stem ending in ντ, the dental τ drops out. But Greek is averse from a σ following ν, and the ν is also dropped. But by way of compensation the preceding vowel is lengthened. The stages are these:

ὀδόντ-σι, ὀδόνσι, ὀδόσι, ὀδοῦσι.
ἱμάντ-σι, ἱμάνσι, ἱμάσι, ἱμᾶσι.

3. As a general guide at this stage we may set down the following scheme of endings for third declension masc. and fem. nouns.

Sing.			Plur.	
N. V.	various	N. V.	-ες
A.	-α	A.	-ας
G.	-ος	G.	-ων
D.	-ι	D.	-σι(ν)

Variations from this scheme will be noted as we come across them.

SIMILARLY DECLINED N.T. NOUNS

	Nom.	English	Gen.
Like φύλαξ	ἡ γυνή (but voc. γύναι)	woman, wife	γυναικός
	ὁ κῆρυξ	herald	κήρυκος
	ἡ σάρξ	flesh	σαρκός
	ὁ φοῖνιξ	date-palm	φοίνικος
Like σάλπιγξ	ἡ μάστιξ	scourge, plague	μάστιγος
Like φλέψ	ὁ Ἄραψ	Arab	Ἄραβος
Like παῖς	ἡ ἐλπίς	hope	ἐλπίδος
	ἡ λαμπάς	torch	λαμπάδος
	ἡ πατρίς	native place	πατρίδος
	ὁ ποῦς	foot	ποδός
	ἡ σφραγίς	seal	σφραγῖδος
Like ἄρχων	ὁ γέρων	old man	γέροντος

No other nouns in the N.T. are declined like ὀδούς and ἱμάς.

QUESTIONS

The Greek question mark is like the English semi-colon ;. Placed at the end of the sentence it indicates that the sentence is interrogative. Greek requires no change in the order of words in an interrogative sentence.

ὁ Κύριος τοὺς ἀποστόλους πέμπει *the Lord sends the apostles.*
ὁ Κύριος τοὺς ἀποστόλους πέμπει; *does the Lord send the apostles?*
ἀγαθός ἐστιν *he is good.*
ἀγαθός ἐστιν; *is he good?*

Many questions are, as in English, introduced by interrogative words, either pronouns (e.g. who? what?) or adverbs (e.g. when? how? where?).

THE INTERROGATIVE PRONOUN, τίς *who?* τί *what?* (See next page)

τίς τοῦτο λέγει; *who says this?*
τίνα πέμψει; *whom will he send?*
τί ἐστι τοῦτο; *what is this?*
τίνος ἐστιν ἡ μάχαιρα; *whose is the sword?*

	Masc. and Fem.	Neut.
Sing. N.	τίς who?	τί what?
A.	τίνα whom?	τί what?
G.	τίνος { of whom? / whose?	τίνος of what?
D.	τίνι to whom?	τίνι to what?
Plur. N.	τίνες who?	τίνα what things?
A.	τίνας whom?	τίνα what things?
G.	τίνων { of whom? / whose?	τίνων of what things?
D.	τίσι(ν) to whom?	τίσι(ν) to what things?

τίς may also be used adjectivally:

τίς γέρων τοῦτο λέγει; *what old man says this?*
τίνα κήρυκα πέμψει; *which herald will he send?*
τίνος πατρίδος ἐστιν; *of what native land is he?*

The most frequent interrogative adverbs are:

ποῦ; *where?*
πότε; *when?*
πῶς; *how?*

ποῦ ἐστιν ἐκεῖνος; *where is he?* (John 7.11).
πότε οὖν ταῦτα ἔσται; *when therefore shall these things be?* (Luke 21.7).
πῶς σταθήσεται ἡ βασιλεία αὐτοῦ; *how shall his kingdom stand?* (Luke 11.18).

σταθήσεται is the 3rd pers. sing. fut. indic. pass. of ἵστημι. See p. 220.

When a question contains a negative careful attention must be paid as to whether the negative used is οὐ or μή.

οὐ introduces a question which expects the answer "Yes".
μή introduces a question which expects the answer "No", or else is very hesitant.

In English this distinction is made by a difference in phraseology. A question like "You will post this letter for me, won't you?" expects the answer "Yes". But "You won't forget me, will you?" expects the answer "No".

In such sentences in Greek a strengthened form of the negative is often used, οὐχί for οὐ, μήτι for μή.

Examples

οὐχὶ καὶ οἱ τελῶναι τὸ αὐτὸ ποιοῦσιν; *do not even the publicans the same?* (Matt. 5.46).

οὐχὶ δώδεκα ὥραί εἰσιν τῆς ἡμέρας; *are there not twelve hours in the day?* (John 11.9).

οὐχ ὑμεῖς μᾶλλον διαφέρετε αὐτῶν; *are not ye of much more value than they?* (Matt. 6.26).

μὴ ὁ νόμος ἡμῶν κρίνει τὸν ἄνθρωπον . . .; *doth our law judge the man . . . ?* (John 7.51).

μήτι ἐγώ εἰμι, Κύριε; *Is it I, Lord?* (Literally, "am I (the one), Lord?") (Matt. 26.22).

μήτι δύναται τυφλὸς τυφλὸν ὁδηγεῖν; *can the blind lead the blind?* (Luke 6.39).

Most of these sentences can be translated into English in other ways: e.g. the last might well be translated "the blind cannot lead the blind, can they?"

Notice that the question may actually receive an answer different from that expected. But this does not affect the use of οὐ and μή, which is determined solely by the answer which is *expected*.

Vocabulary

δοκός, ἡ (2) (N.B. fem.) *beam* (of wood)

νόμος (2) *law*

οὐρανός (2) *heaven*

ὀφθαλμός (2) *eye*

Πειλᾶτος (2) *Pilate*

ἀκρίς, -ίδος, ἡ (3) *locust*

τί; (acc. sing. neut. of τίς;) *why?*

᾽Ιουδαῖος, -α, -ον *Jewish.* Masc. as noun: *a Jew*

πρός (preposition) with acc., *to, towards, at*

κρίνω *judge.* Aor. indic. pass.: ἐκρίθην (for other tenses see p. 185).

ἀποκρίνομαι *answer.* (This is the passive of κρίνω compounded with ἀπό: aor. indic., ἀπεκρίθη.)

προσπίπτω *fall down at* (compound of πίπτω).

EXERCISE 28

Translate into English:

1. οἱ δὲ γέροντες ἤσθιον ἀκρίδας ἐν τῇ ἐρήμῳ. 2. τίς τὸν ἱμάντα τοῦ ὑποδήματός σου ἔλυσεν; [gen. of τὸ ὑπόδημα *sandal*]. 3. ἡ δὲ

γυνὴ προσέπεσεν πρὸς τοὺς πόδας αὐτοῦ. 4. ὁ δὲ ἀπεκρίθη καὶ
εἶπεν αὐτῇ, Τὴν μάστιγά σου θεραπεύσω. 5. τίνι ταῦτα εἶπον;
οὐχὶ σοί; 6. ἐν τῷ οὐρανῷ οὖν τίνος τῶν ἀδελφῶν ἔσται γυνή;

EXERCISE 29

Translate into Greek:

1. Of the woman. 2. For the boys. 3. For the girls. 4. With hope.
5. To my native place. 6. Of the feet. 7. With a seal. 8. With straps.
9. According to the flesh. 10. For the old men. 11. The heralds.
12. Above the palm-trees. 13. With the rulers. 14. Under the feet.
15. Concerning the Arabs. 16. Of a torch. 17. According to my hopes.
18. With scourges. 19. The old man's veins. 20. With the trumpets of
the heralds.

EXERCISE 30

Translate into Greek:

1. The old man has no teeth in his head. 2. Why do you not see the
beam in your eye? 3. Is not this the law and the prophets? 4. Art thou
then the Son of God? 5. How does this man open the eyes of the blind?
6. Lord, dost thou wash my feet? 7. Pilate answered, Am I a Jew?
8. Does a good tree bear bad fruit? 9. Where were the heralds found?
10. How do they put the straps under the boats? 11. When did the
Arabs flee to their native land? 12. The publicans and sinners will not be
in the kingdom of heaven, will they?

Chapter 12

RELATIVE CLAUSES

Up to this point we have been dealing mostly with simple sentences, having only one subject and one verb. But an examination of almost any page of English prose will show that sentences are very frequently more complicated than this, and that they often contain two or more subjects and verbs. Two sentences may be joined together into one, for instance, by a conjunction: e.g. "I went this way, and he went the other": "You say so, but I do not agree". Such sentences are called Compound Sentences. They cause little difficulty.

But frequently the connection between clauses is much more intimate than that provided by a mere conjunction. In the sentence "I wrote to you so that you might know the facts" the connection between the two clauses "I wrote to you" and "you might know the facts" is one of purpose. In the sentence "I shall wait until you come" the connection between the two clauses "I shall wait" and "you come" is one of time. In the sentence "This is the house that Jack built" the connection between the two clauses "This is the house" and "Jack built" is one of relation. The word *that* relates the house which Jack built to the one pointed out by the word *this*. The pronoun *that* is therefore called a *relative pronoun*, and the clause introduced by it a *relative clause*. These three types of sentence are known as Complex Sentences. For the moment we shall only deal with the last type mentioned above, the type which contains a relative clause. Purpose and temporal clauses, as well as others, such as conditional, causal, and consecutive (or result) clauses, will be described later.

One or two more examples in English will help to make the principle of relative clauses clearer:

"The money which you lent me was a great help". Here the main clause is "The money was a great help", and the relative clause is "which you lent me".

"The men who came soon finished the work". The main clause is "The men soon finished the work" and the relative clause "who came".

"The man whom I pointed out ran away". The main clause is "The man ran away" and the relative clause "whom I pointed out".

In English the word *that* is frequently used as the relative pronoun for *which,* and even for *who* and *whom*: "The money that you lent me", "The men that came", "The man that I pointed out". Note also that in English the relative pronoun, when it is the object of the relative clause, is sometimes omitted altogether: "The money you lent me was a great help", "The man I pointed out ran away". It must not be omitted in Greek.

Care must be taken not to confuse the relative pronoun *who, which,* with the interrogative pronoun. In the above examples the words *who, whom,* and *which* do not imply any question.

THE RELATIVE PRONOUN ὅς, ἥ, ὅ

	Masculine	Feminine	Neuter
Sing. N.	ὅς *who, that*	ἥ *who, that*	ὅ *which, that*
A.	ὅν *whom, that*	ἥν *whom, that*	ὅ *which, that*
G.	οὗ *whose, of whom*	ἧς *whose, of whom*	οὗ *of which*
D.	ᾧ *to whom*	ᾗ *to whom*	ᾧ *to which*
Plur. N.	οἵ	αἵ	ἅ
A.	οὕς	ἅς	ἅ
G.	ὧν	ὧν	ὧν
D.	οἷς	αἷς	οἷς

NOTICE: 1. The declension of the relative pronoun follows the endings of the masc. and neut. of the 2nd decl. and the fem. of the 1st, with the exception of the nom. and acc. sing. of the neut. which have no final ν.

2. The accentuation is of great importance in the relative pronoun. The accents must be learned in order to avoid confusion with certain cases of the definite article. Compare ὁ (nom. masc. sing. of the article) and ὅ; ἡ and ἥ; οἱ and οἵ.

THE USE OF THE RELATIVE PRONOUN

The relative pronoun is so called because it relates to a noun or pronoun in another clause of the sentence. This noun or pronoun is called the antecedent, and the clause in which it stands (which is often the main clause of the sentence) is called the antecedent clause. It is called antecedent because it is logically prior to the relative. It may not in fact be *written* first. The clause in which the relative pronoun stands is called the relative clause.

The relative pronoun naturally must agree with the antecedent to which it refers in number and gender. It is also of the same person as its antecedent. Thus in "I who speak am he", *who* is of the 1st pers. In "he who speaks is he", *who* is 3rd pers. But the *case* of the relative pronoun is determined by the way in which it is used in the relative clause. If the relative pronoun stands as subject in the relative clause it will be nominative. If it stands as the direct object of the verb of the relative clause it will be accusative; if as the indirect object it will be dative; and it may be genitive (cf. the use of the English *whose*). A few examples will make all this plain:

'Ραββεί, ὃς ἦν μετὰ σοῦ πέραν τοῦ 'Ιορδάνου, ἴδε οὗτος βαπτίζει (John 3.26) *Rabbi, he that was with thee beyond Jordan, behold the same* (lit. "this man") *baptizeth.*

Here ὅς introduces the relative clause. Its antecedent is οὗτος. (Note that the antecedent may actually be written later than the relative pronoun.) ὅς is singular and masculine because οὗτος is sing. and masc. It is also 3rd pers. Hence ἦν, 3rd pers. imperf. indic. ὅς is nominative not because οὗτος is nominative, but because it is the subject of the relative clause.

καὶ ἰδοὺ ὁ ἀστήρ, ὃν εἶδον ἐν τῇ ἀνατολῇ, προῆγεν αὐτούς (Matt. 2.9) *and lo, the star, which they saw in the east, went before them.*

Here ὅν is singular and masculine because its antecedent ἀστήρ is singular and masculine. It is accusative because it is the object of the verb εἶδον.

Σάρρα ὑπήκουσεν τῷ 'Αβραάμ . . . ἧς ἐγενήθητέ[1] τέκνα (I Pet. 3.6) *Sarah, whose children ye are, obeyed Abraham.*

Here ἧς is singular and feminine because its antecedent Σάρρα is singular and feminine. It is genitive as denoting possession.

τὸ σπέρμα, ᾧ ἐπήγγελται[2] . . . (Gal. 3.19) *the seed, to whom the promise hath been made* . . .

Here ᾧ is singular and neuter because its antecedent σπέρμα is singular and neuter. It is dative, being the indirect object of the verb ἐπήγγελται. The relative pronoun may be linked to its antecedent by means of a

---

1 ἐγενήθητε, 2nd pers. plur. aor. indic. of γίνομαι *become, come to be.*
2 ἐπήγγελται, lit.: *it hath been promised.* 3rd pers. sing. perf. indic. pass. of ἐπαγγέλλω *promise.*

preposition, as in the English "the ship, in which they sailed, was the largest in the world". Here the case of the relative pronoun in Greek is determined by the preposition. Note the following examples:

τίς δέ ἐστιν οὗτος, περὶ οὗ ἀκούω τοιαῦτα; (Luke 9.9) *but who is this, about whom I hear such things?*

οὗ is genitive as being governed by the preposition περί which here requires the genitive.

τινὲς[1] δὲ ἄνδρες . . . ἐπίστευσαν, ἐν οἷς καὶ Διονύσιος ὁ ᾿Αρεοπαγείτης καὶ . . . Δάμαρις (Acts 17.34) *but certain men . . . believed, among whom also (was) Dionysius the Areopagite and . . . Damaris.*

οἷς is dative as being governed by the preposition ἐν which requires the dative.

THE ARTICLE WITH PREPOSITIONAL PHRASES

The definite article in Greek is used with greater elasticity than the English definite article. For example it may be used with a prepositional phrase which thus becomes the equivalent of an adjective:

αἱ γυναῖκες αἱ ἐν τῇ οἰκίᾳ *the women in the house* (literally, "the women the (ones) in the house"). Here ἐν τῇ οἰκίᾳ is treated as an adjective qualifying γυναῖκες.

The article with such a prepositional phrase may often be translated by a relative clause: "the women who were in the house". So also:

Πάτερ ἡμῶν ὁ ἐν τοῖς οὐρανοῖς (Matt. 6.9), *our Father, who art in heaven* (literally, "the (Father) in heaven").
τὴν ἀγάπην τὴν εἰς πάντας τοὺς ἁγίους (Eph. 1.15) *the love which* (ye show) *towards all the saints.*
τὰ πρὸς τὸν Θεόν (Heb. 2.17) *things which concern God.* (R.V. "things pertaining to God".)
οἱ περὶ αὐτόν (Mark 4.10) *they that were about him.*
οἱ παρὰ τὴν ὁδόν (Mark 4.15) *they by the wayside.*

For the use of the article with the participle to which also it can give adjectival force, see p. 163f.
For further uses of the article, see also p. 171f.

[1] τινές, nom. masc. plur. of the indefinite pronoun (here an adjective) τις *certain* (see p. 130).

THE MIDDLE VOICE

The Greek active and passive voices have their counterparts in English, but the middle voice has none. Historically the middle voice is prior to the passive which is a development from it. The middle indicates that the action described is in some way of advantage or significance to the subject. For example νίπτω means "I wash": νίπτω σε *I am washing you*. But in νίπτομαι τὸ πρόσωπον *I am washing my face*, the middle is rightly used, for the action is one which especially concerns the subject.

The middle is not only used to express the idea of an action which is in some way of special advantage to the subject. It also expresses the idea of having a thing done for oneself: e.g.

πάντες εἰς τὸν Μωϋσῆν ἐβαπτίσαντο ἐν τῇ νεφέλῃ καὶ ἐν τῇ θαλάσσῃ (1 Cor. 10.2) *they all submitted to baptism unto Moses in the cloud and in the sea*.

The middle is also used to describe an action which takes place within the subject and does not pass beyond to any external object. An example of this is the following use of the fut. indic. mid. of βουλεύω *I take counsel*.

βουλεύσεται (Luke 14.31) *he will take counsel with himself*, i.e. *he will consider*.

The student must note carefully that the middle voice is not a reflexive. The middle λύεται does not mean "he looses himself". That would require λύει and the reflexive pronoun ἑαυτόν. λύεται means "he looses for himself", i.e. in a way which is of advantage or special significance to himself. In many cases the Greek middle has to be translated into English as though it were an active, but nearly always the student will be able to discern the reason for the Greek use of the middle rather than the active voice.

The greater number of instances of the middle voice in the New Testament, however, is provided by the deponent verbs, which will be described in the next section.

The forms for the middle voice are the same as for the passive in the present, imperfect, perfect and pluperfect tenses. This is true not only of the indicative mood, but of all moods. It is usually a simple matter to decide from the context whether the middle or passive is intended. The future and aorist middle have special forms.

The Middle Voice of λύω (Indicative Mood)[1]

Present λύομαι, etc., *I loose for myself.*
Imperfect ἐλυόμην, etc., *I was loosing for myself.*
Future λύσομαι, λύσῃ, λύσεται, etc., *I shall loose for myself.*
1st aorist

	Singular	Plural
1.	ἐλυσάμην *I loosed for myself*	ἐλυσάμεθα
2.	ἐλύσω	ἐλύσασθε
3.	ἐλύσατο	ἐλύσαντο

Perfect λέλυμαι, etc., *I have loosed for myself.*
Pluperfect (ἐ)λελύμην, etc., *I had loosed for myself.*

Strong Aorist

Verbs which have a strong instead of a weak aorist active also have a strong aorist middle: e.g.

Strong Aorist Indicative Middle of βάλλω

	Singular	Plural
1.	ἐβαλόμην *I put (threw) for myself*	ἐβαλόμεθα
2.	ἐβάλου	ἐβάλεσθε
3.	ἐβάλετο	ἐβάλοντο

The endings of the strong aorist indicative middle are the same as for the imperfect indicative middle, but are added to the augmented *verb stem* (whereas those of the imperfect are added to the *present stem* which is a lengthened form of the verb stem, ἐβαλλόμην *I was throwing for myself*).

DEPONENT VERBS

Certain verbs are deponent. This term comes from the Latin *depono*, "lay aside". They appear to have laid aside and lost the active voice.

[1] The full paradigm of λύω on pp. 294–8 should be studied in order that the relation of the middle to the active and passive may be seen.

The middle voice (and in some cases the passive) is given an active meaning.

SOME COMMON DEPONENT VERBS

Verb	Aorist	
ἅπτομαι *touch*	ἡψάμην	The active ἅπτω means "kindle" (fire or light).
ἄρχομαι *begin*	ἠρξάμην	The active ἄρχω means "rule".
γεύομαι *taste*	ἐγευσάμην	
γίνομαι *become*	ἐγενόμην (strong)	Note verb stem γεν-.
δέχομαι *receive*	ἐδεξάμην	
ἐργάζομαι *work*	ἠργασάμην	
εὐαγγελίζομαι *preach the gospel*	εὐηγγελισάμην	Note the position of augment since the verb is a compound, εὐ-αγγ.
εὔχομαι *pray*	εὐξάμην	The compound προσεύχομαι is more frequent.
πορεύομαι *go, come, journey*	ἐπορευσάμην or ἐπορεύθην	
πυνθάνομαι *ascertain*	ἐπυθόμην (strong)	Note verb stem πυθ-.
χαρίζομαι *grant, forgive*	ἐχαρισάμην	
ψεύδομαι *deceive, lie*	ἐψευσάμην	

It will be noticed that many of these deponent verbs describe actions or states which are closely personal, so that the sense of the middle voice is really retained; e.g. the verbs which describe perception, ἅπτομαι, γεύομαι.

A few deponents have a passive instead of a middle form in the aorist.

> ἀπο-κρίνομαι *answer*: aor. -ἐκρίθην. The classical aorist in the middle form is occasionally found in the N.T. -ἐκρινάμην.
>
> βούλομαι *wish*: aor. ἐβουλήθην.

Some verbs are deponent in certain tenses but not in others; e.g. the very common ἔρχομαι *I come*. The imperf. indic. (not very frequently found) is ἠρχόμην. The other tenses are supplied from other stems:

Future	ἐλεύσομαι *I shall come*	(middle in form).
Strong aorist	ἦλθον *I came*	(active in form).
Perfect	ἐλήλυθα *I have come*	(active in form).

A number of verbs which are active in form in the present tense have a future tense which is deponent. Amongst the commonest are:

-βαίνω *go*;	future: -βήσομαι	(only found as a compound).
γινώσκω *perceive, understand*;	,,	γνώσομαι.
λαμβάνω *take*;	,,	λήμψομαι.
τίκτω *bring forth*;	,,	τέξομαι.
φεύγω *flee*;	,,	φεύξομαι.

Vocabulary

Μαριάμ, Μαρία, *Mary*

ἀναγινώσκω (compound of γινώσκω) *read*

ἀπέρχομαι (compound of ἔρχομαι) *go away, depart*

ἀποκρίνομαι *answer*: takes dat. of person answered (indirect object)

ἐξέρχομαι (compound of ἔρχομαι) *go out, come out*

κατασκευάζω *prepare*

χαρίζομαι *forgive*: takes dat. of person forgiven (indirect object); acc. of the thing forgiven

EXERCISE 31

Translate into English:

1. ἀποστέλλω τὸν ἄγγελόν μου ὃς κατασκευάσει τὴν ὁδόν σου.
2. ὁ Χριστὸς ἔρχεται ᾧ ὑπακούσομεν. 3. οὗτός ἐστιν ὁ κύριος οὗ τοὺς πόδας οἱ δοῦλοι ἔλουσαν. 4. ὃς γὰρ οὐκ ἔστιν καθ' ὑμῶν, ὑπὲρ ὑμῶν ἐστιν. 5. καὶ πρωῒ ἐξῆλθεν καὶ ἀπῆλθεν εἰς ἔρημον τόπον, κἀκεῖ προσηύχετο. 6. ἡ δὲ τέξεται υἱὸν ὃν οἱ ἄνθρωποι οὐ δέξονται.

EXERCISE 32

Translate into Greek:

1. You begin. 2. We shall pray. 3. He ascertained. 4. You (sing.) lie. 5. They tasted. 6. We touch. 7. They were preaching the gospel. 8. You (plur.) become. 9. We were working. 10. You (sing.) received. 11. I anoint my head. 12. He had himself sacrificed. 13. They were considering. 14. He put forward for himself. 15. We journeyed. 16. You (sing.) forgave. 17. He answered. 18. You (plur.) will take. 19. We shall flee. 20. I have arranged for myself.

EXERCISE 33

Translate into Greek:

1. We work the works of God. 2. Peter answered him, Lord, to whom shall we go? 3. The sheep will flee from him because they perceive not his voice. 4. Now shall the ruler of this world be cast out. 5. The hour cometh in which the Son will receive glory. 6. What do you wish? Shall I release Jesus? 7. We shall receive the apostles who preach the gospel. 8. Mary will bring forth a son who will be God with us. 9. He forgave us the sins about which we prayed. 10. The man whose sheep he had loosed was working in the fields. 11. The book which you are reading was written by a holy prophet. 12. Which spear did you put in the place which I found? 13. The woman whom he answered does not understand his words. 14. We hear what you say, but you deceived us.

Chapter 13

THE IMPERATIVE MOOD

Up to this point we have dealt only with the indicative mood of the verb, which is used to express statements, and to ask direct questions. The imperative mood is used to express commands.

The imperative mood has the present, aorist and perfect tenses. The perfect imperative active, however, is not found at all in the New Testament, and the perfect imperative passive is only found once (in Mark 4.39; πεφίμωσο *be silent*). We can, therefore, safely omit it in this book.

The distinction between the present and the aorist imperatives is not one of time. All imperatives must refer to future time, since an order by its very nature refers to a time subsequent to the giving of the order. The distinction is rather between the kinds of action described, the present imperative denoting an action which is to be prolonged or repeated, and the aorist imperative an action which is to be performed once only. This distinction cannot always be expressed neatly in English, and it is often sufficient to translate either a present or an aorist imperative by a simple word of command.

The imperative has no first person. The nearest approach to a command in the first person is the cohortative "let me do it", "let us do it". The Greek translation for such expressions will be given on page 214.

Notice that the aorist imperative has no augment. The augment is the sign of past time, and as we have said above an imperative can only refer to future time.

THE IMPERATIVES OF λύω

ACTIVE VOICE

Present

	Singular	Plural
2	λῦε *loose* (thou)[1]	λύετε *loose* (ye)
3	λυέτω *let him loose*	λυέτωσαν *let them loose*

[1] Or, more strictly, "keep on loosing", "let him keep on loosing", etc.

Aorist

	Singular	Plural
2	λῦσον *loose* (thou)	λύσατε *loose* (ye)
3	λυσάτω *let him loose*	λυσάτωσαν *let them loose*

MIDDLE VOICE

Present

	Singular	Plural
2	λύου *loose* (thou) *for thyself* [1]	λύεσθε *loose* (ye) *for yourselves*
3	λυέσθω *let him loose for himself*	λυέσθωσαν *let them loose for themselves*

Aorist

	Singular	Plural
2	λῦσαι *loose* (thou) *for thyself*	λύσασθε
3	λυσάσθω etc.	λυσάσθωσαν

PASSIVE VOICE

Present

	Singular	Plural
2	λύου *be thou loosed*	λύεσθε *be ye loosed*
3	λυέσθω *let him be loosed*	λυέσθωσαν *let them be loosed*

N.B. The pres. imper. pass. in Greek is the same as the pres. imper. mid.

[1] Or, more strictly, "keep on loosing for thyself", etc.

It is difficult to bring out in English the sense of continuance without cumbrous language; "continue thou to be loosed".

1st Aorist

	Singular	Plural
2	λύθητι *be thou loosed*	λύθητε
3	λυθήτω etc.	λυθήτωσαν

STRONG AORIST IMPERATIVES

Verbs having a strong aorist indicative active form their imperative active and middle from the strong aorist stem (the unaugmented verb stem), to which are added endings as for the present imperative.   Thus βάλλω *I throw, put*; aor. indic. ἔβαλον; verb stem βαλ-.

Strong aor. imper. act.

	Singular	Plural
2	βάλε	βάλετε
3	βαλέτω	βαλέτωσαν

(Compare the pres. imper. act., βάλλε, βαλλέτω, etc.)

Strong aor. imper. mid.

	Singular	Plural
2	βαλοῦ	βάλεσθε
3	βαλέσθω	βαλέσθωσαν

(Compare the pres. imper. mid., βαλλοῦ, βαλλέσθω, etc.)

Verbs having a strong aorist indicative passive form their imperative passive from the stem of the strong aor. pass. to which are added the endings as for the wk. aor. pass., -θι, -τω, -τε, -τωσαν.

Thus: ἀλλάσσω *change* (found usually as a compound) has a str. aor. indic. pass. ἠλλάγην:

Strong aor. imper. pass.

	Singular	Plural
2	ἀλλάγηθι	ἀλλάγητε
3	ἀλλαγήτω	ἀλλαγήτωσαν

NOTE: The -θι of the 2nd pers. sing. is the proper ending. The -τι which is found in the wk. aor. imper. pass. is written for -θι since Greek avoids a repeated θ: λύθητι for λύθηθι.

The imperatives of many other verbs can be found by following the above models; e.g.:

λαμβάνω *take* (str. aor. ἔλαβον).

    Pres. imper. act. 2nd pers. sing.: λάμβανε.

    Str. aor. imper. act. 2nd pers. sing.: λαβέ (for accent see p. 275).

διώκω *pursue*.

    Wk. aor. imper. act. 2nd pers. sing.: δίωξον.

    Wk. aor. imper. pass. 2nd pers. sing.: διώχθητι.

THE IMPERATIVE OF εἰμί *I am*

The imperative of εἰμί *I am*, is as follows:

	Singular	Plural
2	ἴσθι *be* (thou)	ἔστε *be* (ye)
3	ἔστω *let him* or ἤτω (her, it) *be*	ἔστωσαν *let them be*

THE AORIST IMPERATIVES OF -βαίνω AND γινώσκω

-βαίνω and γινώσκω have unthematic strong aorists (see pp. 192–4), -ἔβην, -ἔβης, etc., and ἔγνων, ἔγνως, etc.

The aorist imperatives are:

	Singular	Plural
2	-βηθι or -βα	-βατε
3	-βάτω	-βάτωσαν

	Singular	Plural
2	γνῶθι	γνῶτε
3	γνώτω	γνώτωσαν

Examples of the Use of the Imperative

λέγει αὐτοῖς, Ἔρχεσθε (John 1.39) *He saith unto them, Come.*
γεμίσατε τὰς ὑδρίας (John 2.7) *fill the waterpots.*

(γεμίζω *fill*: aor. indic. act. ἐγέμισα. The aor. imper. is used, since the action is to be performed once only.)

νεκροὺς ἐγείρετε, λεπροὺς καθαρίζετε, δαιμόνια ἐκβάλλετε (Matt. 10.8) *raise the dead, cleanse the lepers, cast out devils.*

(Here present imperatives are used since the actions are to be repeated.)

The difference in meaning between the present and aorist imperatives is well illustrated by St Matthew's and St Luke's versions of the petition for bread in the Lord's Prayer:

St Matthew (6.11) has: τὸν ἄρτον ἡμῶν τὸν ἐπιούσιον δὸς ἡμῖν σήμερον *give us this day our daily bread.*

δός is aorist imperative of δίδωμι *give* (see p. 241), the aorist being used since the verb refers to a single act of giving "this day".

St Luke (11.3) has τὸν ἄρτον ἡμῶν τὸν ἐπιούσιον δίδου ἡμῖν τὸ καθ' ἡμέραν *give us day by day our daily bread.*

δίδου is the present imperative, used since the verb refers to repeated acts of giving "day by day".

φυτεύθητι ἐν τῇ θαλάσσῃ (Luke 17.6) *be thou planted in the sea.*
διαλλάγηθι τῷ ἀδελφῷ σου (Matt. 5.24) *be reconciled to thy brother.*
(2nd pers. sing. str. aor. imper. pass. of διαλλάσσω.)

Prohibitions or Negative Commands

The negative μή is used with the imperative.

μὴ κρίνετε (Matt. 7.1) *judge not.*

μὴ κωλύετε αὐτά (Luke 18.16) *forbid them not.* (κωλύω literally means "prevent".)

Prohibitions are also frequently expressed by μή and the aorist subjunctive, but this construction will be explained later, p. 215.

THIRD DECLENSION NOUNS—*continued*

Nouns with stems ending in the nasal ν

There are two types: (*a*) those which have strong flexion, which is a marked variation in the vowel of the stem as the noun declines, and (*b*) those with no such variation in the stem.

(*a*)	ὁ ποιμήν *shepherd* stem ποιμεν-	ὁ ἡγεμών *leader* stem ἡγεμον-
Sing. N., V.	ποιμήν	ἡγεμών
A.	ποιμένα	ἡγεμόνα
G.	ποιμένος	ἡγεμόνος
D.	ποιμένι	ἡγεμόνι
Plu. N., V.	ποιμένες	ἡγεμόνες
A.	ποιμένας	ἡγεμόνας
G.	ποιμένων	ἡγεμόνων
D.	ποιμέσι(ν)	ἡγεμόσι(ν)

Strong flexion is shown by the variation in the length of the vowel of the stem; η, ε and ω, ο.

Declined liked ποιμήν is   ὁ λιμήν, λιμένος *harbour.*

Declined like ἡγεμών are ὁ βραχίων, βραχίονος *arm.*

                       ὁ γείτων, γείτονος *neighbour.*

                       ἡ εἰκών, εἰκόνος *image.*

                       ὁ κύων, κυνός *dog* (notice that the stem is κυν-).

                       ἡ χιών, χιόνος *snow.*

(b)

		ὁ Ἕλλην Greek stem Ἑλλην–	ὁ αἰών age stem αἰων–
Sing.	N., V.	Ἕλλην	αἰών
	A.	Ἕλληνα	αἰῶνα
	G.	Ἕλληνος	αἰῶνος
	D.	Ἕλληνι	αἰῶνι
Plur.	N., V.	Ἕλληνες	αἰῶνες
	A.	Ἕλληνας	αἰῶνας
	G.	Ἑλλήνων	αἰώνων
	D.	Ἕλλησι(ν)	αἰῶσι(ν)

Note the dat. plur. of these nouns. Where the consonants ν and σ are brought together in inflexion, the ν may be dropped. Usually compensation takes place in the lengthening of the previous vowel if it is short, but this does not occur in the dat. plur. of 3rd decl. nouns of strong flexion (see above). Thus we have dat. plur. ποιμέσι and ἡγεμόσι instead of the expected ποιμῆσι and ἡγεμῶσι.

Declined like Ἕλλην is ὁ μήν, μηνός month.

Declined like αἰών are ὁ ἀμπελών, ἀμπελῶνος vineyard.

ὁ χειμών, χειμῶνος winter.

ὁ χιτών, χιτῶνος tunic.

NOUNS WITH STEMS ENDING IN THE LIQUID ρ

Again there are two types: (a) those with strong flexion, and (b) those with none, or with partial strong flexion.

(a)

		ὁ πατήρ father stem πατερ–	ὁ ἀνήρ man stem ἀνερ–
Sing.	N.	πατήρ	ἀνήρ
	V.	πάτερ	ἄνερ
	A.	πατέρα	ἄνδρα
	G.	πατρός	ἀνδρός
	D.	πατρί	ἀνδρί
Plur.	N., V.	πατέρες	ἄνδρες
	A.	πατέρας	ἄνδρας
	G.	πατέρων	ἀνδρῶν
	D.	πατράσι(ν)	ἀνδράσι(ν)

Strong flexion is shown by the variation in the length of the vowel of the stem, η, ε. It disappears altogether in gen. and dat. sing.

NOTICE: 1. The δ introduced into the flexion of ἀνήρ is simply a helping consonant to facilitate pronunciation. Cf. French *tendre* from Lat. *tener*.

2. The -ασι of the dat. plur. is a survival of a primitive form.

Declined like πατήρ are ἡ γαστήρ, γαστρός *belly, womb.*

ἡ θυγάτηρ, θυγατρός *daughter.*

ἡ μητήρ, μητρός *mother.*

No other nouns are declined like ἀνήρ.

(b)	ὁ ῥήτωρ *orator* stem ῥητορ- (partial strong flexion)	ὁ σωτήρ *saviour* stem σωτηρ-	ἡ χείρ *hand* stem χειρ-	ὁ μάρτυς *witness* stem μαρτυρ-
Sing. N.	ῥήτωρ	σωτήρ	χείρ	μάρτυς
A.	ῥήτορα	σωτῆρα	χεῖρα	μάρτυρα
G.	ῥήτορος	σωτῆρος	χειρός	μάρτυρος
D.	ῥήτορι	σωτῆρι	χειρί	μάρτυρι
Plur. N.	ῥήτορες	σωτῆρες	χεῖρες	μάρτυρες
A.	ῥήτορας	σωτῆρας	χεῖρας	μάρτυρας
G.	ῥητόρων	σωτήρων	χειρῶν	μαρτύρων
D.	ῥήτορσι(ν)	σωτῆρσι(ν)	χερσί(ν)	μάρτυσι(ν)

Notice especially the dat. plur. χερσί(ν) and the nom. sing. μάρτυς not μάρτυρ).

Declined like ῥήτωρ is ὁ ἀλέκτωρ, ἀλέκτορος *cock.*

Declined like σωτήρ is ὁ νιπτήρ, νιπτῆρος *basin.*

ὁ ἀήρ, ἀέρος, *air,* and ὁ ἀστήρ, ἀστέρος *star,* have closer affinities with ῥήτωρ than with σωτήρ, on account of the partial strong flexion. The expected dat. plur. is not, however, found in the N.T. Instead, ἄστροις from τὸ ἄστρον, also meaning *star,* is used.

No other nouns declined like χείρ and μάρτυς are found in the N.T.

Vocabulary

ἐπιστολή, ῆς (1) *letter*
νεφέλη (1) *cloud*
παραβολή, ῆς (1) *parable*

ῥύμη (1) *lane*
συκῆ, ῆς (1) *fig-tree*
πλατεῖα (1) *street*

ὑποκριτής, οῦ (1) *hypocrite*
Αἴγυπτος, ἡ (2) *Egypt*
πῶλος (2) *colt*
ταμεῖον (2) *inner chamber*
Ἰσραήλ, ὁ (indeclinable) *Israel*
αἴρω (aor. act., ἦρα; aor. pass.,
   ἤρθην) *raise*
εἰσέρχομαι (compound of ἔρχομαι) *come in*
ἐκτείνω (compound of τείνω)
   stretch out, forth (aor. ἐξέτεινα)

καταβαίνω (compound of βαίνω)
   go down, come down
προσεύχομαι (compound of εὔχομαι) *pray*
ταράσσω (aor. act., ἐτάραξα; aor.
   pass., ἐταράχθην) *trouble, stir up*
ἐπί (prep.) with dat. *upon*; (in hostile sense) *against, at*
καλῶς (adverb) *rightly, well*
ταχέως (adverb) *quickly*
ὧδε (adverb) *hither, here*

EXERCISE 34
Translate into English:

1. ἐγὼ δὲ λέγω ὑμῖν, προσεύχεσθε ὑπὲρ τῶν ἐχθρῶν ὑμῶν. 2. εἴσελθε εἰς τὸ ταμεῖόν σου καὶ πρόσευξαι τῷ Πατρί σου. 3. ὑποκριτά, ἔκβαλε πρῶτον ἐκ τοῦ ὀφθαλμοῦ σου τὴν δοκόν. 4. ὁ δίκαιος ὑμῶν πρῶτος λίθον ἐπ᾽ αὐτῇ βαλέτω. 5. ἔξελθε ταχέως εἰς τὰς πλατείας καὶ ῥύμας, καὶ τοὺς πτωχοὺς καὶ τυφλοὺς καὶ χωλοὺς εἰσάγαγε ὧδε. 6. κάταβα εἰς τὸν ποταμόν, καὶ βαπτίσθητι.

EXERCISE 35
Translate into Greek:

1. Lead the shepherds here. 2. Do not enter the temple, O Greeks. 3. Let the mother come with her daughters. 4. Listen to the orators. 5. Believe in the Lord. 6. Be thou cast into the sea. 7. Let them write the letters. 8. Work well, and you will have your reward. 9. Pursue these dogs. 10. Let the men be named.

EXERCISE 36
Translate into Greek:

1. Let the house of Israel know the Christ whom God has sent. 2. Draw nigh to the Saviour whose arms are stretched towards you. 3. Let not your heart be troubled. 4. Go into the village and bring the colt. 5. Let the hands of the saints be lifted to their Father in heaven. 6. From the fig-tree learn a parable. 7. And a voice came (ἐγένετο) out of the cloud, This is my beloved Son: hear him. 8. He saith unto the man, Stretch forth thine hand. 9. Take the young child and his mother, and flee into Egypt. 10. Our Father which art in heaven, Hallowed (sanctified) be thy name (ὄνομα), Thy kingdom come.

Chapter 14

THE VERB INFINITE

The moods of the verb which we have so far dealt with, the indicative and the imperative, together with the conjunctive mood which will be dealt with later (see p. 198), comprise what is called the Verb Finite. In these three moods the verb is limited, or defined, by a subject in the nominative case, whether that subject is actually expressed, or left to be understood.

The Verb Infinite, however, consists of those moods in which the verb is not thus limited:

(a) *The infinitive* presents us with the bare idea of the verb, unconnected with any particular subject; e.g. λύειν means "to loose", presenting the idea of the verb without limitation.

(b) *The participle* presents us with the bare idea of the verb as it may be applied, like an adjective, to a suitable noun or pronoun as a description; e.g. ἐσθίων *eating*, presents the idea of the verb "eat" in a general way as it can be applied to any suitable noun or pronoun, *boy, horse*, etc.

(c) *The verbal adjective*, which is much rarer, is an adjective formed by adding -τος to the stem of the verb. Sometimes it is the equivalent of a past passive participle, and sometimes it implies capability. Thus γνωστός means either "known" or "knowable".

We shall now deal with the infinitive, leaving the participle (p. 158ff.) and the verbal adjective (p. 178) until later.

THE INFINITIVE

In English the infinitive is obtained by prefixing the word "to", e.g. "to loose", "to go". This serves as a reminder that in origin the infinitive was a dative or locative case.[1] The Greek present infinitive λύειν originally meant "for loosing" or "in loosing". But, as we shall see, it acquired a much wider range of usage in the developed language.

[1] The locative case was an original case of the noun, used to express "place where" (Latin, *locus*). Vestiges only of the locative case remain in classical Greek and Latin, and in N.T. Greek. In Greek the functions of the locative were assumed largely by the dative, helped with appropriate prepositions.

The Greek use of the infinitive is similar to its use in English. In both languages, for instance, it may be used as the equivalent of a noun. The infinitive, therefore, is a verbal noun. An example or two in English will make this clear:

"To run makes me breathless." Here the infinitive "to run" is the subject of "makes", and is therefore the equivalent of a noun. "I like to run": here the infinitive "to run" is the object of the verb "like", and again, therefore, is the equivalent of a noun.

The English verbal noun ending in -*ing* may be used sometimes instead of the infinitive. The above sentences might equally well be written "running makes me breathless" and "I like running". There is a trap here for the unwary. The word "running" is also the English present participle (as in "the running boy was halted in his stride"): here "running" is not a verbal noun, but a verbal adjective, a different part of speech altogether. The student must be careful not to be misled by the English. He must not translate an English verbal noun by a Greek present participle, but by the infinitive.

The infinitive in Greek, as in English, has tenses. It is also found in all the voices.

THE INFINITIVE MOOD OF λύω

(Refer to pp. 294–8 where λύω is set out in full, in order to see the relation of the infinitives to the other parts of the verb.)

Active Voice	Present	λύειν	*to loose*
	Future	λύσειν	*to be about to loose*
	Aorist	λῦσαι	*to loose* (single action: see p. 101)
	Perfect	λελυκέναι	*to have loosed*
Middle Voice	Present	λύεσθαι	*to loose for oneself*
	Future	λύσεσθαι	*to be about to loose for oneself*
	Aorist	λύσασθαι	*to loose for oneself* (single action)
	Perfect	λελύσθαι	*to have loosed for oneself*
Passive Voice	Present	λύεσθαι	*to be loosed*
	Future	λυθήσεσθαι	*to be about to be loosed*
	Aorist	λυθῆναι	*to be loosed* (single action)
	Perfect	λελύσθαι	*to have been loosed*

NOTICE: 1. The 1st aor. inf. act. and mid. are marked (as in the case of the indicative) by the σ after the stem. The same is true of the fut. infin.

act. and mid. Coalescence takes place when this σ follows a consonant with which this is possible; e.g. διώκω *pursue*: fut. infin. act. διώξειν; aor. infin. act. διῶξαι.

2. The 1st aor. and fut. infin. pass. are marked (as in the case of the indic.) by θη after the stem.

3. The aorist infinitives have no augment. They do not refer to past time.

4. The perfect infinitives show the reduplication, and the 1st perf. infin. act. shows the typical κ after the stem.

THE INFINITIVES OF STRONG TENSES

The Infinitives of the Strong Aorist add -εῖν to the verb stem in the active, -εσθαι in the middle, and -ῆναι in the passive.

The infinitive of the strong future passive adds -ήσεσθαι to the verb stem.

The infinitive of the strong perfect active adds -έναι to the reduplicated verb stem.

These infinitives may be illustrated from βάλλω *throw*, γράφω *write*, and σπείρω *sow*.

Str. aor. infin. act.	βαλεῖν *to throw*
Str. perf. infin. act.	γεγραφέναι *to have written*
Str. aor. infin. middle	βάλεσθαι *to throw for oneself*
Str. fut. infin. pass.	σπαρήσεσθαι *to be about to be sown*
Str. aor. infin. pass.	σπαρῆναι *to be sown*

FINDING THE INFINITIVES OF OTHER VERBS

With little difficulty the student should be able to arrive at the infinitives of other verbs. The infinitives given above will provide the pattern. He must remember especially:

(*a*) That the σ of the aor. and fut. infinitives, active and middle, is liable to coalesce with a preceding consonant: e.g. πέμπω *I send*: 1st aor. infin. act., πέμψαι.

(*b*) That the aor. infin. has no augment; e.g. ἄγω *I lead*: str. aor. indic. act., ἤγαγον; str. aor. infin. act., ἀγαγεῖν.

(*c*) That in the perf. infin. the stem is reduplicated. (If a tense has reduplication in the indic., it has it in all its other moods.)

MEANING OF PRESENT AND AORIST INFINITIVES

The distinction between the present and aorist infinitives is not one of time, but is in the nature of the action. The aorist infinitive λῦσαι does not mean "to have loosed", but "to loose", referring simply to the action without defining its time. The true nature of the Greek aorist tense (Gr. ἀόριστος *undefined*) is perhaps most clearly seen in this use of the aor. infin. The pres. infin. on the other hand is used to refer to an action which is prolonged or repeated. The aor. infin. is more frequently used than the present, which is only employed when attention is to be drawn to the prolonging or repetition of the action.

> ὑπάγω ἁλιεύειν (John 21.3) *I go a fishing* (lit. "to fish"). Here the pres. infin. is used since Peter is referring not to a single occasion but to a prolonged course of action.
>
> Ἡρῴδης θέλει σε ἀποκτεῖναι (Luke 13.31) *Herod would fain kill thee* (lit. "Herod wishes to kill thee"). Here the aor. infin. is used, since a single act of killing is referred to.

THE ARTICULAR INFINITIVE

The infinitive, being used as a noun, is regarded as being neuter, and it may have the neuter article τό.

> ἐμοὶ γὰρ τὸ ζῆν Χριστὸς καὶ τὸ ἀποθανεῖν κέρδος (Philippians 1.21) *for to me to live* (is) *Christ, and to die* (is) *gain.* Here the infinitives ζῆν (see p. 114) and ἀποθανεῖν are nominative, being the subjects of ἐστίν which is to be understood.

The articular infinitive may be used in cases other than the nominative. It is frequently found in the dative case after the preposition ἐν, used in the sense of "in the time that", "during", or "while".

> ἐν δὲ τῷ λαλῆσαι (Luke 11.37) *now as he spake* ... (lit. "in the speaking").

The subject of the infinitive, where it is expressed, is put into the accusative.

> ἐν δὲ τῷ ὑποστρέφειν τὸν Ἰησοῦν (Luke 8.40) *and as Jesus returned* ... (lit. "in (during) the returning (of) Jesus").
>
> ἐν τῷ εἶναι αὐτὸν ἐν μιᾷ τῶν πόλεων (Luke 5.12) *while he was in one of the cities* ... (lit. "in (during) his being in ...").

The articular infinitive may also be used after other prepositions, e.g.

μετὰ δὲ τὸ ἐγερθῆναί με (Matt. 26.32) *but after I am risen again* (lit. "after my being raised"). ἐγερθῆναι is aor. pass. infin. of ἐγείρω: aor. pass. indic., ἠγέρθην. The infin. is here in the accusative after μετά.

THE INFINITIVE AS SUBJECT

οὐκ ἔξεστιν φαγεῖν (Luke 6.4) *it is not lawful to eat.* ἔξεστιν *it is permitted, lawful*, a compound of ἐστίν, is an impersonal verb; i.e. it is not used with a personal subject. In this sentence the infin. φαγεῖν is in reality the subject: "to eat is not lawful".

πειθαρχεῖν δεῖ (Acts 5.29) *we must obey*, or, *it is necessary to obey* (lit. "to obey is necessary"). δεῖ is another impersonal verb. ἔδει *it was necessary.*

δεῖ αὐτὸν εἰς Ἱεροσόλυμα ἀπελθεῖν καὶ πολλὰ παθεῖν (Matt. 16.21) *he must go unto Jerusalem and suffer many things.* Note again that when the subject of the infinitive is expressed it is in the accusative. Here there are two infinitives, both being subjects of the impersonal δεῖ: lit. "(for) him to go ... and to suffer ... is necessary".

Κύριε, καλόν ἐστιν ἡμᾶς ὧδε εἶναι (Matt. 17.4) *Lord, it is good for us to be here* (lit. "(for) us to be here is good"). Note that the καλόν is neuter. The infin. εἶναι as a verbal noun is neuter, and the adjective agreeing with it must be neuter. In turning such an English sentence into Greek the student must beware of translating the *for us* by a dative. Being the subject of the infinitive the pronoun here must be in the accusative.

THE INFINITIVE AS OBJECT

After verbs of ordering and beseeching an infinitive is used as an object in Greek as in English:

ἐκέλευσεν τὸ στράτευμα ἁρπάσαι αὐτόν (Acts 23.10) *he commanded the soldiers to take him by force.*

Here the main verb ἐκέλευσεν has in reality two objects, one being τὸ στράτευμα *the soldiery*, the persons ordered, and the other being the infinitive which represents the *thing* ordered.

παραγγέλλω *order*, however, has the dative of the person ordered.

καὶ παραγγέλλει τῷ ὄχλῳ ἀναπεσεῖν ἐπὶ τῆς γῆς (Mark 8.6) *and he commandeth the multitude to sit down on the ground.*

δέομαι (deponent) *pray, beseech,* has the genitive of person.

δέομαί σου ἐπιβλέψαι ἐπὶ τὸν υἱόν μου (Luke 9.38) *I beseech thee to look upon my son.*

MODAL VERBS

The infinitive also supplies the object to the so-called modal verbs: *will, wish, be able, try, begin.* These are verbs which in themselves present no complete idea, and in Greek as in English need to be completed by an infinitive. E.g. the phrase "they are trying" does not give us a complete idea. We want to know what they are trying. Such verbs are called modal as they make it possible to express a new mode of a verb. Thus, there is no mood of the Greek verb λύω which expresses the idea of being able to loose, or wishing to loose. But the modal verbs δύναμαι *I am able* (see p. 225), βούλομαι or θέλω *I am willing, I wish,* with the infinitive enable this mode or mood to be expressed.

ἐβουλόμην πρότερον πρὸς ὑμᾶς ἐλθεῖν (2 Cor. 1.15) *I wished to come before unto you.*

οὐ δύνασθε Θεῷ δουλεύειν καὶ μαμωνᾷ (Matt. 6.24) *ye cannot serve God and Mammon.*

Note that δύναμαι is followed by the infin. In English *I am able* is followed by the infin., but after *I can* the "to" is omitted.

οὐ θέλετε ἐλθεῖν πρός με (John 5.40) *ye will not* (are not willing to) *come to me.*

THE INFINITIVE EXPRESSING PURPOSE

Purpose may be expressed in English in several ways: e.g. *I sat down in order to think, I sat down to think, I sat down so that I might think,* and (more colloquially) *I sat down so as to think.* These are four different ways of expressing what the purpose of sitting down was.

In Greek also purpose can be expressed in several different ways. At the moment we are only concerned with one, the use of the infinitive (as in the English "I sat down to think"). The other methods are summarized on p. 236.

ὑπάγω ἁλιεύειν (John 21.3) *I go a fishing* (i.e. "I am going to fish").
τότε ὁ ᾿Ιησοῦς ἀνήχθη εἰς τὴν ἔρημον ὑπὸ τοῦ Πνεύματος, πειρασθῆναι ὑπὸ τοῦ διαβόλου (Matt. 4.1) *then was Jesus led up of the Spirit into the wilderness to be tempted of the devil.*

The Infinitive expressing Result

Result or consequence is expressed in English usually in a phrase introduced by "so that" or "so as to"; e.g. *There was a shower of rain so that we
stayed at home. He was so fat as to be ridiculous.*

There is sometimes a little difficulty in deciding whether an English
sentence expresses purpose or consequence; e.g. *They worked so that they
were rewarded.* Does this sentence express purpose or result? A little
thought reveals that we are here told the result or consequence. If the
sentence was intended to convey the purpose for which they worked it
would have been worded differently, "They worked in order to be rewarded", "They worked so that they might be rewarded", or "They
worked for a reward". A moment's thought usually makes clear
whether purpose or consequence is intended. (See also p. 237.)

To express result in Greek the infinitive is used, introduced by ὥστε.
Occasionally the infinitive is used without ὥστε as in Colossians 4.6.
The negative is μή.

> σεισμὸς μέγας ἐγένετο ἐν τῇ θαλάσσῃ, ὥστε τὸ πλοῖον καλύπτεσ-
> θαι ὑπὸ τῶν κυμάτων (Matt. 8.24) *there arose[1] a great tempest in the
> sea, insomuch that (so that) the boat was covered with the waves.*

> ἐθεράπευσεν αὐτόν, ὥστε τὸν κωφὸν λαλεῖν καὶ βλέπειν (Matt.
> 12.22) *he healed him, insomuch that (so that) the dumb man spake and
> saw.*

Note that if the infinitive has a subject it is in the accusative case.

Additional Use of ὥστε

Ὥστε is not always followed by the infinitive. It may be used as an
ordinary conjunction without affecting the construction which follows it.
It then has the meaning "so then", "therefore", "accordingly", and the
negative is οὐ.

> Τὸ σάββατον διὰ τὸν ἄνθρωπον ἐγένετο, καὶ οὐχ ὁ ἄνθρωπος
> διὰ τὸ σάββατον· ὥστε κύριός ἐστιν ὁ Υἱὸς τοῦ ἀνθρώπου καὶ
> τοῦ σαββάτου (Mark 2.27–28) *the sabbath was made for man, and
> not man for the sabbath: therefore the Son of man is Lord also of the
> sabbath.* (In such a sentence there is not so much stress on the
> consequential connection between the ὥστε clause and the pre-
> ceding clause.)

---

1 The examples on this page illustrate some of the different shades of meaning of
the verb γίνομαι *become, be, come to pass.* The aorist ἐγένετο may be translated
"became", "was", "came to pass", "arose", "came into being", "was made".

Vocabulary

ζωή, ῆς (1) *life*

θρίξ, τριχός, ἡ (see p. 9) (3) *hair*

Δαυείδ (indeclinable) *David*

ἀποδέχομαι (compound of δέχομαι) *welcome*

διαφημίζω *spread abroad*

δύναμαι, -ασαι, -αται, -άμεθα, -ασθε, -ανται; infin., δύνασθαι (see p. 225) *can, am able*

ἐκμάσσω *wipe*

ἥκω (pres. with perf. meaning) *have come*

καλύπτω (aor. act., ἐκάλυψα; aor. pass., ἐκαλύφθην) *cover*

καλῶ (aor. act., ἐκάλεσα; see p. 117) *call*

νίπτω *wash*

σώζω or σῴζω *save*

ὑποστρέφω (compound of στρέφω) *return*

εἶναι (infin. of εἰμί *I am*) *to be*

οὐκέτι (μηκέτι where the negative required is μή) *no longer*

οὔπω (μήπω) *not yet*

φανερῶς *openly, clearly*

EXERCISE 37

Translate into English:

1. οὐκ ἦλθον καλέσαι δικαίους ἀλλὰ ἁμαρτωλούς. 2. ὁ δὲ ἤρξατο διαφημίζειν τὸν λόγον (the matter), ὥστε μηκέτι τὸν Ἰησοῦν δύνασθαι φανερῶς εἰς πόλιν (city) εἰσελθεῖν. 3. Δαυεὶδ εἰσῆλθεν εἰς τὸν οἶκον τοῦ Θεοῦ καὶ τοὺς ἄρτους ἔφαγεν, οὓς οὐκ ἔξεστιν φαγεῖν. 4. ἐν δὲ τῷ ὑποστρέφειν τὸν Ἰησοῦν ἀπεδέξατο αὐτὸν ὁ ὄχλος. 5. ὁ Ἰησοῦς λέγει τῇ μητρί, Οὔπω ἥκει ἡ ὥρα μου. 6. καὶ ἤρξατο νίπτειν τοὺς πόδας τῶν μαθητῶν καὶ ἐκμάσσειν.

EXERCISE 38

Translate into Greek:

1. It is not lawful to bring Greeks into the temple. 2. It was necessary for the shepherds to guard their sheep. 3. We commanded the witnesses to be brought from the prison. 4. O Saviour, we beseech thee to hear us. 5. I am willing to be baptized. 6. The father wished to send his daughter to the village. 7. You will begin to read this book. 8. He arranged the tunic so that his hand was covered. 9. While they were walking (use πορεύομαι), the crowd began to shout. 10. The Son of man must come in the glory of the Father.

EXERCISE 39

Translate into Greek:

1. It is good for thee to enter into life lame.   2. Can a devil open the eyes of the blind?   3. I did not come to judge the world but to save the world.   4. She began to wipe his feet with the hairs of her head.   5. And he commanded them not (μή, see p. 239) to speak of these things.   6. He went into the house to eat bread.   7. And while he was coming a crowd was gathering.[1]   8. Is it lawful to heal men on the Sabbath?   9. The Greeks have good orators, but the Jews have good leaders.   10. The apostles so (οὕτως) preached that the Greeks believed.

[1] Consider carefully what voice should be used.

Chapter 15

THIRD DECLENSION NOUNS: STEMS ENDING IN A VOWEL

The 3rd decl. nouns previously dealt with have all had stems ending in a consonant. There are also several types which have stems ending in a vowel or diphthong:

		Stems ending in ι (changing to ε before vowels and in dat. plur.)	Stems ending in υ (changing to ε before vowels and in dat. plur.)
		ἡ πόλις *city* stem πολι-	ὁ πῆχυς *ell, cubit* stem πηχυ-
Sing.	N.	πόλις	πῆχυς
	V.	πόλι	—
	A.	πόλιν	πῆχυν
	G.	πόλεως	πήχεως or πήχεος
	D.	πόλει	πήχει
Plur.	N., V.	πόλεις	πήχεις
	A.	πόλεις	πήχεις
	G.	πόλεων	πηχῶν or πήχεων
	D.	πόλεσι(ν)	πήχεσι(ν)

The irregular accent of the gen. sing. and plur. is apparently due to εω having been pronounced as one syllable. Thus the accent is really paroxytone. See p. 277.

These nouns in -ις, -εως must not be confused with nouns in -ις which have dental stems, e.g. ἐλπίς, ἐλπίδος *hope*; χάρις, χάριτος *grace*.

3rd decl. nouns with stems ending in ι are feminine.

Some N.T. nouns with stems ending in ι and declining like πόλις:

ἄλυσις *chain*	κρίσις *judgement*
ἀνάστασις *resurrection*	κωμόπολις *country town*
ἀπόκρισις *answer*	ὄψις *face, appearance*
ἄφεσις *forgiveness, remission*	πίστις *faith*
βρῶσις *food*	πόσις *drink*
δύναμις *power*	πρόθεσις *setting forth, purpose*
ἔκστασις *amazement, trance*	πρόφασις *pretext*
ζήτησις *questioning, examination*	πώρωσις *hardening*
θλίψις *tribulation*	ῥύσις *flowing, issue*
κατοίκησις *dwelling*	συνείδησις *conscience*
κίνησις *movement, a moving*	ὑπάντησις *meeting*
κοίμησις *rest, a reclining*	

No other nouns declined like πῆχυς are found in the N.T.

	Stems ending in υ (unchanging)	Stems ending in ευ	Stems ending in ου
	ὁ ἰχθύς *fish* stem ἰχθυ-	ὁ βασιλεύς *king* stem βασιλευ-	ὁ βοῦς *ox* stem βου-
Sing. N.	ἰχθύς	βασιλεύς	βοῦς
V.	ἰχθύ	βασιλεῦ	βοῦ
A.	ἰχθύν	βασιλέα	βοῦν
G.	ἰχθύος	βασιλέως	βοός
D.	ἰχθύϊ	βασιλεῖ	βοΐ
Plur. N., V.	ἰχθύες	βασιλεῖς	βόες
A.	ἰχθύας	βασιλεῖς	βόας
G.	ἰχθύων	βασιλέων	βοῶν
D.	ἰχθύσι(ν)	βασιλεῦσι(ν)	βουσί(ν)

NOTICE: The stem of nouns like βασιλεύς changes to ε before a vowel. The accusative plural ends like the nominative in -εῖς (contracted from έ-ες) instead of in έας.

The stem of βασιλεύς originally ended with digamma (βασιλεϝ-) which, like σ, disappeared between two vowels.

Declined like ἰχθύς are ὁ βότρυς, βότρυος *bunch of grapes*
    ἡ ἰσχύς, ἰσχύος *strength*
    ἡ ὀσφύς, ὀσφύος *loins*
    ἡ ὀφρύς, ὀφρύος *brow*
    ἡ στάχυς, στάχυος *ear of corn*
    ἡ ὗς, ὑός *sow*

Declined like βασιλεύς are ὁ ἁλιεύς, ἁλιέως *fisherman*
    ὁ ἀρχιερεύς, ἀρχιερέως *high-priest*
    ὁ γονεύς, γονέως *father* (plur. *parents*)
    ὁ γραμματεύς, γραμματέως *scribe*
    ὁ ἱερεύς, ἱερέως *priest*

Declined like βοῦς are ὁ νοῦς, νοός *mind*
    ὁ πλοῦς, πλοός *voyage*
    ὁ χοῦς, χοός *dust*

(In classical Greek these three nouns (but not βοῦς) belonged to the 2nd decl. with genitive in -οῦ.)

CONTRACTED VERBS

In verbs whose stems end in the vowels α, ε, or ο contraction takes place when this vowel coalesces with the first vowel of the ending. Thus φιλέομεν (the uncontracted form of the 1st pers. plur. pres. indic. act. of φιλῶ) contracts to φιλοῦμεν.

These three types we shall deal with together. This may seem to the student like taking one bite at three cherries, but there is much to be said for this course. These verbs have much in common, and the three sets of rules for contraction are most conveniently learned together. These rules are here summarized, and the student should learn them by heart. It should be noted that these verbs have forms which are subject to contraction in the present tense (all voices and moods) and the imperfect (all voices); but in no others.

Stems ending in α

1. α and ε or η become long α. Thus: τιμά-ετε becomes τιμᾶτε; τιμά-ητε (a subjunctive form) also becomes τιμᾶτε.
2. α and ο or ω become ω. Thus: τιμά-ομεν becomes τιμῶμεν; τιμά-ω becomes τιμῶ.
3. ι is subscript (i.e. written underneath). Thus: τιμά-εις becomes τιμᾷς.
4. υ disappears. Thus: τιμά-ουσι becomes τιμῶσι.

Stems ending in ε

1. ε and ε become ει.   Thus: φιλέ-ετε becomes φιλεῖτε.
2. ε and ο become ου.   Thus: φιλέ-ομεν becomes φιλοῦμεν.
3. ε followed by a long vowel or diphthong disappears.   Thus: φιλέ-ω becomes φιλῶ; φιλέ-ουσι becomes φιλοῦσι.

Stems ending in ο

1. ο and a long vowel become ω.   Thus: δηλό-ητε (a subjunctive form) becomes δηλῶτε.
2. ο and a short vowel become ου.   Thus: δηλό-ομεν becomes δηλοῦμεν; ἐδήλο-ε becomes ἐδήλου.
3. ο before ου disappears.   Thus: δηλό-ουσι becomes δηλοῦσι.
4. Any combination of ο with an ι becomes οι.   Thus: δηλό-εις becomes δηλοῖς.

The inflection of the contracted verbs is very regular.   The student has only to apply the rules of contraction where the final vowel of the stem and the vowel of the ending come together.   The endings are those with which he should already be familiar in λύω.

THE PARADIGMS OF τιμάω *honour*; φιλέω *love*; AND δηλόω *show*

The uncontracted forms are given in brackets.

Present Indicative Active

Sing.	1	τιμῶ (ά-ω)	φιλῶ (έ-ω)	δηλῶ (ό-ω)
	2.	τιμᾷς (ά-εις)	φιλεῖς (έ-εις)	δηλοῖς (ό-εις)
	3.	τιμᾷ (ά-ει)	φιλεῖ (έ-ει)	δηλοῖ (ό-ει)
Plur.	1.	τιμῶμεν (ά-ομεν)	φιλοῦμεν (έ-ομεν)	δηλοῦμεν (ό-ομεν)
	2.	τιμᾶτε (ά-ετε)	φιλεῖτε (έ-ετε)	δηλοῦτε (ό-ετε)
	3.	τιμῶσι(ν) (ά-ουσι)	φιλοῦσι(ν) (έ-ουσι)	δηλοῦσι(ν) (ό-ουσι)

Imperfect Indicative Active

Sing.	1.	ἐτίμων (α-ον)	ἐφίλουν (ε-ον)	ἐδήλουν (ο-ον)
	2.	ἐτίμας (α-ες)	ἐφίλεις (ε-ες)	ἐδήλους (ο-ες)
	3.	ἐτίμα (α-ε)	ἐφίλει (ε-ε)	ἐδήλου (ο-ε)
Plur.	1.	ἐτιμῶμεν (ά-ομεν)	ἐφιλοῦμεν (έ-ομεν)	ἐδηλοῦμεν (ό-ομεν)
	2.	ἐτιμᾶτε (ά-ετε)	ἐφιλεῖτε (έ-ετε)	ἐδηλοῦτε (ό-ετε)
	3.	ἐτίμων (α-ον)	ἐφίλουν (ε-ον)	ἐδήλουν (ο-ον)

Present Imperative Active

Sing.	2	τίμα (α-ε)	φίλει (ε-ε)	δήλου (ο-ε)
	3.	τιμάτω (α-έτω)	φιλείτω (ε-έτω)	δηλούτω (ο-έτω)
Plur.	2.	τιμᾶτε (ά-ετε)	φιλεῖτε (έ-ετε)	δηλοῦτε (ό-ετε)
	3.	τιμάτωσαν (α-έτωσαν)	φιλείτωσαν (ε-έτωσαν)	δηλούτωσαν (ο-έτωσαν)

Present Indicative Middle and Passive

Sing. 1. τιμῶμαι (ά-ομαι)   φιλοῦμαι (έ-ομαι)   δηλοῦμαι (ό-ομαι)
2. τιμᾶσαι[1] (ά-εσαι)   φιλῇ (έ-η)   δηλοῖ (ό-η)
3. τιμᾶται (ά-εται)   φιλεῖται (έ-εται)   δηλοῦται (ό-εται)
Plur. 1. τιμώμεθα (α-όμεθα)   φιλούμεθα (ε-όμεθα)   δηλούμεθα (ο-όμεθα)
2. τιμᾶσθε (ά-εσθε)   φιλεῖσθε (έ-εσθε)   δηλοῦσθε (ό-εσθε)
3. τιμῶνται (ά-ονται)   φιλοῦνται (έ-ονται)   δηλοῦνται (ό-ονται)

Imperfect Indicative Middle and Passive

Sing. 1. ἐτιμώμην (α-όμην)   ἐφιλούμην (ε-όμην)   ἐδηλούμην (ο-όμην)
2. ἐτιμῶ (ά-ου)   ἐφιλοῦ (έ-ου)   ἐδηλοῦ (ό-ου)
3. ἐτιμᾶτο (ά-ετο)   ἐφιλεῖτο (έ-ετο)   ἐδηλοῦτο (ό-ετο)
Plur. 1. ἐτιμώμεθα (α-όμεθα)   ἐφιλούμεθα (ε-όμεθα)   ἐδηλούμεθα (ο-όμεθα
2. ἐτιμᾶσθε (ά-εσθε)   ἐφιλεῖσθε (έ-εσθε)   ἐδηλοῦσθε (ό-εσθε))
3. ἐτιμῶντο (ά-οντο)   ἐφιλοῦντο (έ-οντο)   ἐδηλοῦντο (ό-οντο)

Present Imperative Middle and Passive

Sing. 2. τιμῶ (ά-ου)   φιλοῦ (έ-ου)   δηλοῦ (ό-ου)
3. τιμάσθω (ά-εσθω)   φιλείσθω (ε-έσθω)   δηλούσθω (ο-έσθω)
Plur. 2. τιμᾶσθε (ά-εσθε)   φιλεῖσθε (έ-εσθε)   δηλοῦσθε (ό-εσθε)
3. τιμάσθωσαν   φιλείσθωσαν   δηλούσθωσαν
(α-έσθωσαν)   (ε-έσθωσαν)   (ο-έσθωσαν)

ACCENTUATION

It must not be assumed that where contraction takes place the resulting diphthong or vowel is always accented with the circumflex. At first this appears to be the case if attention is paid only to the pres. indic. act. But note e.g. the accent of the imperf. ἐτίμων. The correct accent may be found in the following way:

(a) Write out the uncontracted form of the verb, and accent it according to the ordinary rules (see p. 274); e.g. τιμάομεν, φιλέομαι.

(b) Place a grave accent over any syllable which does not already possess an accent. (In Greek any syllable with no written accent is regarded as having a grave accent.) Thus: τὶμάὸμὲν, φὶλέὸμαὶ.

(c) When two syllables come together in contraction in such a way that a grave accent follows an acute, a circumflex results. Thus τιμάὸμεν gives τιμῶμεν, φιλέομαι gives φιλοῦμαι.

(d) When two syllables come together in contraction in such a way that an acute accent follows a grave, the acute accent remains. Thus τιμάόμεθα gives τιμώμεθα and ἐφιλὲόμην gives ἐφιλούμην.

[1] τιμᾷ in classical Greek.

The Other Tenses of the Contracted Verbs

There is no contraction of stem and termination in any tenses but the present and imperfect, for in all the other tenses a consonant precedes the termination. In the future and aorist with very few exceptions the final vowel of the stem is lengthened, α into η, ε into η, and o into ω. The appropriate endings as in λύω are then added.

In the perfect the final vowel of the stem is similarly lengthened, reduplication takes place, and the endings as in λύω are added.

Fut. indic. act.	τιμήσω	φιλήσω	δηλώσω
Fut. indic. mid.	τιμήσομαι	φιλήσομαι	δηλώσομαι
Fut. indic. pass.	τιμηθήσομαι	φιληθήσομαι	δηλωθήσομαι
Aor. indic. act.	ἐτίμησα	ἐφίλησα	ἐδήλωσα
Aor. indic. mid.	ἐτιμησάμην	ἐφιλησάμην	ἐδηλωσάμην
Aor. indic. pass.	ἐτιμήθην	ἐφιλήθην	ἐδηλώθην
Aor. imper. act.	τιμῆσον	φιλῆσον	δηλῶσον
Aor. imper. mid.	τιμῆσαι	φιλῆσαι	δηλῶσαι
Aor. imper. pass.	τιμήθητι	φιλήθητι	δηλώθητι
Perf. indic. act.	τετίμηκα	πεφίληκα	δεδήλωκα
Perf. indic. mid.	τετίμημαι	πεφίλημαι	δεδήλωμαι
Perf. indic. pass.	τετίμημαι	πεφίλημαι	δεδήλωμαι

The Infinitives of the Contracted Verbs

Pres. infin. act.	τιμᾶν	φιλεῖν	δηλοῦν
Pres. infin. mid. and pass.	τιμᾶσθαι	φιλεῖσθαι	δηλοῦσθαι
Aor. infin. act.	τιμῆσαι	φιλῆσαι	δηλῶσαι
Aor. infin. mid.	τιμήσασθαι	φιλήσασθαι	δηλώσασθαι
Aor. infin. pass.	τιμηθῆναι	φιληθῆναι	δηλωθῆναι

NOTICE: The pres. infin. termination was originally -εεν: hence τιμαεεν = τιμᾶν; δηλοεεν = δηλοῦν (no *iota subscript*).

We give here some lists of contracted verbs, including all those which appear in St Mark's Gospel. It is not suggested that the student should at this stage learn all these by heart. They are given here in order to show how extensive is this class of verb.

The uncontracted form of the 1st pers. sing. of the pres. ind. act. is given. The few irregularities met with in this class of verb are noted.

STEMS IN α

ἀγαπάω *love*

ἀναπηδάω (compound) *leap up*

-ἀντάω *meet*. Compounds ἀπαντάω, ὑπαντάω, also mean *meet*; aor. act. ὑπήντησα

βλαστάω *sprout, spring up*

βοάω *cry, shout*

γελάω *laugh*. Irreg. fut. γελάσω. Compound καταγελάω *laugh to scorn*

δαπανάω *spend*

διαπεράω (compound) *cross over*

ἐρωτάω *ask, question*. Compound ἐπερωτάω *question*

κλάω *break*. Compound κατακλάω *break in pieces*

μεριμνάω *be anxious*. Compound προμεριμνάω *be anxious beforehand*

ὁρμάω *rush*

πεινάω *hunger*. Irreg. fut. πεινάσω; aor. act. ἐπείνασα

πλανάω *lead astray*; pass., *be led astray, err*. Compound ἀποπλανάω *lead astray*

σιωπάω *be silent*

σπάω *draw*. Compound διασπάω *break asunder*

τελευτάω *die*

τιμάω *honour*. Compound ἐπιτιμάω *rebuke*

τολμάω *dare*

χαλάω *loosen, let down*

ὁράω *see* is to be included in this class. Imperf. act. ἑώρων; perf. act. ἑόρακα or ἑώρακα. But the stem ὁρα- is not found in the fut. or aor., and these tenses are supplied from other stems Ϝειδ- and ὀπ- (hence called suppletives): str. aor. act. εἶδον *I saw*; fut. ὄψομαι (deponent); fut. pass. ὀφθήσομαι; aor. pass. ὤφθην

DEPONENTS

αἰτιάομαι *accuse*

ἐμβριμάομαι (compound) *be moved with anger*

θεάομαι *behold*. α pure (see below): aor. ἐθεασάμην; perf. τεθέαμαι

ἰάομαι *heal*. α pure (see below): aor. ἰασάμην; aor. pass. ἰάθην *I was healed*

καταράομαι *curse*

μοιχάομαι *commit adultery*

α *pure*:

Where the final α of the stem follows ε, ι, or ρ it lengthens in the future, aorist and perfect into ᾱ not η (α pure—cf. its occurrence in 1st decl. nouns, p. 27). The following are examples, but none of them happens to occur in St Mark. (But see several deponents above.)

ἀγαλλιάω *exult, rejoice greatly.* Aor. act. ἠγαλλίασα; aor. pass. ἠγαλλιάθην
ἐάω *allow.* Fut. ἐάσω; aor. act. εἴασα (note irreg. augment)
κοπιάω *toil.* Perf. act. κεκοπίακα

Two verbs usually given in the dictionaries as ζάω (contracting to ζῶ), *live* and χράομαι (contracting to χρῶμαι), *use*, in reality have stems ending in η, ζήω, χρήομαι.
Parts not found in the N.T. are bracketed.

Present Indicative

Sing.	1.	ζῶ	(χρῶμαι)
	2.	ζῇς	(χρᾶσαι)
	3.	ζῇ	(χρᾶται)
Plur.	1.	ζῶμεν	χρώμεθα
	2.	ζῆτε	(χρᾶσθε)
	3.	ζῶσι(ν)	χρῶνται

Imperfect Indicative

Sing.	1.	ἔζων	(ἐχρώμην)
	2.	(ἔζης)	(ἐχρῶ)
	3.	(ἔζη)	(ἐχρᾶτο)
Plur.	1.	(ἐζῶμεν)	(ἐχρώμεθα)
	2.	ἐζῆτε	(ἐχρᾶσθε)
	3.	(ἔζων)	ἐχρῶντο

Present infinitive   ζῆν (χρᾶσθαι).
Present imperative   χρῶ.
Present participle (see chapter 20)   ζῶν, ζῶσα, ζῶν.
                                      χρώμενος, χρωμένη, χρώμενον.
Aorist indicative   ἔζησα, ἐχρησάμην.

STEMS IN ε

ἀγανακτέω *be indignant*
ἀγνοέω *be ignorant*
ἀγρυπνέω *be wakeful, watchful*

ἀδημονέω *be troubled*
ἀθετέω *reject*
αἰτέω *ask* (of requests).   Compound παραιτέομαι (dep.) *beg, beg off*
ἀκολουθέω *follow*.   Compounds ἐπακολουθέω *follow after*; παρα-
κολουθέω *follow up closely*; συνακολουθέω *accompany*
ἀπιστέω *be faithless*
ἀποδημέω (compound) *go abroad*
ἀπορέω *be perplexed*
ἀποστερέω (compound) *deprive*
ἀσθενέω *be sick*
βλασφημέω *blaspheme*
βοηθέω *help*
γαμέω *marry*.   Aor. ἐγάμησα and ἔγημα both found
γονυπετέω *fall down before, kneel*
γρηγορέω *be awake, watch*
διακονέω *minister*.   Fut. διακονήσω; aor. διηκόνησα as though the verb
    were a compound δια-κονέω.   Actually it is not a compound, the
    stem being derived from the noun διάκονος.
δοκέω *suppose, seem*.   Compound εὐδοκέω *think it good, be well pleased*
δωρέω *bestow*
ἐλεέω *pity*.   Fut. ἐλεήσω; aor. ἠλέησα; perf. act. not found; perf. pass.
    ἠλέημαι; aor. pass. ἠλεήθην (ἐλεάω is also found in pres. indic.)
ἐξουθενέω or ἐξουδενέω *set at nought, despise*.   Derived from οὐδέν
    nothing
ἐνειλέω (compound) *roll in, wind in*
εὐκαιρέω *to have leisure*
εὐλογέω *bless*.   Compound κατευλογέω *bless fervently*
εὐχαριστέω *give thanks*
ζητέω *seek*.   Compound συνζητέω *discuss, dispute*
θαμβέω *astonish*.   Compound ἐκθαμβέω *astonish, terrify*
θαρρέω, or θαρσέω *be confident, take heart*
θεωρέω *behold*
θορυβέω *make a noise, trouble*
θροέω *make an outcry*.   In N.T. pass. only found, *to be troubled*
κακολογέω *speak ill of, abuse*
καρτερέω *be stedfast*.   Compound προσκαρτερέω *continue stedfastly*
κατηγορέω *accuse*.   Fut. κατηγορήσω; aor. act. κατηγόρησα
κινέω *move*
κληρονομέω *inherit*
κρατέω *be strong, take hold*
λαλέω *talk*.   Compound συνλαλέω or συλλαλέω *talk with*

λατομέω *hew* (stones)

λυπέω *grieve.* Compound συνλυπέω or συλλυπέω. In pass., *be moved with grief, sympathize*

μαρτυρέω *bear witness.* Compounds, καταμαρτυρέω *bear witness against*; ψευδομαρτυρέω *bear false witness*

μετρέω *measure*

μισέω *hate*

νοέω *understand.* Compound μετανοέω *repent*

οἰκοδομέω *build.* Fut. οἰκοδομήσω: aor. act. ῳκοδόμησα. But the aor. pass. shows no augment, οἰκοδομήθην. So also perf. pass., οἰκοδόμημαι.

ὁμολογέω *agree, confess.* Compound ἐξομολογέω *profess*; in mid., *confess, give praise*

πενθέω *mourn*

περιπατέω (compound) *walk*

ποιέω *make, do.* Compound κακοποιέω *do evil*

προσκυνέω (compound) *worship*

πωλέω *sell*

συνεργέω *work together*

τηρέω *guard, keep.* Compounds, παρατηρέω *watch closely*; συντηρέω *keep safe*

ὑμνέω *sing, sing praise*

ὑστερέω *come short of, be in want of, fail*

φιλέω *love, kiss.* Compound καταφιλέω *kiss fondly*

φορέω *carry, wear.* Compound καρποφορέω *bear fruit*

φρονέω *think, be mindful of*

φωνέω *call*

χωρέω *make room, have a place, hold.* Compound ἀναχωρέω *withdraw*

ὠφελέω *help*

αἱρέω *take*, is to be included in this class. But the fut. and aor. act. are supplied from another root ἑλ-. Fut. act., ἑλῶ (but αἱρήσω is found in contemporary Greek); str. aor. act. -εῖλον (infinitive ἑλεῖν); aor. pass. ᾑρέθην; perf. pass. -ᾕρημαι.

The compound καθαιρέω *take down, destroy*, is found in St Mark.

DEPONENTS

ἀρνέομαι *deny.* Compound ἀπαρνέομαι, also means "deny"

ἡγέομαι *lead, think.* Compound διηγέομαι *describe*

ὀρχέομαι *dance*

φοβέομαι *fear.* A passive deponent; aor. ἐφοβήθην

In the following the final ε of the stem does not lengthen in the fut., aor., and perfect. Of these only καλέω, its compounds, and συντελέω are found in St Mark.

αἰνέω *praise.* Fut., αἰνέσω; aor. act., ᾔνεσα
ἀρκέω *suffice.* Fut. ἀρκέσω; aor. act., ἤρκεσα
καλέω *call.* Fut., καλέσω; aor. act., ἐκάλεσα; there is a variation of stem in other tenses: perf. act. κέκληκα; perf. pass., κέκλημαι; aor. pass., ἐκλήθην. Compounds παρακαλέω *beseech, comfort*; προσκαλέω (in mid.) *call to oneself*; συνκαλέω *call together*
τελέω *finish.* Fut., –τελέσω; aor. act., ἐτέλεσα; perf. act., τετέλεκα; perf. pass., τετέλεσμαι; aor. pass., ἐτελέσθην. Compound συντελέω *finish, accomplish*

In verbs with stems of one syllable ending in ε, the contraction of εε and εει into ει takes place, but εω and εο are left uncontracted; e.g.:

πλέω *I sail* δέομαι (deponent) *I entreat*

		Present indicative	Present indicative
Sing.	1.	πλέω	δέομαι
	2.	πλεῖς	δέῃ
	3.	πλεῖ	δεῖται
Plur.	1.	πλέομεν	δεόμεθα
	2.	πλεῖτε	δεῖσθε
	3.	πλέουσι(ν)	δέονται

		Imperfect indicative	Imperfect indicative
Sing.	1.	ἔπλεον	ἐδεόμην
	2.	ἔπλεις	ἐδέου
	3.	ἔπλει	ἐδεῖτο
Plur.	1.	ἐπλέομεν	ἐδεόμεθα
	2.	ἐπλεῖτε	ἐδεῖσθε
	3.	ἔπλεον	ἐδέοντο

The principal N.T. verbs in this class are:

δέω *bind.* Fut. δήσω; aor. act. ἔδησα; but perf. act. δέδεκα; perf. pass. δέδεμαι; aor. pass. ἐδέθην

δέομαι (deponent) *entreat.*　Aor. ἐδεήθην
πλέω *sail.*　Aor. act. ἔπλευσα
πνέω *breathe.*　Aor. act. ἔπνευσα
ῥέω *flow.*　Fut. ῥεύσω
-χέω *pour.*　Several irregularities: fut. -χεῶ; aor. act. -ἔχεα; perf. pass.
　-κέχυμαι; aor. pass. -ἐχύθην.　Compound ἐκχέω *pour out*

The stem of these verbs originally ended with digamma, πλεϝ, δεϝ-,
etc., which, like σ, disappeared between two vowels.　But in the case of
an original ϝ contraction of εω, εο, εου, εοι, and εη does not take place.

Stems in o

ἀκυρόω *invalidate*
βεβαιόω *confirm*
ζημιόω *damage.*　In pass., *lose*
θανατόω *put to death*
κεφαλιόω *wound in the head*
κοινόω *make common*
κολοβόω *cut off*
μορφόω *form.*　Compound μεταμορφόω *transform*
ὁμοιόω *liken.*　Fut. ὁμοιώσω; aor. act. ὡμοίωσα; aor. pass. ὡμοιώθην;
　perf. pass. has both -ὡμοίωμαι and -ὁμοίωμαι
πληρόω *fill, fulfil*
σκηνόω *dwell.*　Compound κατασκηνόω *dwell*
σταυρόω *crucify.*　Compound συνσταυρόω *crucify together with*
φανερόω *make manifest*
φραγελλόω *scourge*

The following are common, although not found in St Mark:

δικαιόω *declare righteous, justify*
ἐλευθερόω *make free*
ταπεινόω *humble*
τελειόω *finish, make perfect*
τυφλόω *blind*
ὑψόω *exalt, lift up*

Vocabulary

γενεά, -ᾶς (1) *generation*
Μωυσῆς, -έως, ὁ *Moses.* Acc. -ῆν
　and -έα, dat. -ῆ and -εῖ

ὄφις, -εως, ὁ (the only masc. noun
　of this type) *serpent*
παράδοσις, -εως, ἡ *tradition*

ἀγοράζω *buy*

ἀκολουθέω (takes dat.) *follow*

ἀποκτείνω *kill* (aor. ἀπέκτεινα).

ἀριθμέω *number*

δέομαι (takes gen. of person) *en-treat, beseech*·

κατηγορέω (takes gen. of person) *accuse*

διὰ τί; *why?* (literally, "on account of what?")

καθώς *as, even as* (usually balanced by οὕτως, *so*, in the corresponding clause)

Translate into English:

1. καὶ ἐπερωτῶσιν αὐτὸν Διὰ τί οὐ περιπατοῦσιν οἱ μαθηταί σου κατὰ τὴν παράδοσιν τῶν πρεσβυτέρων; 2. ὁ δὲ εὐθὺς ἐλάλησεν μετ᾽αὐτῶν, καὶ λέγει αὐτοῖς Θαρσεῖτε, μὴ φοβεῖσθε. 3. παρετηροῦντο δὲ αὐτὸν οἱ γραμματεῖς καὶ οἱ Φαρισαῖοι. 4. τίνι οὖν ὁμοιώσω τοὺς ἀνθρώπους τῆς γενεᾶς ταύτης; 5. τὰ ἔργα τοῦ Θεοῦ ἐφανερώθη ἐν αὐτῷ. 6. ἐδεήθην τῶν μαθητῶν σου ἐκβαλεῖν τὸ δαιμόνιον.

Translate into Greek:

1. I will let down the nets. 2. The scripture was fulfilled. 3. Whom seekest thou? 4. What shall I do? 5. He called them, and they followed him. 6. His disciples asked him this. 7. You accused the king. 8. We were building the city. 9. They were made free by the power of the resurrection. 10. Some will be humbled, but others will be exalted.

Translate into Greek:

1. The fishermen were selling fish and buying bread. 2. Parents were put to death by children. 3. Honour thy father and thy mother. 4. Her daughter was healed by faith. 5. And they call the blind man, and say, Be of good comfort; he calleth thee. 6. They were afraid to ask him. 7. The bunches of grapes have been numbered by the high-priest. 8. Ye seek to kill me, because my word hath no place in you. 9. And as Moses lifted up the serpent in the wilderness, even so must the Son of man be lifted up. 10. They dwelt among the sheep and oxen.

Chapter 16

THIRD DECLENSION NEUTER NOUNS

There are three main types. As with neuter nouns of the 2nd declension, the same form is used for the nom., voc., and acc. both in the singular and in the plural.

		(1) *Nom. in -α and* *stem in -ατ-*	(2) *Nom. in -ας and* *stem in -ατ-*	(3) *Nom. in -ος and* *stem in -ες-*
		τὸ γράμμα *letter* Stem γραμματ-	τὸ κέρας *horn* Stem κερατ-	τὸ γένος *race* Stem γενες-
Sing.	N., V., A. G. D.	γράμμα γράμματος γράμματι	κέρας κέρατος κέρατι	γένος γένους γένει
Plur.	N., V., A. G. D.	γράμματα γραμμάτων γράμμασι(ν)	κέρατα κεράτων κέρασι(ν)	γένη γενῶν γένεσι(ν)

NOTICE: In the dat. plur. in classes (1) and (2) the τ is dropped before the σι. See p. 7. Contraction takes place in nouns of class (3). The gen. sing. in its full form is γένεσος. The ς drops out between two vowels (see p. 9) leaving γένεος which contracts (see p. 110) to γένους. So in the dat. sing., the ς drops out of γένεσι leaving γένει. In the plural γένη is for γένεα (see p. 10).

Declined like γράμμα, -τος are:

αἷμα *blood*   πνεῦμα *spirit*
βρῶμα *food*   ῥῆμα *word, saying*
θέλημα *will*   σπέρμα *seed*
νόσημα *sickness*   στόμα *mouth*
ὄνομα *name*   σῶμα *body*

Declined like κέρας, κέρατος are:

ἅλας *salt* (but in Mark 9.50, ἅλα is found as an acc.)

πέρας *end*

τέρας *marvel*

φῶς, φωτός *light*, is similarly declined

Declined like γένος, γένους are:

ἔθνος *nation*	πλῆθος *crowd*
ἔθος *custom*	σκέλος *leg*
εἶδος *appearance, form*	σκεῦος *vessel*
ἔτος *year*	σκότος *darkness*
μέρος *part*	στῆθος *breast*
ὄρος *mountain* (gen. plu. ὀρῶν	ψεῦδος *falsehood*
or ὀρέων).	

OTHER THIRD DECLENSION NOUNS

There are in the N.T. a number of third declension nouns which cannot be exactly placed in any of the classes hitherto dealt with. Some of them are given here, together with the gen. sing., from which the other cases can usually be discovered easily. Irregularities are noted.

ἡ γυνή *woman, wife*. Gen. sing. γυναικός; voc. sing. γύναι: declined otherwise as φύλαξ.

ἡ ἔρις *strife*. Gen. sing. ἔριδος; acc. sing. ἔριν. For the nom. plur. both ἔρεις and ἔριδες are found. Declension otherwise like ἐλπίς.

ἡ θρίξ *hair*. Gen. sing. τριχός. This unexpected genitive is the result of the operation of the rule (see p. 9) by which if a syllable begins with an aspirate and the next syllable also begins with an aspirate, the first aspirate is dropped. The gen. sing. of θρίξ in its original form is θριχός (stem θριχ-); but the first aspirate is dropped, giving us τριχός (stem τριχ-). The rest of the declension is as φύλαξ, but dat. plur. θριξίν

ἡ κλείς *key*. Gen. sing. κλειδός; acc. sing. κλεῖν and κλεῖδα; nom. plur. κλεῖς and κλεῖδες. Declension otherwise like ἐλπίς.

ἡ ναῦς *ship*. Found once only in N.T. in acc. sing.: ναῦν (Acts 27.41). (The declension of ναῦς, gen. sing. νεώς, is irregular in classical Greek. The usual N.T. word for "ship" is πλοῖον.)

ἡ νύξ *night*. Gen. sing. νυκτός. Rest of declension as φύλαξ, but dat. plur. νυξί(ν).

ὁ πούς *foot*. Gen. sing. ποδός. Declension as ἐλπίς. Dat. plur. ποσί(ν),

ἡ χάρις *grace, favour*.   Gen. sing. χάριτος; acc. sing. χάριν and χάριτα. Declension otherwise like ἐλπίς.

ἡ ὠδίν *travail*.   Gen. sing. ὠδῖνος.   Declension like αἰών.

Among the *neuters* are:

τὸ γάλα *milk*.   Gen. sing. γάλακτος.

τὸ γῆρας *old age*.   Found only once (Luke 1.36) in dat. γήρει.

τὸ γόνυ *knee*.   Gen. sing. γόνατος.   Declined like γράμμα.

τὸ δάκρυ *tear*.   Dat. plur. δάκρυσι(ν) is found; but in other cases the 2nd decl. neuter form δάκρυον is used.

τὸ ἥμισυ *half*.   Gen. sing. ἡμίσους.   For the plural ἡμίση, ἡμίσεα, and ἡμίσια are found in different N.T. manuscripts.

τὸ κρέας *flesh*.   Nom. and acc. plur. κρέα.

τὸ οὖς *ear*.   Gen. sing. ὠτός.   Otherwise declined like γράμμα.

τὸ πῦρ *fire*.   Gen. sing. πυρός.

τὸ ὕδωρ *water*.   Gen. sing. ὕδατος.   Otherwise declined like γράμμα.

τὸ φρέαρ *well*.   Gen. sing. φρέατος.   Otherwise declined like γραμμα.

The 2nd decl. neuter σάββατον, *Sabbath, week*, has a 3rd decl. dat. plur. σάββασι(ν).

VERBS WITH STEMS ENDING IN CONSONANTS

(*a*) GUTTURAL STEMS

Verbs in -ω whose stems end in consonants follow λύω closely.   But note:

1. Changes take place when the consonant with which the stem ends is followed by σ, θ, μ, and τ.   (See below, and also p. 6.)

2. Some of these verbs have certain strong tenses.   They will be noted in the verb lists given.

3. Irregularities are found in these verbs.   They too will be noted in the verb lists.

STEMS ENDING IN A GUTTURAL: γ, κ, χ

A guttural and σ coalesce to ξ (= κς).

The 1st pers. sing. fut. indic. act. of ἄγω is ἄξω (ἄγ-σω).

The 1st pers. sing. fut. indic. act. of ἄρχομαι is ἄρξομαι (ἄρχ-σομαι).

The guttural is χ when followed by θ; i.e. it acquires an aspirate.

The 1st pers. sing. aor. indic. pass. of τίκτω (stem τεκ) is ἐτέχθην.

The 1st pers. sing. aor. indic. pass. of ἄγω is ἤχθην.

Present	Compounds	English	Future	Aorist	Perfect	Perf. pass.	Aor. pass.
ἄγω		lead	ἄξω	ἤγαγον		ἦγμαι	ἤχθην
	ἀπάγω	lead away					
	ἐπισυνάγω	gather together					
	ἐξάγω	lead out					
	παράγω	pass by					
	περιάγω	go about					
	προάγω	go before					
	συνάγω	gather together					
	ὑπάγω	depart					
ἀνοίγω¹		open	ἀνοίξω	⎧ ἤνοιξα ⎨ ἀνέῳξα ⎩ ἠνέῳξα	ἀνέῳγα	⎧ ἀνέῳγμαι ⎨ ἠνέῳγμαι ⎩ ἤνοιγμαι	⎧ ἀνεῴχθην ⎨ ἠνεῴχθην ⎩ ἠνοίχθην
	διανοίγω	open					
ἄρχω		rule					
ἄρχομαι	(Mid. of ἄρχω)	begin	ἄρξομαι	ἠρξάμην			
δέχομαι	(dep.)	receive	δέξομαι	ἐδεξάμην	δέδεγμαι		ἐδέχθην
	παραδέχομαι	receive					
	προσδέχομαι	wait for					
διώκω		pursue, persecute					
	καταδιώκω	follow					
εὔχομαι		pray	-εὔξομαι	εὐξάμην			
	προσεύχομαι	pray					
ἔχω²		have	ἕξω	ἔσχον	ἔσχηκα		
	ἀνέχομαι (Mid.)	endure					
	ἀπέχω						
	ἐνέχω	set oneself against					
	παρέχω	provide, show					
ἥκω	(has perfect significance)	have come	ἥξω	ἧξα			
-λέγω³		gather	-λέξω	-ἔλεξα		λέλεγμαι	
	ἐκλέγομαι (Mid.)	choose					
πλέκω		entwine, plait		ἔπλεξα	πέπλεχα		-ἐπλάκην
πνίγω		choke					
	συνπνίγω	choke					
στήκω	(an -ω form sometimes used in N.T. for ἵστημι)	stand					
τρώγω		eat	τρώξω	ἔτρωξα			
φεύγω		flee	φεύξομαι (dep.)	ἔφυγον	πέφευγα		

¹ ἀνοίγω. A compound of οἴγω with ἀνά. The simplex is not found in N.T. Notice the variation of the augment. In ἤνοιξα the prefix is augmented. In ἀνέῳξα the verb is doubly augmented, ε-ῳξα. In ἠνέῳξα there are really three augments, a double augment to the stem, and an augment to the prefix. Similar variation in augment is found in the perf. pass. and aor. pass.

² ἔχω. The original verb stem is σεχ-. In the fut. a rough breathing replaces the initial σ, ἕξω. In the aor. the stem is shortened to σχ-, ἔσχον. In the perf. this shortened stem is lengthened to σχη-, ἔσχηκα.

³ -λέγω gather. Not to be confused with (though it is in fact the same word) λέγω say, for which see below, p. 124.

The guttural is γ when followed by μ.

The 1st pers. sing. perf. indic. pass. of πλέκω is πέπλεγμαι.
The 1st pers. sing. perf. indic. pass. of ἄγω is ἦγμαι.

The guttural is κ when followed by τ.

The 3rd pers. sing. perf. indic. pass. of πλέκω is πέπλεκται.
The 3rd pers. sing. perf. indic. pass. of ἄγω is ἦκται.

Verbs of this class have a strong perfect active (i.e. without the κ of the ending, which would be inconvenient, since the stem ends in a guttural); e.g. πέφευγα from φεύγω. The guttural is, however, frequently aspirated, e.g. πέπλεχα from πλέκω.

The usual endings of the principal parts of verbs with guttural stems, therefore, are:

Pres.	Fut.	Wk. aor. act.	Perf. act.	Perf. pass.	Wk. aor. pass.
-γω -κω -χω	-ξω	-ξα	-χα	-γμαι	-χθην

The table on p. 123 lists verbs with guttural stems found in St Mark's Gospel. The compounds which appear in St Mark are also listed. Note particularly those which have a str. aor. act. If no part is given it means either that it is formed regularly, or that it is not found in the N.T.

Some verbs have a stem ending in a guttural in the present, whilst other tenses are supplied from quite different stems. Thus τρέχω I run has no aorist from a stem τρεχ-. Instead ἔδραμον is employed from a stem δραμ-. Ἔρχομαι I come and λέγω I say both very common verbs, have their principal parts supplied from different stems.

Present	Compounds	English	Future	Aor. act.	Perf. act.	Perf. pass.	Aor. pass.
ἔρχομαι		come, go	ἐλεύσομαι	ἦλθον	ἐλήλυθα		
	ἀπέρχομαι	go away					
	διέρχομαι	go through					
	εἰσέρχομαι	go into					
	ἐξέρχομαι	go out					
	παρέρχομαι	pass by					
	προσέρχομαι	approach					
	συνεισέρχομαι	enter together					
λέγω		say	ἐρῶ	εἶπον	εἴρηκα	εἴρημαι	⌈ -ἐλέχθην ⌊ ἐρρέθην
	διαλέγομαι (Mid.)	discuss, argue					
	προλέγω	say beforehand, declare					
τρέχω		run		ἔδραμον			
	ἐπισυντρέχω	run together again					
	περιτρέχω	run about					
	προστρέχω	run to					

Guttural Stems lengthened in Present by Consonantal ι

We shall find that in the present tense a verb stem is frequently lengthened by the addition of a letter or a syllable (see p. 190). Sometimes this is done by the addition of a y sound, known as consonantal ι (i.e. iota pronounced as a consonant). A guttural followed by this y becomes σσ or ζ.

Thus: φυλακ-ι-ω becomes φυλάσσω *guard*.
πραγ-ι-ω becomes πράσσω *do* (common in N.T. but not found in St Mark).
κραγ-ι-ω becomes κράζω *cry*.

The following is a list of the verbs with guttural stems found in St Mark which have present stems lengthened in this way:

Present	Compounds	English	Fut.	Aor. act.	Perf. act.	Perf. pass.	Aor. pass.
ἁρπάζω		seize, snatch away	ἁρπάσω	ἥρπασα	ἥρπακα		{ ἡρπάγην { ἡρπάσθην
	διαρπάζω	plunder		[as though with dental stem]			
κηρύσσω		proclaim	κηρύξω	ἐκήρυξα	κεκήρυχα	κεκήρυγμαι	ἐκηρύχθην
κράζω		cry	κράξω	ἔκραξα	κέκραγα		
	ἀνακράζω	cry out					
ὀρύσσω		dig		ὤρυξα			ὠρύγην
	ἐξορύσσω	dig out					
παίζω		play	-παίξω	-ἔπαιξα			-ἐπαίχθην
	ἐμπαίζω	mock at					
πατάσσω		strike	πατάξω	ἐπάταξα			
-πλήσσω		strike		-ἔπληξα			{ ἐπλήγην { -ἐπλάγην
	ἐκπλήσσω	astonish					
ῥήσσω		rend, break	ῥήξω	ἔρηξα or ἔρρηξα			
	διαρήσσω	break asunder					
σπαράσσω		tear		ἐσπάραξα			
	συνσπα-ράσσω or συσπα-ράσσω	convulse					
-στεγάζω		roof (not found in N.T.)					
	ἀπο-στεγάζω	unroof					
στενάζω		groan		ἐστέναξα			
	ἀνα-στενάζω	sigh deeply					
ταράσσω		trouble, stir up		ἐτάραξα		τετάραγμαι	ἐταράχθην
τάσσω		arrange	τάξομαι (Mid.)	ἔταξα	τέταχα	τέταγμαι	ἐτάχθην
	ἀποτάσσω	take leave of					
	ἐπιτάσσω	command					
	προσ-τάσσω	command					
-τινάσσω		shake		-ἐτίναξα			
	ἐκτινάσσω	shake off					
φυλάσσω		guard	φυλάξω	ἐφύλαξα			

PERFECT MIDDLE AND PASSIVE OF VERBS WITH GUTTURAL
STEMS

Remember that a guttural and σ coalesce to ξ,
the guttural is χ when followed by θ,
the guttural is γ when followed by μ,
the guttural is κ when followed by τ.

The 3rd pers. plur. ending -νται cannot follow a consonant, and therefore a periphrastic construction is substituted (periphrastic, "roundabout"
from περιφράζω, "to express by circumlocution"); the 3rd pers. plur.
of εἰμί is used with the plural of the middle or perfect passive participle
(see p. 173–5).

Thus the perfect middle and passive indicative of ἄγω and κηρύσσω
are:

Sing.	1.	ἦγμαι	κεκήρυγμαι
	2.	ἦξαι	κεκήρυξαι
	3.	ἦκται	κεκήρυκται
Plur.	1.	ἤγμεθα	κεκηρύγμεθα
	2.	ἦχθε	κεκήρυχθε
	3.	ἠγμένοι εἰσί(ν)	κεκηρυγμένοι εἰσί(ν)

Vocabulary

ἄκανθα (1) *thorn*

ἡδονή, ῆς (1) *pleasure*

'Ιουδαία (1), fem. of adj. 'Ιουδαῖος,
-α, ον *Judaea*

μέριμνα (1) *care*

ῥίζα (1) *root*

σφαγή, -ῆς (1) *slaughter*

χαρά, ᾶς (1) *joy*

διάβολος (2) *devil.* Masc. of adj.
διάβολος, -ον

πλοῦτος (2) *riches*

στέφανος (2) *crown*

συνέδριον (2) *council* (usually of the
Sanhedrin)

ἀρχαῖος, -α, -ον *ancient, of old time*

ἄφωνος, -ον *dumb*

πλεῖστος, -η, -ον (superlative of
πολύς: see p. 151) *most, very
great*

δώδεκα (indeclinable adjective)
twelve

συνέρχομαι *come together*

φονεύω *murder, slay*

παρά (preposition), with acc. *beside, by*

πρός (preposition), with acc. *to, towards*

ὡς (adverb) *as, just as*

EXERCISE 43

Translate into English:

1. καὶ εἶπεν αὐτῇ, Διὰ τοῦτον τὸν λόγον ὕπαγε, ἐξελήλυθεν ἐκ τῆς
θυγατρός σου τὸ δαιμόνιον. 2. καὶ ἀπήγαγον τὸν Ἰησοῦν πρὸς τὸν

ἀρχιερέα, καὶ συνέρχονται οἱ ἀρχιερεῖς καὶ οἱ πρεσβύτεροι καὶ οἱ
γραμματεῖς. 3. ἀπεκρίθη αὐτοῖς ὁ Ἰησοῦς, Οὐκ ἐγὼ ὑμᾶς τοὺς
δώδεκα ἐξελεξάμην; καὶ ἐξ ὑμῶν εἷς (one) διαβολός ἐστιν. 4. καὶ
πάλιν ἤρξατο διδάσκειν παρὰ τὴν θάλασσαν· καὶ συνάγεται πρὸς
αὐτὸν ὄχλος πλεῖστος. 5. καὶ ἐξεπλήσσοντο ἐπὶ τῇ διδαχῇ αὐτοῦ,
ὅτι ἐν ἐξουσίᾳ ἦν ὁ λόγος αὐτοῦ. 6. ὡς πρόβατον ἐπὶ σφαγὴν ἤχθη,
καὶ ὡς ἀμνός ἐστιν ἄφωνος, οὕτως οὐκ ἀνοίγει τὸ στόμα αὐτοῦ.

EXERCISE 44

Translate into Greek:

1. We shall pray in the Spirit. 2. They waited for him. 3. He will do
the will of God. 4. He cast out the unclean spirit. 5. In his mouth there
is no falsehood. 6. The women wiped the feet of the Twelve with their
hair. 7. I have the keys of death. 8. You have been led to the well.
9. The apostle was received with favour. 10. They plaited a crown of
thorns.

EXERCISE 45

Translate into Greek:

1. Thou hast come from God and shalt go to God. 2. It was said by
them of old time, Thou shalt do no murder. 3. The high priests and
scribes were gathered together, and led him away into their council.
4. And again he prayed and said the same words. 5. Let them (that are)
in Judaea flee to the mountains. 6. These receive the word with joy, but
they have no root. 7. They are choked by cares and riches and pleasures.
8. Which of the prophets did your fathers not persecute? 9. I send thee to
open the eyes of the blind and to turn men from darkness to light.
10. The disciples were troubled and cried out for (ἀπό) fear.

Chapter 17

PRONOUNS

POSSESSIVE PRONOUNS

Possessive pronouns, as they are called in many grammars, are in reality adjectives since they always qualify a noun, expressed or unexpressed. They denote possession, e.g. *my, our, thy, your*. The Greek possessives are all of the second and first declension type (see pp. 36–37).

1st Person: *My*: ἐμός, ἐμή, ἐμόν.

           Our: ἡμέτερος, α, ον (α pure).

2nd Person: *Thy* (or *your*, singular): σός, σή, σόν.

           Your (plural): ὑμέτερος, α, ον (α pure).

There is no Greek possessive for the third person, *his, her, its, their*. These are translated by the genitive of the personal pronoun (see p. 47) αὐτοῦ, αὐτῆς, αὐτοῦ, αὐτῶν.

The genitive of the personal pronouns of the 1st and 2nd persons may, of course, be used instead of the possessives. "My father" may be translated either by ὁ ἐμὸς πατήρ (or ὁ πατὴρ ὁ ἐμός) or by ὁ πατήρ μου.

The gender of the possessive is determined by the noun with which it agrees, and not by the sex of the person to whom the pronoun refers. Thus "our mother" is ἡ ἡμέτερα μητήρ (or ἡ μητὴρ ἡ ἡμέτερα), ἡμέτερα being fem. sing., even though the word "our" may refer to men. This is felt to be a difficulty by some beginners. But it must be remembered that the so-called possessive pronoun is an adjective, and therefore agrees in number, case, and gender with *the noun it qualifies*.

Possessive pronouns, like other adjectives, may be used predicatively (see p. 40), when the English translation will be *mine, ours, thine, yours*: τὸ ἀργύριόν ἐστιν ἐμόν *the silver is mine*.

Examples

τὸ ἐμὸν ὄνομα (Matt. 18.20) *my name*.

ἡ χαρὰ ἡ ἐμή (John 3.29) *my joy*.

ἐν τῷ σῷ ὀφθαλμῷ (Matt. 7.3) *in thine own eye*.

τῆς ἡμετέρας θρησκείας (Acts 26.5) *of our religion*.

ὁ καιρὸς ὁ ὑμέτερος (John 7.6) *your time*.

Note also τὸ σόν *that which is thine*; τὰ σά *thy goods*; τὸ ὑμέτερον *that which is your own*; τὸ ἡμέτερον *that which is our own, our own property.*

REFLEXIVE PRONOUNS

Reflexive pronouns are those which refer back to (or "bend back upon", from Latin *reflecto, bend back*) the subject of the clause. In the sentences, "You will hurt *yourself*"; "I cannot see *myself* doing that", the words in italics are reflexive pronouns. It will be obvious that there can be no nominative or vocative cases of a reflexive pronoun, but all the oblique cases are found, accusative, genitive (e.g. of myself) and dative (e.g. to myself).

The reflexive pronoun must be carefully distinguished from the definite pronoun used for emphasis, αὐτός *self* (see p. 47), which is, of course, used frequently in the nominative: "I myself saw the accident".

1st person sing. myself		2nd person sing. thyself		3rd person sing. himself, herself, itself		
Masc.	*Fem.*	*Masc.*	*Fem.*	*Masc.*	*Fem.*	*Neut.*
A. ἐμαυτόν	ἐμαυτήν	σεαυτόν	σεαυτήν	ἑαυτόν	ἑαυτήν	ἑαυτό
G. ἐμαυτοῦ	ἐμαυτῆς	σεαυτοῦ	σεαυτῆς	ἑαυτοῦ	ἑαυτῆς	ἑαυτοῦ
D. ἐμαυτῷ	ἐμαυτῇ	σεαυτῷ	σεαυτῇ	ἑαυτῷ	ἑαυτῇ	ἑαυτῷ

αὐτόν, αὐτήν, αὐτό is also found for third person singular.

The plural of all three persons is the same:

1st, 2nd, and 3rd persons plural ourselves, yourselves, themselves		
Masc.	*Fem.*	*Neut.*
A. ἑαυτούς	ἑαυτάς	ἑαυτά
G. ἑαυτῶν	ἑαυτῶν	ἑαυτῶν
D. ἑαυτοῖς	ἑαυταῖς	ἑαυτοῖς

Examples

οὐδὲ ἐμαυτὸν ἠξίωσα (Luke 7.7) *neither thought I myself worthy.*

ἐγὼ ἀπ᾽ ἐμαυτοῦ λαλῶ (John 7.17) *I speak from myself.*

σὺ περὶ σεαυτοῦ μαρτυρεῖς (John 8.13) *thou bearest witness of thyself.*

ἄλλους ἔσωσεν, σωσάτω ἑαυτόν (Luke 23.35) *he saved others; let him save himself.*

ἄφες τοὺς νεκροὺς θάψαι τοὺς ἑαυτῶν νεκρούς (Luke 9.60) *leave the dead to bury their own dead.*

μαρτυρεῖτε ἑαυτοῖς (Matt. 23.31) *ye witness to yourselves.*

ἐν ἑαυτοῖς στενάζομεν (Rom. 8.23) *we groan within ourselves.*

THE INDEFINITE PRONOUN

The indefinite pronoun τις is the equivalent of the English *some, any, somebody, anybody, certain, a man*. Its declension is the same as that of the interrogative τίς, apart from accentuation.

		Masc. and Fem.	Neut.
Sing.	N.	τις	τι
	A.	τινά	τι
	G.	τινός	τινός
	D.	τινί	τινί
Plur.	N.	τινές	τινά
	A.	τινάς	τινά
	G.	τινῶν	τινῶν
	D.	τισί(ν)	τισί(ν)

Τις is an enclitic. It throws its accent back on to the preceding word (see p. 273). But if the preceding word has an acute accent on the last syllable but one, it receives no accent from the enclitic. In this case the indefinite pronoun loses its accent altogether if it is the monosyllable τις or τι; if it is a disyllable the accent is retained as shown above.

The indefinite pronoun may stand by itself, being used as a pronoun proper. E.g.

πῶς δύναταί τις εἰσελθεῖν ... (Matt. 12.29) *how can anybody* (or "a man") *enter ....*

εἴδομέν τινα ... (Luke 9.49) *we saw a certain man* (or "some one").

οὐ χρείαν εἶχεν ἵνα τις μαρτυρήσῃ (John 2.25) *he had no need that any should bear witness.*

The indefinite pronoun may also be used adjectivally. In this case it usually follows the noun with which it agrees, and no article is used:

ὑπήντησεν ἀνήρ τις ἐκ τῆς πόλεως (Luke 8.27) *there met* (him) *a certain man out of the city.*

ἐγένετο . . . ἱερεύς τις ὀνόματι Ζαχαρίας (Luke 1.5) *there was a certain priest by name Zacharias.*

μετὰ δὲ ἡμέρας τινὰς . . . (Acts 24.24) *after some* (or "certain") *days.*

THE RECIPROCAL PRONOUN

The reciprocal pronoun ἀλλήλους translates the English "one another", "each other". There is no nominative case. No fem. or neut. forms are found in the New Testament.

	Masc.
A.	ἀλλήλους
G.	ἀλλήλων
D.	ἀλλήλοις

ἀφορίσει αὐτοὺς ἀπ᾽ ἀλλήλων (Matt. 25.32) *he will separate them from one another.*

καὶ ἔλεγον πρὸς ἀλλήλους (Mark 4.41) *and they said to each other.*

VERBS WITH STEMS ENDING IN CONSONANTS

(b) DENTAL STEMS

When the stem of a verb ends in a dental consonant (δ, τ, or θ) the following points are to be noticed (see p. 7):

A dental drops out before σ and κ.

The 1st pers. sing. fut. indic. act. of πείθω *persuade*, is πείσω (for πείθ-σω).

A dental followed by θ, μ, or τ becomes σ.

The 1st pers. sing. perf. indic. pass. of πείθω is πέπεισμαι (for πέπειθ-μαι).

The 1st pers. sing. aor. indic. pass. of πείθω is ἐπείσθην (for ἐπείθ-θην).

We have already noticed (p. 125) that some verbs have a lengthening of the verb stem in the present. Most of the verbs with dental stems form their present by adding to the stem the y sound known as "consonantal ι" (cf. p. 125). A dental followed by this y usually becomes ζ. ἁγιαδ-ι-ω becomes ἁγιάζω. ἁγνιδ-ι-ω becomes ἁγνίζω. These two verbs may serve as models of a large number of verbs in -αζω and -ιζω.

Present	Stem	English	Future	Aor. act.	Perf. act.	Perf. pass.	Aor. pass.
ἁγιάζω	ἁγιαδ-	sanctify	ἁγιάσω	ἡγίασα	ἡγίακα	ἡγίασμαι	ἡγιάσθην
ἁγνίζω	ἁγνίδ-	purify	ἁγνίσω	ἥγνισα	ἥγνικα	ἥγνισμαι	ἡγνίσθην

There is a group of verbs like κόπτω *cut* which the student might think to have stems ending in a dental. The τ, however, is an addition in the present to the verb stem κοπ-. These verbs will be treated under the heading of labial stems (p. 143).

πίπτω *fall*. The verb stem of this irregular verb ends in a dental, πετ-. In the present the stem is lengthened by reduplication with ι as a helping vowel; thus πι-πετ-ω. The ε of the stem drops out for ease of pronunciation, giving us πίπτω. The future, which is deponent, and the aorist are from another stem πεσ-, and the perfect from yet another πτω-: πίπτω, πεσοῦμαι, ἔπεσον, πέπτωκα.

πίπτω has a compound which is common in the N.T., ἀναπίπτω *recline* (at a meal).

Opposite is a list of the verbs with dental stems which appear in St Mark's Gospel. They are given here so that the student may see how extensive and how useful this class of verb is. Most of the verbs in this class are in -άζω and -ίζω, and follow the model of ἁγιάζω and ἁγνίζω (see above). But some verbs with dental stems show certain irregularities. The following must be noted:

ἐργάζομαι (deponent) *work*, has aor. ἠργασάμην; perf. εἴργασμαι; and aor. pass. -εἰργάσθην.

εὐαγγελίζω *bring good tidings*, mid. *proclaim good tidings*, especially in the Christian sense, *preach the gospel*; pass. *have good tidings proclaimed to one*; aor. εὐηγγέλισα; perf. pass. εὐηγγέλισμαι; aor. pass. εὐηγγελίσθην.

καθαρίζω *cleanse*, has fut. καθαριῶ (see p. 134); aor. ἐκαθέρισα; perf. pass. κεκαθέρισμαι. Notice the ε after the θ in both aor. act. and perf. pass. This is a mistaken second augment, as though the verb

Verb	Compounds	English	Verb	Compounds	English
ορράζω		buy	καυματίζω		burn, scorch up
αλάζω		shout, wail	κολαφίζω		buffet
ίζω		salt, season	κοπάζω		grow weary, abate
αγκάζω		compel			
αθεματίζω		anathematize, curse	κτίζω		create
			λογίζομαι		reckon, suppose,
οκεφαλίζω		behead	(dep.)	διαλογίζομαι	consider, reason
ρτίζω		fit	μερίζω		divide
	καταρτίζω	furnish, complete		διαμερίζω	distribute
πάζομαι (dep.)		greet, salute	ὀνειδίζω		reproach
ιμάζω		dishonour	ὁρκίζω		make swear, adjure
ρίζω		foam (at mouth)			
πτίζω		baptize	πειράζω		try, tempt
σανίζω		torture	ποτίζω		give to drink
στάζω		carry	ῥαντίζω		sprinkle
μίζω		give in marriage	σκανδαλίζω		cause to stumble
ιζω		fill	-σκευάζω		prepare
μονίζομαι (dep.)		be possessed		κατασκευάζω	prepare
			-σκιάζω		shade
μάζω		tame		ἐπισκιάζω	overshadow
ιμάζω		test, prove	σκορπίζω		scatter
	ἀποδοκιμάζω	reject		διασκορπίζω	scatter abroad
ξάζω		glorify	σκοτίζω		darken
ιζω		draw near	σμυρνίζω		mingle with myrrh
γκαλίζομαι (dep.)		take into one's arms	σπλαγχνίζομαι (dep.)		feel pity
υσιάζω		exercise authority over	στυγνάζω		have a gloomy face
	κατεξουσιάζω	exercise authority over	σχίζω		cleave, divide
μάζω		prepare	τρίζω		grind, gnash
μάζω		wonder	φημίζω		spread a report
	ἐκθαυμάζω	wonder greatly		διαφημίζω	spread abroad
άζω		suckle	χορτάζω		fill, satisfy (with food)
τίζω		clothe	χωρίζω		separate

were a compound with κατά, which it is not. It is connected with the adjective καθαρός *clean*.

καθίζω *set, appoint*, is used more frequently in the intransitive sense, *sit*: fut. καθίσω; aor. ἐκάθισα; perf. κεκάθικα.

σώζω or σῴζω *save*: fut. σώσω; aor. ἔσωσα; perf. σέσωκα; perf. pass. σέσωσμαι or σέσωμαι; aor. pass. ἐσώθην.

πείθω *persuade*: pass. *be persuaded, obey*, with dat. Fut. πείσω; aor. ἔπεισα; str. perf. πέποιθα, with present meaning, *I trust*; perf. pass. πέπεισμαι; aor. pass. ἐπείσθην.

σπεύδω *hasten*: aor. ἔσπευσα.

καθεύδω *sleep*. Frequent in imperfect of continuous action, ἐκάθευδεν *he was asleep*.

CONTRACTED FUTURE

A few verbs in this class have a contracted future. They are:

Present	English	Fut. act.	Aor. act.	Perf. act.	Perf. pass.
γνωρίζω	make known	γνωριῶ[1]	ἐγνώρισα	ἤλπικα	
ἐλπίζω	hope	ἐλπιῶ	ἤλπισα		
καθαρίζω	cleanse	καθαριῶ	ἐκαθέρισα		κεκαθέρισμαι
μακαρίζω	bless	μακαριῶ			

This future appears to be a survival from classical Greek. In Attic Greek, verbs with stems of more than one syllable ending in ιδ had a contracted future. The σ, having caused the loss of the preceding dental, was dropped as coming between two vowels; ε was substituted, and this ε contracted with the vowel of the personal ending. The steps in this process in arriving at the future of ἐλπίζω, for instance, would be ἐλπίδ-σω, ἐλπίσω, ἐλπίω, ἐλπιέω, ἐλπιῶ.

The conjugation of the fut. ind. act. in full is: ἐλπιῶ
ἐλπιεῖς
ἐλπιεῖ
ἐλπιοῦμεν
ἐλπιεῖτε
ἐλπιοῦσι(ν)

As will be clear from the list on p. 133 the majority of N.T. verbs in -ίζω have fut. in -ίσω.

PERFECT MIDDLE AND PASSIVE OF VERBS WITH DENTAL
STEMS

Remember that a dental drops out before σ. The dental becomes σ when followed by θ, μ, or τ.

The periphrastic 3rd pers. plur. is found, as in the case of guttural stems (see p. 126). The perfect indicative mid. and pass. of ἁγιάζω, which may provide a model is:

Sing.	1. ἡγίασμαι	Plur.	1. ἡγιάσμεθα
	2. ἡγίασαι		2. ἡγίασθε
	3. ἡγίασται		3. ἡγιασμένοι εἰσί(ν)

1 In Col. 4.9 γνωριῶ and γνωρίσω are found as variants in different MSS.

Vocabulary

Ἡρῴδης (1) *Herod*

ἥλιος (2) *sun*

κόπος (2) *trouble*

Τιμόθεος (2) *Timothy*

ἔντιμος, ον *prized, precious, dear*

ἕξ (indecl. adj.) *six*

ἐμβαίνω (compound of βαίνω; aor. infin. ἐμβῆναι) *go into, embark*

κακόω *afflict*

μέλλω, followed by infin., *to be about* (to . . .). Imperf. ἤμελλον; fut. μελλήσω

παραδώσουσιν 3rd pers. plur. fut. act. of παραδίδωμι (see p. 243) *they will deliver, betray*

πάρειμι (compound of εἰμί) *be at hand, be present*

ποταπός, ή, όν (interrog. pronoun) *of what kind?*

οὔτε . . . οὔτε (conjunctions) *neither . . . nor*

πάντοτε (adverb) *always*

EXERCISE 46

Translate into English:

1. ὁ καιρὸς ὁ ἐμὸς οὔπω πάρεστιν, ὁ δὲ καιρὸς ὁ ὑμέτερος πάντοτέ ἐστιν ἕτοιμος. 2. τί αὐτῇ κόπους παρέχετε; καλὸν ἔργον ἠργάσατο ἐν ἐμοί. 3. ἑκατοντάρχου δέ τινος δοῦλος ἤμελλεν τελευτᾶν, ὃς ἦν αὐτῷ ἔντιμος. 4. οἱ δὲ ἄνθρωποι ἐθαύμασαν καὶ εἶπον Ποταπός ἐστιν οὗτος, ὅτι καὶ οἱ ἄνεμοι καὶ ἡ θάλασσα αὐτῷ ὑπακούουσιν; 5. καὶ εὐθὺς ἠνάγκασεν τοὺς μαθητὰς αὐτοῦ ἐμβῆναι εἰς τὸ πλοῖον. 6. καὶ τότε σκανδαλισθήσονται πολλοὶ καὶ ἀλλήλους παραδώσουσιν καὶ μισήσουσιν ἀλλήλους.

EXERCISE 47

Translate into Greek:

1. Who will save me from this death? 2. Yours is the kingdom of God. 3. In those days the sun shall be darkened. 4. There are six days in which men ought to work. 5. Love worketh no evil. 6. Save me from (ἐκ) this hour. 7. God was glorified in his works. 8. Separate yourselves from evil. 9. We trust in the Lord. 10. He cleansed himself.

EXERCISE 48

Translate into Greek:

1. In the resurrection they neither marry nor are given in marriage, but are as the angels in heaven. 2. We have hoped in God who is the Saviour

of men. 3. There was there a certain disciple by name Timothy, the son of a Jewish woman. 4. I declared unto them thy name, and will declare it. 5. God has called us to preach the gospel to them. 6. He hastened to come down from the tree. 7. They will not be persuaded because their hearts are not sanctified. 8. They dishonour themselves who cause the little ones to stumble. 9. And some of them were persuaded and baptized. 10. Herod the king put forth his hands to afflict certain of the faithful.

Chapter 18

PREPOSITIONS—*continued*

The prepositions dealt with on p. 30 were followed by one case only of the noun or pronoun. Those dealt with on p. 62ff. were found with the acc. and gen. There are three prepositions which may govern the acc., gen., or dat. They are as follows:

Preposition	Before vowel	Before aspirate	Meaning with acc.	Meaning with gen.	Meaning with dat.
ἐπί	ἐπ'	ἐφ'	towards, upon, against	upon, in the presence of, in the time of	on, at, on account of, with a view to
παρά	παρ'	παρ'	by the side of, along, beside, by, contrary to, in comparison with, than	from beside, from, by	at the side of, near, with, in the sight of
πρός	πρός	πρός	to, towards, at, with	in the interest of	at, near

Examples

ἐπί *with acc.* often, but by no means always, includes the idea of movement towards:

ἔρχονται ἐπὶ τὸ μνῆμα (Mark 16.2) *they come towards the sepulchre.*

ἐπὶ τὸ δῶμα (Luke 5.19) *upon (i.e., up on to) the housetop.*

ἔρχεται ἡ ὀργὴ τοῦ Θεοῦ ἐπὶ τοὺς υἱοὺς τῆς ἀπειθείας (Eph. 5.6) *the wrath of God cometh upon the sons of disobedience.*

βασιλεία ἐπὶ βασιλείαν (Matt. 24.7) *kingdom against kingdom.*

ἐπί *with gen.* rather gives the idea of position without movement:

ἐπὶ τῶν νεφελῶν (Matt. 24.30) *on the clouds.*

But occasionally the idea of motion is present:

ὡς ἄνθρωπος βάλῃ τὸν σπόρον ἐπὶ τῆς γῆς (Mark 4.26) *as a man might throw seed on to the ground.*

ἐπὶ σοῦ (Acts 23.30) *in thy presence.*
ἐγένετο ἐπὶ Κλαυδίου (Acts 11.28) *it came to pass in the days of Claudius.*

ἐπί *with the dat.* denotes rest on or at a place:

ἐπὶ πίνακι (Matt. 14.8) *on a charger.*
ἐπὶ τῇ πηγῇ (John 4.6) *at the well.*
οὐκ ἐπ' ἄρτῳ μόνῳ (Luke 4.4) *not for (or* on account of) *bread alone.*
ἐπὶ ἔργοις ἀγαθοῖς (Eph. 2.10) *for (or* with a view to) *good works.*

παρά *with the acc.* denotes movement to or along the side of. The root
meaning of παρά is "beside".

παρὰ τὴν θάλασσαν (Matt. 4.18) *by the sea.*
παρὰ τὴν ὁδόν (Mark 4.4) *by the wayside.*
παρὰ τὸν νόμον (Acts 18.13) *contrary to the law.*
μὴ ὑπερφρονεῖν παρ' ὃ δεῖ φρονεῖν (Rom. 12.3) *not to think more
highly than* (in comparison with what) *he ought to think.*
διαφορώτερον παρ' αὐτοὺς κεκληρονόμηκεν ὄνομα (Heb. 1.4) *he
hath obtained a more excellent name than they.*

παρά *with the gen.* denotes movement from the side of:

παρὰ τῶν ἀρχιερέων (Mark 14.43) *from the high priests.*
τὸ Πνεῦμα τῆς ἀληθείας ὃ παρὰ τοῦ Πατρὸς ἐκπορεύεται (John
15.26) *the Spirit of truth which proceedeth from the Father.*

Sometimes παρά with the gen. denotes agent, in a sense approximating
to that of ὑπό with the gen. But all N.T. instances are quotations from
the Septuagint.

παρὰ Κυρίου ἐγένετο αὕτη (Mark 12.11) *this was done by the Lord.*

παρά *with the dat.* denotes rest at the side of:

παρ' αὐτῷ ἔμειναν (John 1.39) *they stayed with* (beside) *him.*
παρὰ δὲ Θεῷ πάντα δυνατά (Matt. 19.26) *but with God all things are
possible.*
εὗρες γὰρ χάριν παρὰ τῷ Θεῷ (Luke 1.30) *for thou hast found favour
with* (in the sight of) *God.*

πρός *with the acc.* usually means *to* or *towards*, and is used with persons,
whilst εἰς is more often used with things. But there is no hard and
fast rule about this.

σὺ ἔρχῃ πρός με; (Matt. 3.14) *dost thou come to me?*

It is sometimes used after verbs of speaking:

λέγει ἡ μήτηρ τοῦ 'Ιησοῦ πρὸς αὐτόν (John 2.3) *the mother of Jesus saith to him.*

It is also used of close proximity, *at* or *with*:

πρὸς τὴν ῥίζαν τῶν δένδρων (Matt. 3.10) *at the root of the trees.*
καὶ οὐκ εἰσὶν αἱ ἀδελφαὶ αὐτοῦ ὧδε πρὸς ἡμᾶς; (Mark 6.3) *and are not his sisters here with us?*

It is also used of time:

πρὸς ἑσπέραν ἐστίν (Luke 24.29) *it is towards evening.*

πρός *with the gen.* occurs only once in the N.T.:

τοῦτο γὰρ πρὸς τῆς ὑμετέρας σωτηρίας ὑπάρχει (Acts 27.34) *for this is for* (in the interest of) *your safety.*

πρός *with the dat.* occurs only six times in the N.T.:

πρὸς τῷ ὄρει (Mark 5.11) *near the mountain.*
πρὸς τῇ θύρᾳ (John 18.16) *at* (close to) *the door.*

Improper Prepositions

Some prepositions which, unlike those we have already described, are never used as prefixes in compound verbs, are styled improper prepositions. The reason for this description is not clear. There are some forty of them. Two or three of the less common are used with the dative case, the rest with the genitive. Those which occur in St Mark are as follows. They are all used with the genitive case:

ἔμπροσθεν *in front of, before, in the presence oj*
ἕνεκα, ἕνεκεν *because of*
ἔξω *outside, out of*
ἔξωθεν *outside, out of*
ἔσω *within*
ἕως *until, as far as*
ὀπίσω *behind, after*
πέραν *beyond*
πλήν *except*
χωρίς *apart from, without.*

Other improper prepositions found commonly in the N.T. are:

ἄχρι, ἄχρις *until, as far as*
ἐνώπιον *before, in the presence of*
ἐπάνω *above*
μεταξύ *between*
μέχρι, μέχρις *until, as far as.*

These also are used with the genitive case.

DEMONSTRATIVE PRONOUNS AND THEIR CORRELATIVES

Declined like οὗτος (see p. 48) are the demonstrative pronouns τοιοῦτος, τοιαύτη, τοιοῦτο *such*; τοσοῦτος, τοσαύτη, τοσοῦτο *so great, so much* (plur. *so many*); and τηλικοῦτος, τηλικαύτη, τηλικοῦτο *so great*. In the declension of these words, however, the initial τ of the oblique cases of οὗτος is omitted; thus the acc. masc. sing. of τοιοῦτος is τοιοῦτον, not τοιτοῦτον. Τοιοῦτον, τοσοῦτον, τηλικοῦτον are found as alternatives in the neut. sing.

... τὸν Θεὸν τὸν δόντα ἐξουσίαν τοιαύτην τοῖς ἀνθρώποις (Matt. 9.8) ... *God who had given such power unto men.*

οὐδὲ ἐν τῷ Ἰσραὴλ τοσαύτην πίστιν εὗρον (Luke 7.9) *not even in Israel have I found so great faith.*

ὃς ἐκ τηλικούτου θανάτου ἐρύσατο ἡμᾶς (2 Cor. 1.10) *who delivered us out of so great a death.*

In English the words "such", "so great", "as much", etc. are often balanced in a subordinate clause by the word "as": e.g. *I will show you such things as you have never seen; I have not so much* (or *as much*) *energy as I used to have.* In these examples "as" is a relative word introducing the subordinate (relative) clause, and relating it to the words "such" and "so" in the main clauses.

Similarly in Greek demonstrative pronouns have correlatives. For example, ἐκεῖνος and οὗτος are frequently found with their correlative ὅς, ἥ, ὅν (see pp. 81ff.).

With τοιοῦτος may be found the relative pronouns οἷος, οἵα, οἷον, *o, what kind, as*; ὁποῖος, ὁποία, ὁποῖον, *of what kind, as,* or the relative adverb ὡς *as.*

οἷοί ἐσμεν τῷ λόγῳ ... τοιοῦτοι ... τῷ ἔργῳ (2 Cor. 10.11) *what we are in word ... such (are we) also in deed.*

πάντας . . . γενέσθαι τοιούτους ὁποῖος καὶ ἐγώ εἰμι (Acts 26.29)
all . . . might become such as I also am.

τοιοῦτος ὢν ὡς Παῦλος πρεσβύτης (Philem. 9) *being such a one as
Paul the aged.*

With τοσοῦτος may be found the relative pronoun ὅσος, ὅση, ὅσον
as much as, as, or the consecutive adverb ὥστε *so as to, as.*

τοσούτῳ μᾶλλον ὅσῳ βλέπετε ἐγγίζουσαν τὴν ἡμέραν (Heb.
10.25) *so much the more as ye see the day drawing nigh.* (τοσούτῳ
and ὅσῳ are datives of difference, "more by so much".)

ἄρτοι τοσοῦτοι ὥστε χορτάσαι ὄχλον τοσοῦτον (Matt. 15.33) *so
many loaves as to fill so great a multitude.*

TIME AND PLACE

TIME

English expresses time by using a preposition followed by the noun of
time, e.g.:

(*Extent of time*) "for two hours"; "throughout the night"; (*Time
within which*) "it happened in the night"; "he went out during the
second act; (*Time at which,* i.e. point of time) "on the first day"; "at
midnight".

Greek expressses time by placing the noun of time in the appropriate
case, the accusative for extent of time, the genitive for time within which
(though the genitive has other temporal uses as well) and the dative for
point of time.

Extent of Time. The accusative case is used:

ἔμειναν οὐ πολλὰς ἡμέρας (John 2.12) *they remained not many days.*
ποιήσας τε μῆνας τρεῖς (Acts 20.3) *and when he had spent three
months.*

But in the following the dative appears to be used of extent of time:

χρόνῳ ἱκανῷ οὐκ ἐνεδύσατο ἱμάτιον (Luke 8.27) *for a long time he
had worn no clothes.*

Time within which. The genitive case is used:

ἵνα μὴ γένηται ἡ φυγὴ ὑμῶν χειμῶνος (Matt. 24.20) *that your
flight be not in winter.*

οὗτος ἦλθεν πρὸς αὐτὸν νυκτός (John 3.2) *this man came to him by night* (i.e. during the night).

νηστεύω δὶς τοῦ σαββάτου (Luke 18.12) *I fast twice in the week*.

νυκτὸς καὶ ἡμέρας (Mark 5.5) *night and day* (i.e. during the night and during the day).

Time at which. The dative case is used:

εὐθὺς τοῖς σάββασιν (Mark 1.21) *straightway on the Sabbath day*.

τῇ πρώτῃ ἡμέρᾳ τῶν ἀζύμων (Mark 14.12) *on the first day of unleavened bread*.

The accusative is also used (but rarely) to denote point of time:

ἐχθὲς ὥραν ἑβδόμην ἀφῆκεν αὐτὸν ὁ πυρετός (John 4.52) *yesterday at the seventh hour the fever left him*.

Prepositions are, however, sometimes employed in expressing time: ἐν with dat. may express both time within which and time at which:

ἐγένετο δὲ ἐν ταῖς ἡμέραις ἐκείναις (Luke 2.1) *now it came to pass in those days*.

ἐν τοῖς σάββασιν (Luke 13.10) *on the Sabbath day*.

Other prepositions, as we have already seen (pages 63f.), may have a temporal meaning:

δία with gen., *throughout*:

δι᾽ ὅλης νυκτὸς κοπιάσαντες (Luke 5.5) *having toiled throughout the whole night*.

The accusative of extent of time would equally well express this: νυκτὰ ὅλην κοπιάσαντες.

μετά with acc., *after*:

μετὰ δὲ τρεῖς μῆνας (Acts 28.11) *and after three months*.

πρό with gen., *before*:

πρὸ τοῦ κατακλυσμοῦ (Matt. 24.38) *before the flood*.

PLACE

Extent of Space or distance is expressed by the accusative case:

εἰς κώμην ἀπέχουσαν σταδίους ἑξήκοντα ἀπὸ Ἰερουσαλήμ (Luke 24.13) *to a village which was sixty furlongs from Jerusalem* (literally, "a village distant from Jerusalem sixty furlongs").

ἀπὸ τῆς γῆς ἐπαναγαγεῖν ὀλίγον (Luke 5.3) *to put out a little from the land.* (ὀλίγον is the acc. neut. of the adjective ὀλίγος, and may be used for extent of space or time, *a little way, for a short while.*)

ὅστις σε ἀγγαρεύσει μίλιον ἕν ... (Matt. 5.41) *and whosoever shall compel thee to go one mile. . . .*

Route. Twice in St Luke a genitive is used to express route, or "the way by which":

ποίας εἰσενέγκωσιν αὐτόν (Luke 5.19) *by what (way) they might bring him in.*

ἐκείνης ἤμελλεν διέρχεσθαι (Luke 19.4) *he was about to go through by that (way).*

With both ποίας and ἐκείνης in these sentences ὁδοῦ is to be understood.

Place where. To express "place where", the preposition ἐν *in*, and παρά *beside*, with the dative are usual. But a locative dative without a preposition is found in such phrases as καθαροὶ τῇ καρδίᾳ (Matt. 5.8) *pure in heart*; Χριστοῦ παθόντος σαρκί (1 Pet. 4.1) *as Christ suffered* (genitive absolute, see p. 178f.) *in the flesh.*

The student will by now be familiar with the common prepositions which have a locative meaning, εἰς, ἐκ, ἀπό, etc.

VERBS WITH STEMS ENDING IN CONSONANTS

(c) LABIAL STEMS

When the stem of a verb ends in a labial consonant (β, π, φ) the following points are to be noted (see p. 7):

A labial and ς coalesce to give ψ:

The 1st pers. sing. fut. indic. act. of κλέπτω (stem κλεπ-) is κλέψω.

A labial followed by θ becomes φ, i.e. it acquires an aspirate:

The 1st pers. sing. aor. indic. pass. of καλύπτω (stem καλυπ-) is ἐκαλύφθην.

A labial followed by μ becomes μ.

The 1st pers. sing. perf. indic. pass. of γράφω is γέγραμμαι.

A labial followed by τ takes the form π.

The 3rd pers. sing. perf. indic. pass. of τρίβω is τέτριπται.

Present	Compounds	English	Future	Aor. act.	Perf. act.	Perf. pass.	Aor. pass
ἀλείφω		anoint	ἀλείψω	ἤλειψα			ἠλείφθην
ἅπτω (not in Mark)		kindle, light		ἧψα			ἥφθην
ἅπτομαι (deponent)		lay hold of		ἡψάμην			
βάπτω		dip	βάψω	ἔβαψα		βέβαμμαι	
	ἐμβάπτω	dip in					
βλάπτω		hurt		ἔβλαψα			
βλέπω		see, look at	βλέψω	ἔβλεψα			
	ἀναβλέπω	look up, re-cover sight					
	διαβλέπω	see clearly					
	ἐμβλέπω	look at					
	περιβλέπω	look about					
γράφω		write	γράψω	ἔγραψα	γέγραφα	γέγραμμαι	ἐγράφην
	ἐπιγράφω	write upon, inscribe					
θλίβω		press, afflict				τέθλιμμαι	
	συνθλίβω	press together					
καλύπτω		cover	καλύψω	ἐκάλυψα		κεκάλυμμαι	ἐκαλύφθην
	ἀποκαλύπτω (not in Mark)	reveal					
	περικαλύπτω	cover around					
κλέπτω		steal	κλέψω	ἔκλεψα			
κόπτω		beat, cut	κόψω	ἔκοψα			ἐκόπην
	ἀποκόπτω	cut off					
	κατακόπτω	cut in pieces					
κύπτω		stoop		ἔκυψα			
λείπω		leave	λείψω	ἔλιπον (strong)		λέλειμμαι	ἐλείφθην
	ἐγκαταλείπω	leave behind, forsake					
	καταλείπω	leave behind, forsake					
νίπτω		wash		ἔνιψα			
πέμπω		send	πέμψω	ἔπεμψα			ἐπέμφθην
-ράπτω		sew					
	ἐπιράπτω	sew upon					
σέβομαι (deponent)		worship					
στίλβω		shine					
στρέφω		turn	στρέψω	ἔστρεψα		ἔστραμμαι	ἐστράφην
	ἐπιστρέφω	turn towards					
	καταστρέφω	overturn					
-τρέπω		turn		-ἔτρεψα			-ἐτράπην
	ἐπιτρέπω	turn to, en-trust, permit					
-τρίβω		rub	-τρίψω	-ἔτριψα		-τέτριμμαι	-ἐτρίβην
	συντρίβω	break in pieces, crush, bruise					
τύπτω		strike					

Notice also the following which do not occur in St Mark:

κρύπτω		hide	κρύψω	ἔκρυψα	κέκρυφα	κέκρυμμαι	ἐκρύβην
ῥιπτέω (classical ῥίπτω)		throw		ἔριψα ⎫ ἔρριψα ⎭		ἔριμμαι ⎫ ἔρριμμαι ⎭	
τρέφω[1]		nurture, feed		ἔθρεψα		τέθραμμαι	ἐτράφην

[1] The stem of this verb is θρεφ-. The first aspirate disappears when the second is present as in the pres. indic. The first aspirate remains when the second is absent as in the aor. ἔθρεψα Cf. τριχός for θριχός gen. of θρίξ hair (see p. 121).

In this class occurs the group of verbs mentioned on p. 132 whose present stems are lengthened by the addition of τ, as τύπ-τ-ω: imperf. ἔτυπτον. Note, however, that πίπτω (stem πετ-) is not of this class. See p. 132.

On page 144 is a list of verbs with labial stems which occur in St Mark. Where principal parts are not given they either do not occur in St Mark, or, if they do, show no irregularity. Notice that wherever the perf. indic. act. occurs it is strong. Notice also the frequency of str. aor. passives (ending in -ην and not -θην).

PERFECT MIDDLE AND PASSIVE OF VERBS WITH LABIAL STEMS

The periphrastic 3rd pers. plur. is required. See p. 126. The perf. indic. mid. and pass. of κρύπτω will serve as a model:

Sing.	1.	κέκρυμμαι
	2.	κέκρυψαι
	3.	κέκρυπται
Plur.	1.	κεκρύμμεθα
	2.	κέκρυφθε
	3.	κεκρυμμένοι εἰσί(ν).

Vocabulary

ἐργάτης (1) *workman*
δάκτυλος (2) *finger*
κτίσις, εως (3) *creation*
'Αβιάθαρ, (ὁ) indecl. *Abiathar*
νήπιος, α, ον *childish* (as noun, *babe*)
συνετός, ή, όν *wise, prudent*
ἀφορίζω *separate*
ἐκπορεύομαι (dep.) *proceed*
ἐπιτρέπω (with dat.) *permit, allow*

ἔφαγον (strong aor.) *I ate* (suppletive of ἐσθίω)
καταψύχω *cool*
κολλάω *unite*. Pass.: *cleave to* (with dat.)
οὐδέποτε (adv.) *never*
πρῶτον (adv.) *first, at first* (acc. neut. of πρῶτος)
δύο *two* (see p. 170)

EXERCISE 49

Translate into English:

1. καὶ ἐκείνην τὴν ὥραν τὸ δαιμόνιον συνεσπάραξεν αὐτόν, καὶ ἔπεσεν ἐπὶ τῆς γῆς. 2. ὁ δὲ Δαυεὶδ εἰσῆλθεν εἰς τὸν οἶκον τοῦ Θεοῦ ἐπὶ 'Αβιάθαρ ἀρχιερέως καὶ τοὺς ἄρτους ἔφαγεν, οὓς οὐκ ἔξεστιν φαγεῖν. 3. ἐν ταῖς ἡμέραις ἐκείναις πέμψω ὑμῖν τὸ Πνεῦμα τῆς

ἀληθείας ὃ παρὰ τοῦ Πατρὸς ἐκπορεύεται. 4. τοιαύτη γὰρ θλῖψις
ἔσται οἵα οὐ γέγονεν ἀπ᾽ ἀρχῆς κτίσεως ἕως τοῦ νῦν. 5. ἕνεκα τούτου
καταλείψει ἄνθρωπος τὸν πατέρα καὶ τὴν μητέρα καὶ κολληθήσεται
τῇ γυναικὶ αὐτοῦ. 6. ἐξομολογοῦμαί σοι, Πάτερ, ὅτι ἔκρυψας ταῦτα
ἀπο σοφῶν καὶ συνετῶν, καὶ ἀπεκάλυψας αὐτὰ νηπίοις.

EXERCISE 50

Translate into Greek:

1. The Lord will not forsake us in our affliction. 2. In the days of
Herod the king the word of God came to John. 3. They sat down by the
river. 4. The book will be written in two months. 5. We must lay hold
of eternal life. 6. How was so great a multitude fed? 7. These are the
men whom the apostles anointed. 8. Dip your finger in water and cool
my brow. 9. You were sent to us for your people's sake. 10. He did
not teach them without a parable.

EXERCISE 51

Translate into Greek:

1. They asked him to stay with them; and he stayed there two days.
2. The nations will be gathered together before him, and he will separate
them one from another. 3. I have never seen such a house as these
workers have built. 4. This saying was spread abroad until this day.
5. They took so many fish that their nets were breaking.[1] 6. The evil
spirit harmed him no longer. 7. The truth which evil men have hidden
will be revealed in those days. 8. We have been afflicted because of
our sins. 9. Lord, allow me first to go and bury my father. 10. First
he washed his hands, and then he washed their feet.

---

[1] Consider carefully which voice should be used.

Chapter 19

INDECLINABLE NOUNS

Hebrew and Aramaic names which occur in the N.T. are often not declined, i.e. there is no change in the ending for the various cases. This is always so with those which end in a consonant which Greek does not tolerate at the end of a word (see p. 9), e.g. Ἀβραάμ, Δαυείδ, Ἰσραήλ, Ἰωσήφ, Ἰερουσαλήμ. The genealogies in Matt. 1 and Luke 3 provide many instances.

A few place names ending in α and η are also indeclinable, e.g. Κανά *Cana*; Ῥαμά *Ramah*; Βηθφαγή *Bethphage*; Ναζαρά *Nazareth* (alternative forms of this place name are Ναζαρέτ, Ναζαρέθ, Ναζαράθ, all indeclinable).

HEBREW NAMES

Many Hebrew names are indeclinable (see above). But others are declined according to the Greek declensions.

1st *Declension*: those which end in α, η (fem.), ας, and ης (masc.). E.g. Ἄννα, Μαρία, Σαλώμη, Ἰούδας, Ἰωάνης.

Note, however, that an indeclinable Μαριάμ is found as an alternative to Μαρία.

Ἰωάνης has dative Ἰωάνει.

2nd *Declension*: those which end in ος, e.g. Λάζαρος, Μαθθαῖος.

Ἱεροσόλυμα, (an alternative for the indeclinable Ἱερουσαλήμ, which is fem.) is a neut. plur. noun of the 2nd decl.: gen., Ἱεροσολύμων; dat., Ἱεροσολύμοις. But in Matt. 2.3 it is 1st decl. feminine.
Σόδομα also is 2nd decl. neut. plur.

3rd *Declension*: those which end in ων, e.g. Σίμων, -ωνος; Σολομών, -ῶνος.

Μωυσῆς *Moses*, is declined on the analogy of βασιλεύς, but 1st decl. alternatives appear in acc. and dat.

> N.   Μωυσῆς
> A.   Μωυσέα or Μωυσῆν
> G.   Μωυσέως
> D.   Μωυσεῖ or Μωυσῇ

'Ιωσῆς *Joses*, has gen. 'Ιωσῆτος.
For the declension of 'Ιησοῦς *Jesus*, see p. 32.

ADJECTIVES EMPLOYING THE THIRD DECLENSION

Some adjectives have 3rd declension forms in the masc. and neut. while the fem. is of the 1st declension.

(*a*) Stems in -αντ:

πᾶς, πᾶσα, πᾶν *all, every*

	Masculine	*Feminine*	*Neuter*
Sing. N.	πᾶς	πᾶσα	πᾶν
A.	πάντα	πᾶσαν	πᾶν
G.	παντός	πάσης	παντός
D.	παντί	πάσῃ	παντί
Plur. N.	πάντες	πᾶσαι	πάντα
A.	πάντας	πάσας	πάντα
G.	πάντων	πασῶν	πάντων
D.	πᾶσι(ν)	πάσαις	πᾶσι(ν)

This adjective is very common. The only other adjective of this type is ἅπας, ἅπασα, ἅπαν, which is a strengthened form of πᾶς. But certain participles are so declined, namely the wk. aor. part. act. of verbs in -ω (see p. 161), and ἱστάς the pres. part. act. of ἵστημι and στάς its str. aor. part. act.

πᾶς is used with a noun without the article meaning "every" in the singular (πᾶν δένδρον *every tree*) and "all" in the plural (πρὸς πάντας ἀνθρώπους *to all men*). With the article (before or after it) it usually means "whole"; πᾶσα ἡ ἀγέλη *the whole herd*; τὸν πάντα χρόνον *the whole time*. It may also stand alone: πάντες *all men*; πάντα *all things*.

(b) Stems in -οντ:

ἑκών, ἑκοῦσα, ἑκόν *willing*

	Masculine	Feminine	Neuter
Sing. N.	ἑκών	ἑκοῦσα	ἑκόν
A.	ἑκόντα	ἑκοῦσαν	ἑκόν
G.	ἑκόντος	ἑκούσης	ἑκόντος
D.	ἑκόντι	ἑκούσῃ	ἑκόντι
Plur. N.	ἑκόντες	ἑκοῦσαι	ἑκόντα
A.	ἑκόντας	ἑκούσας	ἑκόντα
G.	ἑκόντων	ἑκουσῶν	ἑκόντων
D.	ἑκοῦσι(ν)[1]	ἑκούσαις	ἑκοῦσι(ν)[1]

[1] For ἑκόντσι(ν).

ἑκών is found only twice in the N.T. The only other adjective so declined is ἄκων, ἄκουσα, ἄκον *unwilling* (once only in N.T.). But the pres. part. act. of uncontracted verbs in -ω, the fut. part. act., and the str. aor. part. act. of -ω verbs are similarly declined (see pp. 159ff.).

(c) Stems in -υ (the stem changes to ταχε- before a vowel, and in the dat. plur.):

ταχύς, ταχεῖα, ταχύ *swift*

	Masculine	Feminine	Neuter
Sing. N.	ταχύς	ταχεῖα	ταχύ
A.	ταχύν	ταχεῖαν	ταχύ
G.	ταχέως	ταχείας	ταχέως
D.	ταχεῖ	ταχείᾳ	ταχεῖ
Plur. N.	ταχεῖς	ταχεῖαι	ταχέα
A.	ταχεῖς	ταχείας	ταχέα
G.	ταχέων	ταχειῶν	ταχέων
D.	ταχέσι(ν)	ταχείαις	ταχέσι(ν)

Adjectives similarly declined are:

βαθύς, εῖα, ύ *deep*
βαρύς, εῖα, ύ *heavy*
βραδύς, εῖα, ύ *slow*
βραχύς, εῖα, ύ *short*
γλυκύς, εῖα, ύ *sweet*
εὐθύς, εῖα, ύ *direct, straight*

θῆλυς, εια, υ *female*
ὀξύς, εῖα, ύ *sharp*
πλατύς, εῖα, ύ *broad* (fem. used as noun, *street*)
πραΰς, εῖα, ΰ *gentle, meek*
τραχύς, εῖα, ύ *rough*

(*d*) Stems in -ν:

μέλας, μέλαινα, μέλαν *black*

		Masculine	Feminine	Neuter
Sing.	N.	μέλας	μέλαινα	μέλαν
	V.	μέλαν	μέλαινα	μέλαν
	A.	μέλανα	μέλαιναν	μέλαν
	G.	μέλανος	μελαίνης	μέλανος
	D.	μέλανι	μελαίνῃ	μέλανι
Plur.	N.,V.	μέλανες	μέλαιναι	μέλανα
	A.	μέλανας	μελαίνας	μέλανα
	G.	μελάνων	μελαινῶν	μελάνων
	D.	μέλασι(ν)	μελαίναις	μέλασι(ν)

No other adjective of this type is found in the N.T.

(*e*) The very common and irregular adjective μέγας, μεγάλη, μέγα *great*, uses two stems, μεγα- and μεγαλο-:

		Masculine	Feminine	Neuter
Sing.	N.,V.	μέγας	μεγάλη	μέγα
	A.	μέγαν	μεγάλην	μέγα
	G.	μεγάλου	μεγάλης	μεγάλου
	D.	μεγάλῳ	μεγάλῃ	μεγάλῳ
Plur.	N.,V.	μεγάλοι	μεγάλαι	μεγάλα
	A.	μεγάλους	μεγάλας	μεγάλα
	G.	μεγάλων	μεγάλων	μεγάλων
	D.	μεγάλοις	μεγάλαις	μεγάλοις

No other adjective of this type appears in the N.T.

(*f*) The common and irregular adjective πολύς, πολλή, πολύ *much* (plur. *many*) also uses two stems πολυ- and πολλο-.

		Masculine	Feminine	Neuter
Sing.	N.	πολύς	πολλή	πολύ
	A.	πολύν	πολλήν	πολύ
	G.	πολλοῦ	πολλῆς	πολλοῦ
	D.	πολλῷ	πολλῇ	πολλῷ
Plur.	N.	πολλοί	πολλαί	πολλά
	A.	πολλούς	πολλάς	πολλά
	G.	πολλῶν	πολλῶν	πολλῶν
	D.	πολλοῖς	πολλαῖς	πολλοῖς

No other adjective of this type appears in the N.T.

δαιμόνια πολλὰ ἐξέβαλεν (Mark 1.34) *he cast out many devils.*
μετὰ δὲ πολὺν χρόνον (Matt. 25.19) *and after a long time.*

The masc. plur. is also used substantively (i.e. as a noun):

καὶ συνήχθησαν πολλοί (Mark 2.2) *and many were gathered together.*

THIRD DECLENSION ADJECTIVES OF TWO TERMINATIONS

There are also 3rd decl. adjectives of two terminations, the masc. and fem. being the same:

(*a*) Stems in -ν:

ἄφρων, ἄφρον *foolish*; ἄρσην, ἄρσεν *male*

		Masc. and Fem.	Neut.	Masc. and Fem.	Neut.
Sing.	N.	ἄφρων	ἄφρον	ἄρσην	ἄρσεν
	A.	ἄφρονα	ἄφρον	ἄρσενα	ἄρσεν
	G.	ἄφρονος	ἄφρονος	ἄρσενος	ἄρσενος
	D.	ἄφρονι	ἄφρονι	ἄρσενι	ἄρσενι
Plur.	N.	ἄφρονες	ἄφρονα	ἄρσενες	ἄρσενα
	A.	ἄφρονας	ἄφρονα	ἄρσενας	ἄρσενα
	G.	ἀφρόνων	ἀφρόνων	ἀρσένων	ἀρσένων
	D.	ἄφροσι(ν)	ἄφροσι(ν)	ἄρσεσι(ν)	ἄρσεσι(ν)

Declined like ἄφρων is σώφρων, σῶφρον *sober*. No other adjectives declined like ἄρσην appear in the N.T.

Declined also like ἄφρων, but with certain alternatives, is the comparative adjective in -ων; e.g. μείζων, μεῖζον *greater*.

		Masculine and Feminine	Neuter
Sing.	N.	μείζων	μεῖζον
	A.	μείζονα and μείζω	μεῖζον
	G.	μείζονος	μείζονος
	D.	μείζονι	μείζονι
Plur.	N.	μείζονες and μείζους	μείζονα and μείζω
	A.	μείζονας and μείζους	μείζονα and μείζω
	G.	μειζόνων	μειζόνων
	D.	μείζοσι(ν)	μείζοσι(ν)

The alternative forms are derived from a different stem, in -ος.   In the masc. and fem. acc. sing. and the neut. nom. and acc. plur. μείζω is from μείζοσα the σ being dropped between two vowels, and the μείζοα contracting to μείζω (see p. 10).   The masc. and fem. nom. and acc. plur. μείζους is from μείζοσες, the σ being dropped and the οε contracting to ου.

Like μείζων are declined the other comparative adjectives in -ων, for a list of which see p. 158.

(*b*) Stems in -ες (compare the nouns like γένος, stem γενες-, p. 120):

ἀληθής, ἀληθές true

		Masc. and Fem.	Neut.
Sing.	N.	ἀληθής	ἀληθές
	V.	ἀληθές	ἀληθές
	A.	ἀληθῆ	ἀληθές
	G.	ἀληθοῦς	ἀληθοῦς
	D.	ἀληθεῖ	ἀληθεῖ
Plur.	N., V.	ἀληθεῖς	ἀληθῆ
	A.	ἀληθεῖς	ἀληθῆ
	G.	ἀληθῶν	ἀληθῶν
	D.	ἀληθέσι(ν)	ἀληθέσι(ν)

The masc. and fem. acc. sing. and the neut. nom. and acc. plur. ἀληθῆ, is from ἀληθέσα, the σ being dropped, and έα contracting to ῆ.

The gen. sing. ἀληθοῦς is from ἀληθέσος, becoming ἀληθέος which contracts to ἀληθοῦς.

The dat. sing. ἀληθεῖ is from ἀληθέσι.

The masc. and fem. nom. and acc. plur. ἀληθεῖς is from ἀληθέσες, the σ dropping out, and έε contracting to εῖ.

LIKE ἀληθής ARE DECLINED:

ἀσθενής, ές *weak, sick*   συγγενής, ές (dat. plur. συγγεν-
εὐγενής, ές *well-born, noble*   εῦσι(ν)), *akin, natural* (masc. as
θεοσεβής, ές *god-fearing*   noun, *kinsman*)
μονογενής, ές *only-begotten*   ὑγιής, ές *whole, healthy*
                              ψευδής, ές *false*

πλήρης, ες *full*, may be included in this class, but in some instances it appears to be indeclinable.

VERBS WITH STEMS ENDING IN CONSONANTS

(d) LIQUID STEMS

Greek tends to avoid a σ immediately following a liquid (λ and ρ). In the fut. of verbs with liquid stems ε is inserted between the liquid and the σ (see p. 9). Then the σ, as usual between two vowels, is dropped, and the ε contracts with the vowel of the ending. Thus the fut. of στέλλω *send* (stem στελ-), στέλσω, becomes στελέσω, στελέω, and finally στελῶ.

In the wk. aor. act. the σ is dropped, and a lengthening of the stem takes place in compensation. Thus ἔστελσα becomes ἔστειλα.

θ, μ, and τ, when added to a liquid stem, cause no change. Thus ἠγέρθην (1st pers. sing. aor. indic. pass. of ἐγείρω); ἤγγελμαι (1st pers. sing. perf. indic. pass. of ἀγγέλλω); ἤγγελται (3rd pers. sing. of the same).

The present stems of these verbs are usually lengthened by the addition of the y sound (consonantal ι). In λ stems this generally has the effect of doubling the λ. Thus στελιω becomes στέλλω. In ρ stems the vowel of the stem is lengthened. Thus ἀριω becomes αἴρω *raise*.

The future tenses in full of στέλλω, active and middle, are:

		Active	*Middle*
Sing.	1.	στελῶ	στελοῦμαι
	2.	στελεῖς	στελῇ
	3.	στελεῖ	στελεῖται
Plur.	1.	στελοῦμεν	στελούμεθα
	2.	στελεῖτε	στελεῖσθε
	3.	στελοῦσι(ν)	στελοῦνται

The following is a list of verbs with liquid stems which occur in St Mark:

Present	*Compounds*	English	*Fut.*	*Aor. act.*	*Perf. act.*	*Perf. pass.*	*Aor. pass*
ἀγγέλλω		announce	ἀγγελῶ	ἤγγειλα		ἤγγελμαι	-ἠγγέλην
	ἀπαγγέλλω	announce, report					
	ἐπαγγέλλω	promise, profess (often mid.)					
	παραγγέλλω	command					
αἴρω		raise, take up	ἀρῶ	ἦρα	ἦρκα	ἦρμαι	ἤρθην
	ἀπαίρω	take away					
βάλλω		throw, put	βαλῶ	ἔβαλον[2]	βέβληκα	βέβλημαι	ἐβλήθην
	ἀμφιβάλλω	throw around, cast (a net)					
	ἀποβάλλω	throw off					
	ἐκβάλλω	cast out					
	ἐπιβάλλω	throw upon, put upon					
	περιβάλλω	throw around, clothe					
βούλομαι (dep.)		wish					ἐβουλήθην I wishe
δέρω		beat		ἔδειρα			ἐδάρην[1]
ἐγείρω		arouse	ἐγερῶ	ἤγειρα		ἐγήγερμαι	ἠγέρθην
	διεγείρω	arouse thoroughly					
θέλω		will	θελήσω[3]	ἠθέλησα			
μέλλω		intend, be about to	μελλήσω[3]				
μέλω[4]		be a care					
σκύλλω		trouble, vex				ἔσκυλμαι	
σπείρω		sow		ἔσπειρα		ἔσπαρμαι	ἐσπάρην[1]
-στέλλω		send	-στελῶ	-ἔστειλα	-ἔσταλκα	-ἔσταλμαι	-ἐστάλην
	ἀποστέλλω	send away					
	διαστέλλομαι (mid.)	command					

Present	Compounds	English	Fut.	Aor. act.	Perf. act.	Perf. pass.	Aor. pass.
τέλλω		meaning of simplex is uncertain	-τελοῦμαι⁵	-ἔτειλα	-τέταλκα	-τέταλμαι	
	ἀνατέλλω	cause to rise, rise					
	ἐντέλλομαι (mid.)	command					
	ἐξανατέλλω	spring up					
-ίλλω		pluck					
φέρω⁶		bring	οἴσω	ἤνεγκα	-ἐνήνοχα		ἠνέχθην
	ἀναφέρω	lead up					
	ἀποφέρω	carry off					
	ἐκφέρω	carry out					
	παραφέρω	take away					
	περιφέρω	carry about					
	προσφέρω	bring to, offer					
‚αίρω⁷		rejoice	χαρήσομαι				ἐχάρην¹ I rejoiced

1 Notice these str. aorists passive. Future passives of these verbs, where they occur, are formed from these str. aor. pass. stems, e.g. δαρήσομαι *I shall be beaten.*

2 ἔβαλον is the only str. aor. act. in this class.

3 Note that θέλω and μέλλω in the fut. and aor. depart from the usual liquid stem future forms.

4 μέλω is frequently found in the 3rd pers. sing. pres. indic. act., μέλει, used impersonally, "it matters", e.g. οὐ μέλει σοι; (Mark 4.38) *does it not matter to you?*

5 -τέλλω has a deponent (middle) future.

6 φέρω, imperfect ἔφερον, has the liquid stem φερ-. The other principal parts are supplied from different stems, οἰ- and ἐνεγκ-.

7 χαίρω has fut. and aor. deponent, with strong passive forms.

Vocabulary

δῶρον (2) *gift*
θερισμός (2) *harvest*
ζιζάνιον (2) *darnel, tare*
ἱμάτιον (2) *garment*
κοράσιον (2) *maiden*
κράββατος (2) *pallet, bed*
σεισμός (2) *earthquake*
σῖτος (2) *wheat*
ἀμπελών, ῶνος, ὁ (3) *vineyard*
κῦμα, ατος, τό (3) *wave*
λαῖλαψ, απος, ἡ (3) *storm*
πίναξ, ακος, ὁ (3) *dish, charger*
σπέρμα, ατος, τό (3) *seed*
μισθωτός, ή, όν *hired.* As noun, *hireling*

ὑψηλός, ή, όν *high*
ἀναπίπτω (compound of πίπτω: see p. 132) *recline, sit down*
ἐπαγγέλλομαι (mid.) *promise,* with dat. of person
δημοσίᾳ (adv.) *publicly*
λάθρᾳ (adv.) *secretly*
ὅτε (conjunction) *when*
πόσος, η, ον (interrog. adj.) *how much?* (plur. *how many?*)
πότε (interrog. adv.) *when?*
διὰ τί or τί *why?*
ἑπτά *seven*

EXERCISE 52

Translate into English:

1. ὁ δὲ εἶπεν αὐτοῖς Τί ὑμῖν ἐνετείλατο Μωυσῆς; 2. ἔφερον πρὸς αὐτὸν πάντας τοὺς ἀσθενεῖς. 3. καὶ ἐγένετο λαῖλαψ μεγάλη ἀνέμου καὶ τὰ κύματα ἐπέβαλλεν εἰς τὸ πλοῖον, ὥστε ἤδη γεμίζεσθαι τὸ πλοῖον. 4. τοῖς ἀγγέλοις αὐτοῦ ἐντελεῖται περὶ σοῦ καὶ ἐπὶ χειρῶν ἀροῦσίν σε. 5. πολλοὺς τῶν υἱῶν Ἰσραὴλ ἐπιστρέψει ὁ Ἰωάνης ἐπὶ τὸν Κύριον. 6. ὅτε δὲ ἤγγισεν ὁ καιρὸς τοῦ θερισμοῦ, ἀπέστειλεν τοὺς δούλους αὐτοῦ εἰς τὸν ἀμπελῶνα.

EXERCISE 53

Translate into Greek:

1. God promised Abraham a son. 2. I will lift up my eyes unto the hills. 3. The devils were cast out by the power of the Lord. 4. The way was straight, but their feet were slow. 5. Take up thy bed and walk. 6. The disciples rejoiced when they saw the Lord. 7. Elders must be sober. 8. A hireling does not care about the sheep. 9. We have offered our gifts in the temple. 10. The good seed has been sown in the fields.

EXERCISE 54

Translate into Greek:

1. Jesus asked them, How many loaves have ye? They said, Seven. And he commanded the multitude to sit down on the ground. 2. They beat us publicly and cast us into prison; and do they now thrust us out secretly? 3. After six days Jesus led them up into a high mountain. 4. John's head was brought on a charger to the maiden, who brought it to her mother. 5. Now shall the ruler of this world be cast out, and I shall be lifted up. 6. There was a great earthquake, and the sun became black. 7. Write therefore in a book the things which you are about to behold. 8. Did you not sow wheat in your field? When will it spring up? 9. A colt was brought to Jesus, and the disciples put their garments on it. 10. They that were whole wanted to bring the sick to Christ.

Chapter 20

THE COMPARISON OF ADJECTIVES

The comparison of adjectives is a process by which an adjective assumes
a different ending to denote a greater or less degree in quantity or quality;
e.g. the adjective "bright" may be given the suffix -er, "brighter", or
-est, "brightest", to express increasing degrees of brightness. We are
thus able to *compare* one thing with another. There are three degrees of
comparison; positive, comparative, and superlative; thus, "a good man"
(positive degree), "a better man" (comparative), "the best man" (superlative).

In Greek the principal rule is that -τερος is added to the masculine stem
of the positive adjective to form the comparative, and -τατος to form the
superlative. Thus δίκαιος, stem δικαιο- *righteous*, has:

> comparative, δικαιότερος, α, ον *more righteous* (declined like ἅγιος),
> superlative, δικαιότατος, η, ον *most righteous*, *very righteous* (de-
> clined like ἀγαθός).

ἀκριβής, stem ἀκριβες- *exact*, has:

> comparative, ἀκριβέστερος, α, ον *more exact*,
> superlative, ἀκριβέστατος, η, ον *most exact*, *very exact*

If the final ο of the stem is preceded by a short vowel the ο in the comparative and superlative is lengthened to ω. Thus σοφός, stem σοφο-
wise, has:

> comparative, σοφώτερος, α, ον *wiser*,
> superlative, σοφώτατος, η, ον *wisest*, *very wise*.

Stems ending in -ον have the comparative in -εστερος and the superlative in -εστατος. Thus δεισιδαίμων, stem δεισιδαιμον- *religous*, has:

> comparative, δεισιδαιμονέστερος.

The comparative can also express the sense of "rather" or "somewhat".
So in Acts 17.22 the meaning of δεισιδαιμονεστέρους ὑμᾶς is "you (are)
quite religious" (or possibly if the comparative stands for the superlative
(see p. 158) "you are very religious").

The superlative form δεισιδαιμονέστατος is not found in the N.T.

Some adjectives drop the final vowel of the stem and add -ιων for the comparative, and -ιστος for the superlative. ἡδύς *sweet*, which is not found in the N.T., however, affords a regular example of this formation: stem ἡδυ-, comparative ἡδίων, superlative ἥδιστος (of which the neut. plur. is found in the N.T. as an adverb). All the N.T. examples of this type introduce some irregularity. It will be noticed that the positive adjective with which these comparatives and superlatives are usually associated is frequently from a different stem.

Positive	English	Comparative	Superlative
ἀγαθός	good	κρείσσων, κρείττων, βελτίων	κράτιστος
κακός	bad	ἥσσων, ἥττων, χείρων	
μέγας	great	μείζων, μειζότερος	μέγιστος
μικρός	small	μικρότερος, ἐλάσσων, ἐλάττων *smaller, less*	ἐλάχιστος *least*
πολύς	much (plur. *many*)	πλείων, πλέων	πλεῖστος

The superlative degree is much less frequent in the N.T. than in classical Greek, and the comparative often does duty for it; e.g. ὁ δὲ μικρότερος ἐν τῇ βασιλείᾳ τοῦ Θεοῦ (Luke 7.28) *he that is least in the kingdom of God*.

The declension of μείζων, p. 152, provides the pattern for the declension of comparatives ending in -ων.

The noun or pronoun with which the comparison is made is put in the genitive:

μείζων αὐτοῦ *greater than he.*

εὐγενέστερός ἐστι τοῦ ἀδελφοῦ *he is more noble than his brother.*

Or the comparative may be followed by the particle ἤ, *than*, and a noun or pronoun in the same case.

εὐγενέστερός ἐστιν ἢ ὁ ἀδελφός *he is more noble than his brother.*

τομωτέροις λόγοις ἢ τούτοις *with sharper words than these.*

THE PARTICIPLE

Together with the infinitive the participle belongs to the verb infinite (see p. 98). But whereas the infinitive is a verbal noun, the participle is adjectival. It is used to attach the idea of a verb to a noun (or pronoun), as in "the *frightened* boys", "the *returning* warriors". Here *frightened* and *returning* are the equivalents of adjectives, qualifying the nouns.

But an adjective may also be attached to a noun in a different way, and a way which emphasizes its verbal rather more than its adjectival character; e.g. "The boys, *seeing* the headmaster approach, returned to their seats". In this sentence *seeing* is adjectival to the extent that it "qualifies" *the boys*. It adds a descriptive phrase to the noun *boys*, and it agrees with *boys*, as would be evident if the English participle were inflected. Yet it is verbal in that it emphasizes the action of seeing, and as a verb takes an object. The participial clause is, in fact, the equivalent of a clause with a main verb, and the sentence might, without change of meaning, be written "The boys saw the headmaster approach and returned to their seats". The sentence could also be expressed "When they saw the headmaster approach, the boys . . ." or "Since they saw the headmaster approach, the boys . . .". We shall see that a Greek participle can often best be translated into English by a temporal or a causal clause.

Being verbal, the participle has tense and voice. The English verb does not provide a full scheme of participles, and auxiliary verbs have to be used to complete the number. The participles of the verb "to do" may be set out as follows:

	Active voice	Passive voice
Present	doing	being done
Future	being about to do	being about to be done
Past	having done	done

Being adjectival, the participle in Greek is fully inflected.

Present Participle Active

The pres. part. act. is declined like the adjective ἑκών (see p. 149). -ων is added to the present stem. The fem. ending is -ουσα, and the neuter -ον. Thus: λύων, λύουσα, λῦον *loosing*; βάλλων, βάλλουσα, βάλλον *throwing*.

The stem of the pres. part. act. for masc. and neut. is λυοντ-. The dat. plur., masc. and neut. is from λύοντσι(ν) (cf. dat. plur. of ἄρχων, ἄρχοντος, p. 75). There is little danger of confusion with the similar 3rd pers. plur. pres. indic. act., since the context invariably makes clear which is intended.

	Masculine	Feminine	Neuter
Sing. N., V.	λύων	λύουσα	λῦον
A.	λύοντα	λύουσαν	λῦον
G.	λύοντος	λυούσης	λύοντος
D.	λύοντι	λυούσῃ	λύοντι
Plur. N., V.	λύοντες	λύουσαι	λύοντα
A.	λύοντας	λυούσας	λύοντα
G.	λυόντων	λυουσῶν	λυόντων
D.	λύουσι(ν)	λυούσαις	λύουσι(ν)

The pres. part. act. of all verbs in -ω may be found by adding -ων, -ουσα, -ον to the present stem. They are all declined like λύων.

The rules of contraction, as described above, p. 109, apply in the pres. participles act. of verbs with stems ending in α, ε, and ο.

Pres. part. act. of τιμῶ (τιμά-ω):

Nom. sing.	τιμῶν, τιμῶσα, τιμῶν
Gen. sing.	τιμῶντος, τιμώσης, τιμῶντος
Dat. plur.	τιμῶσι(ν), τιμώσαις, τιμῶσι(ν)

Pres. part. act. of φιλῶ (φιλέ-ω):

Nom. sing.	φιλῶν, φιλοῦσα, φιλοῦν.
Gen. sing.	φιλοῦντος, φιλούσης, φιλοῦντος
Dat. plur.	φιλοῦσι(ν), φιλούσαις, φιλοῦσι(ν)

Pres. part. act. of δηλῶ (δηλόω):

Nom. sing.	δηλῶν, δηλοῦσα, δηλοῦν
Gen. sing.	δηλοῦντος, δηλούσης, δηλοῦντος
Dat. plur.	δηλοῦσι(ν), δηλούσαις, δηλοῦσι(ν)

The *pres. part.* of εἰμί *I am*, is as follows:

Nom. sing.	ὤν, οὖσα, ὄν
Gen. sing.	ὄντος, οὔσης, ὄντος
Dat. plur.	οὖσι(ν), οὔσαις, οὖσι(ν)

FUTURE PARTICIPLE ACTIVE

-ων is added to the stem after the σ has been inserted. Thus λύ-σ-ων *being about to loose*; stem: λυσοντ-; fem.: λύσουσα; neut.: λῦσον. The declension is the same as for the pres. part.

In verbs in which the σ disappears in the fut. indic. (those with stems ending in liquids, p. 154, nasals, p. 184, and dentals, p. 134, which have contracted fut. indic.) the σ is naturally not found in the fut. part. Thus στέλλω: fut. indic. στελῶ; fut. part. στελῶν, στελοῦσα, στελοῦν *being about to send* (pres. part. act. στέλλων). ἐλπίζω: fut. indic. ἐλπιῶ; fut. part. ἐλπιῶν, ἐλπιοῦσα, ἐλπιοῦν *being about to hope*.

WEAK AORIST PARTICIPLE ACTIVE

-ας is added to the stem after the insertion of the σ characteristic of the aorist. Thus λύ-σ-ας *having loosed*;[1] stem: λυσαντ-; fem.: λύσασα; neut.: λῦσαν.

The declension is the same as that of πᾶς, πᾶσα, πᾶν.

		Masculine	Feminine	Neuter
Sing.	N., V.	λύσας	λύσασα	λῦσαν
	A.	λύσαντα	λύσασαν	λῦσαν
	G.	λύσαντος	λυσάσης	λύσαντος
	D.	λύσαντι	λυσάσῃ	λύσαντι
Plur.	N., V.	λύσαντες	λύσασαι	λύσαντα
	A.	λύσαντας	λυσάσας	λύσαντα
	G.	λυσάντων	λυσασῶν	λυσάντων
	D.	λύσασι(ν)	λυσάσαις	λύσασι(ν)

Note that there is no augment in the aorist participle. Where the σ before the ending is not found in the weak aorist active indicative, neither is it found in the participle: e.g. the aor. indic. act. of ἀγγέλλω is ἤγγειλα, and the aor. act. part. is ἀγγείλας; the aor. indic. act. of ἀποκτείνω is ἀπέκτεινα and the aor. act. part. is ἀπόκτεινας. In effect, therefore, the wk. aor. part. act. of a verb may be found by adding a final ς to the 1st pers. sing. of the wk. aor. indic. act., and omitting the augment. Consider the following:

	Wk. aor. indic. act.	Aor. part. act.
βλέπω	ἔβλεψα	βλέψας *having looked*
αἴρω	ἦρα	ἄρας *having raised*
φέρω	ἤνεγκα	ἔνεγκας *having brought*
σημαίνω	ἐσήμανα	σήμανας *having signified*

[1] "Having loosed" is usually given as the nearest English equivalent. But see pp. 165–6 on the time sequence in participles.

STRONG AORIST PARTICIPLE ACTIVE

In those verbs which have a strong aorist active the aorist participle is formed by adding -ών to the verb stem. Thus βάλλω, str. aor. indic. ἔβαλον; aor. part. act. βαλών (cf. pres. part. act. βάλλων). Stem, βαλοντ-; fem., βαλοῦσα; neut., βαλόν. The declension is the same as that of λύων, but the accent is oxytone (i.e. acute on the last syllable; see p. 275).

Consider the following:

	Str. aor. indic. act.	Aor. part. act.
ἄγω	ἤγαγον	ἀγαγών having led
ἔρχομαι	ἦλθον	ἐλθών having come
λέγω	εἶπον	εἰπών having said
πίπτω	ἔπεσον	πεσών having fallen
λαμβάνω	ἔλαβον	λαβών having taken
μανθάνω	ἔμαθον	μαθών having learned

PERFECT PARTICIPLE ACTIVE

-ώς is added to the reduplicated stem after the κ which is characteristic of the perfect has been inserted: λελυ-κ-ώς having loosed; stem, λελυκοτ-; fem., λελυκυῖα; neut., λελυκός. The accent is oxytone (see pp. 275–6).

		Masculine	Feminine	Neuter
Sing.	N.	λελυκώς	λελυκυῖα	λελυκός
	A.	λελυκότα	λελυκυῖαν	λελυκός
	G.	λελυκότος	λελυκυίας	λελυκότος
	D.	λελυκότι	λελυκυίᾳ	λελυκότι
Plur.	N.	λελυκότες	λελυκυῖαι	λελυκότα
	A.	λελυκότας	λελυκυίας	λελυκότα
	G.	λελυκότων	λελυκυιῶν	λελυκότων
	D.	λελυκόσι(ν)	λελυκυίαις	λελυκόσι(ν)

In verbs which have a strong perfect the κ is not, of course, found in the participle; γράφω I write: perf. indic. γέγραφα I have written; perf. part.

act. γεγραφώς, γεγραφυῖα, γεγραφός. The declension is as for λελυκώς.

There is some evidence for a gen. and dat. sing. fem. in -ης, -ῃ. Acts 5.2 has συνειδυίης in the best manuscripts. (This is the gen. sing. fem. of the participle of συνοῖδα, a compound of οἶδα *I know*, which is a perfect, the present of which does not exist. The participle is εἰδώς, εἰδυῖα, εἰδός.)

The perfect participle active is much less frequently used than the aorist. It retains the specifically perfect reference to a completed state, whilst the aor. part. most frequently merely draws attention to the fact that its action was prior to the action of the main verb of the sentence. λελυκώς means "being in a state of having loosed", though such a translation would be cumbrous.

THE USE OF THE PARTICIPLE

The participle in Greek is used in much the same way as in English, but it is used more frequently, and with greater flexibility.

The negative with the participle is μή. The N.T. has a few instances of οὐ with the participle, but most of these exceptions are to be explained by the fact that the negative is separated from the participle, or is felt to attach to some noun in the clause rather than to the participle.

The Greek participle, broadly speaking, may be said to be used in two ways, adjectivally and adverbially.

THE ADJECTIVAL USE OF THE PARTICIPLE

The participle may be used to qualify a noun after the manner of an adjective. Used in this way it generally has the article. It is often best translated by a relative clause; e.g.:

ὁ υἱός σου οὗτος ὁ καταφαγών σου τὸν βίον (Luke 15.30) *this thy son who hath devoured thy living.*

πᾶς ὁ ποιῶν τὴν ἁμαρτίαν (John 8.34) *every one that committeth sin.*

Occasionally the article is omitted:

τίς γυνὴ δραχμὰς ἔχουσα δέκα... (Luke 15.8) *what woman who has* (or "having") *ten pieces of silver...*

The participle is frequently found with the article, but with no preceding noun, although such a noun may easily be understood:

ὁ πιστεύων εἰς ἐμέ (John 12.44) *he that believeth in me* (where ὁ πιστεύων is the equivalent of ὁ ἄνθρωπος ὁ πιστεύων).

οἱ ἰσχύοντες . . . οἱ κακῶς ἔχοντες (Mark 2.17) *they that are whole . . . they that are sick* (the equivalent of οἱ ἄνθρωποι οἱ ἰσχύοντες etc.).

τῷ τύπτοντί σε (Luke 6.29) *to him that striketh thee.*

The participle with the article thus practically becomes a noun: οἱ κακῶς ἔχοντες above virtually means "the sick". So ὁ βαπτίζων of Mark 1.4 is the equivalent of ὁ βαπτιστής. The student will find this useful in translating English into Greek. If the required Greek noun is not known, a participle may often be found to serve; e.g. ὁ σπείρων (Mark 4.3) *the sower.*

THE ADVERBIAL USE OF THE PARTICIPLE

A participle, while belonging to and agreeing with a noun, may be used to add something to the idea of the main verb. In this case the participle has no article with it:

ταῦτα εἰπών Ἰησοῦς ἐταράχθη (John 13.21) *having said* (or "when he had said") *these things, Jesus was troubled.*

Here the participial clause, ταῦτα εἰπών, further defines the main verb, ἐταράχθη, and therefore has the function of an adverb. In this example the participial clause is the equivalent of a temporal clause, "when he had said these things". A participial clause, however, may be the equivalent of other types of adverbial clauses, causal, concessive, and conditional.

Causal meaning in the Participle:

νυνὶ δὲ μηκέτι τόπον ἔχων ἐν τοῖς κλίμασι τούτοις . . . πορεύομαι εἰς Ἱερουσαλήμ (Rom. 15.23–25) *but now having* (or "since I have") *no more any place in these regions . . . I go unto Jerusalem.*

Concessive meaning in the Participle

Καίπερ *although*, may be written with the participle to give it a concessive sense. But the participle may have a concessive sense even without καίπερ:

θέλων κληρονομῆσαι τὴν εὐλογίαν ἀπεδοκιμάσθη . . . καίπερ μετὰ δακρύων ἐκζητήσας αὐτήν (Heb. 12.17) *when he desired to inherit the blessing, he was rejected . . . though he sought it diligently with tears.*

Even if the particle καίπερ were not written here the participle
ἐκζητήσας would have to be given a concessive sense. Cf. the following:

ἐλεύθερος γὰρ ὢν ἐκ πάντων πᾶσιν ἐμαυτὸν ἐδούλωσα (1 Cor.
9.19) *for though I was free from all men, I brought myself under bondage
to all.*

Conditional meaning in the Participle

Here the participle is the equivalent of an if-clause:

ἡ γὰρ... γυνὴ τῷ ζῶντι ἀνδρὶ δέδεται νόμῳ (Rom. 7.2) *the
woman... is bound by law to the husband if he is alive.*

Here the participle ζῶντι might, perhaps, equally well be translated
temporally, *as long as* (or "while") *he is alive.* The next verse, however,
begins with a conditional clause, "But if the husband die", and suggests
the conditional sense of the participle in verse 2.

It is frequently a matter of nice choice to decide the type of adverbial
clause which best translates a Greek participle. Nor must it be forgotten
that there are occasions when an English participle also best suits the context: e.g.

ἐβαπτίζοντο... ἐξομολογούμενοι τὰς ἁμαρτίας αὐτῶν (Mark
1.5) *they were baptized... confessing their sins.*
καὶ ἐκήρυσσεν λέγων... (Mark 1.7) *and he preached, saying...*

TIME SEQUENCE IN THE PARTICIPLES

The broad distinction between the present and aorist participles is that
the pres. part. describes an action thought of as contemporaneous with the
action of the main verb, whilst the aor. part. describes an action which
precedes that of the main verb:

καὶ οἱ μαθηταὶ αὐτοῦ ἤρξαντο ὁδὸν ποιεῖν τίλλοντες τοὺς
στάχυας (Mark 2.23) *and his disciples began to make their way
plucking the ears of corn.*

The pres. part. τίλλοντες is used because the action of plucking was
contemporaneous with "began to make their way". It is to be noted
carefully that the pres. part. does not necessarily refer to present time. It
is present only in relation to the main verb. The actual time to which the
participle refers, past, present, or future, is governed by the main verb

of the sentence.   In the above sentence it is in past time because ἤρξαντο is past.

καὶ κράξας καὶ πολλὰ σπαράξας ἐξῆλθεν (Mark 9.26) *and having cried out, and torn him much, he came out.*

Here the two aor. parts., κράξας and σπαράξας, refer to the action of the evil spirit before it came out.   Again it is to be noted that the aor. part. does not always refer to past time.   It is past in relation to the main verb. The actual time to which the participle refers is governed by the main verb: e.g. in the following the time is future:

ἀναστὰς πορεύσομαι πρὸς τὸν πατέρα μου (Luke 15.18) *I will arise and go to my father* (or "having arisen I will go . . .").

Here the aor. part. (of ἀνίστημι, see pp. 221–4) is used because the action of arising is conceived of as prior to that of πορεύσομαι.

Other uses of the participles to be noted are:

Present Participle

There being no imperfect participle, the pres. part. is used to convey the particular shades of meaning of the imperfect: repetition, incompleteness, and continuance (see p. 24).   The student must be ready to identify such pres. participles, e.g. τοὺς σωζομένους in Acts 2.47, "those that were being saved".

Aorist Participle

Where the main verb is in past time an aor. part. is sometimes used to describe an action contemporaneous with it, where, in accordance with what has been said above, we should expect a pres. part.   The best known example of this is the frequent ἀποκριθεὶς εἶπεν *he answered and said.* Here it is obvious that the answering does not precede the saying; the two actions are contemporaneous, and indeed the participle ἀποκριθείς refers to the same action as εἶπεν.

Perfect Participle

The perfect participle always represents the particular shade of meaning of the perfect, the completeness (at the time referred to in the sentence) of an action begun previously:

εὗρεν . . . τὸ δαιμόνιον ἐξεληλυθός (Mark 7.30) *she found . . . the devil gone out.*

The part. here is the neut. of ἐξεληλυθώς, participle of the perfect ἐξελήλυθα, associated with (ἐξ)έρχομαι. The use of the perf. part. here expresses the idea that the devil had gone out of the child before the return of the mother, and the child was still free from it.

Future Participle

The future participle naturally represents future time, e.g. οἱ λύσοντες *those who will loose, those that are about to loose.*

But the future participle can also express purpose:

ἐληλύθει προσκυνήσων (Acts 8.27), he had come to worship (literally "being about to worship").

ἐλεημοσύνας ποιήσων εἰς τὸ ἔθνος μου παρεγενόμην (Acts 24.17) *I came to bring* (do) *alms to my nation.*

Vocabulary

πορνή (1) *harlot*
ἄργυρος (2) *silver*
βίος (2) *life, living, livelihood*
κόκκος (2) *grain*
μόσχος (2) *calf*
πειρασμός (2) *trial, temptation*
σημεῖον (2) *sign*
χρυσός (2) *gold*
σίναπι, εως, τό (3) *mustard*
σιτευτός, ή, όν *fattened*

ἀκολουθέω (with dat.) *follow*
ἀναβαίνω *ascend*
ἐλέγχω *convict*
εὑρίσκω, fut. εὑρήσω *find*
κατεσθίω (compound of ἐσθίω; suppletive aor. ἔφαγον) *devour*
κατοικέω *dwell*
ἄλλος, η, ο (adj.) *another, other*
πόρρωθεν (adv.) *from afar*

EXERCISE 55

Translate into English:

1. ὁ πιστεύων εἰς ἐμὲ οὐ πιστεύει εἰς ἐμὲ ἀλλὰ εἰς τὸν πέμψαντά με. 2. καὶ ἐλθόντες πρὸς τοὺς μαθητὰς εἶδον ὄχλον πολὺν περὶ αὐτοὺς καὶ γραμματεῖς συνζητοῦντας πρὸς αὐτούς. 3. ὅτε ὁ υἱός σου οὗτος ὁ καταφαγών σου τὸν βίον μετὰ πορνῶν ἦλθεν, ἔθυσας αὐτῷ τὸν σιτευτὸν μόσχον. 4. μετὰ ταῦτα εἶδον ἄλλον ἄγγελον καταβαίνοντα ἐκ τοῦ οὐρανοῦ, ἔχοντα ἐξουσίαν μεγάλην. 5. ὁμοία ἐστὶν ἡ βασιλεία τῶν οὐρανῶν κόκκῳ σινάπεως, ὃν λαβὼν ἄνθρωπος ἔσπειρεν ἐν τῷ ἀγρῷ αὐτοῦ. 6. οἱ δὲ μαθηταὶ ἰδόντες αὐτὸν ἐπὶ τῆς θαλάσσης περιπατοῦντα ἐταράχθησαν.

EXERCISE 56

Translate into Greek, using the participle as much as possible:

1. All things are possible to him that believes. 2. Every one that exalteth himself shall be abased. 3. Gold is more precious than silver. 4. When they heard these things they were baptized. 5. Blessed is he who keeps these sayings. 6. He that is least in the kingdom of heaven is greater than John. 7. The Son of man came eating and drinking. 8. He that hateth me hateth my Father also. 9. And when he comes he will convict the world. 10. We have more sheep than you.

EXERCISE 57

Translate into Greek, using the participle as much as possible:

1. You shall see the angels of God ascending and descending upon the Son of man. 2. He that doeth righteousness shall have greater honour. 3. They came to the city and stayed there six months. 4. He would not receive the messengers although they had come from far. 5. When they did not find him there they departed. 6. They followed Jesus as he was walking by the sea. 7. When the Son of man comes shall he find faith on the earth? 8. The Lord took a loaf and blessed and brake it. 9. All who saw the signs which he did besought him to remain. 10. I will keep thee from the hour of temptation which is about to test them that dwell upon the earth.

Chapter 21

NUMERALS

The numerals are of three types: *Cardinals*, which are adjectives denoting number, *one, two, a hundred*, etc., *Ordinals*, which are adjectives denoting order, *first, second, hundredth*, etc., and *Adverbials*, which denote occurrence, *once, twice, thrice*. For the higher adverbial numerals English resorts to the phrase *four times, a hundred times*, etc.

The following numerals occur in the N.T.:

	Cardinals	*Ordinals*	*Adverbials*
1	εἷς, μία, ἕν *one*	πρῶτος, η, ον *first*	ἅπαξ *once*
2	δύο	δεύτερος, α, ον	δίς
3	τρεῖς. τρία	τρίτος, η, ον	τρίς
4	τέσσαρες, τέσσαρα	τέταρτος, η, ον	τετράκις
5	πέντε	πέμπτος, η, ον	πεντάκις
6	ἕξ	ἕκτος, η, ον	
7	ἑπτά	ἕβδομος, η, ον	ἑπτάκις
8	ὀκτώ	ὄγδοος, η, ον	
9	ἐννέα	ἔνατος, η, ον	
10	δέκα	δέκατος, η, ον	
11	ἕνδεκα	ἑνδέκατος, η, ον	
12	δώδεκα *or* δεκαδύο	δωδέκατος, η, ον	
14	δεκατέσσαρες, α	τεσσαρεσκαιδέκατος, η, ον	
15	δεκαπέντε	πεντεκαιδέκατος, η, ον	
16	δέκα ἕξ		
18	δέκα ὀκτώ *or* δέκα καὶ ὀκτώ		
20	εἴκοσι(ν)		
30	τριάκοντα		
40	τεσσαράκοντα *or* τεσσεράκοντα		
50	πεντήκοντα	πεντηκοστός, ή, όν	
60	ἑξήκοντα		
70	ἑβδομήκοντα		ἑβδομηκοντάκις
80	ὀγδοήκοντα		
90	ἐνενήκοντα		
100	ἑκατόν		
200	διακόσιοι, αι,α		
300	τριακόσιοι, αι, α		
400	τετρακόσιοι, αι, α		
500	πεντακόσιοι, αι, α		

Cardinals		Ordinals	Adverbials
600	ἑξακόσιοι, αι, α		
1,000	χίλιοι, αι, α		
2,000	δισχίλιοι, αι, α		
3,000	τρισχίλιοι, αι, α		
4,000	τετρακισχίλιοι, αι, α		
5,000	πεντακισχίλιοι, αι, α		
	or χιλιάδες πέντε		
7,000	ἑπτακισχίλιοι, αι, α		
	or χιλιάδες ἑπτά		
10,000	μύριοι, αι, α, or		
	δέκα χιλιάδες		
12,000	δώδεκα χιλιάδες		
20,000	εἴκοσι χιλιάδες		
50,000	μυριάδες πέντε		
100,000,000	μυριάδες μυριάδων		

Those which do not occur in the N.T. may be easily supplied. Thus:

"thirteen" is δεκατρεῖς or τρεῖς καὶ δέκα.

"twentieth", εἰκοστός, ή, όν; "hundredth", ἑκατοστός, ή, όν.

"six times", ἑξάκις; "eight times", ὀκτάκις; "nine times", ἐνάκις; "ten times", δεκάκις; "twenty times", εἰκοσάκις; "a hundred times", ἑκατοντάκις; "a thousand times", χιλιάκις; "ten thousand times", μυριάκις.

The first four cardinal numbers are declined as follows:

	Masc.	Fem.	Neut.		Masc., Fem., and Neut.
N.	εἷς one	μία	ἕν	N.	δύο two
A.	ἕνα	μίαν	ἕν	A.	δύο
G.	ἑνός	μιᾶς	ἑνός	G.	δύο
D.	ἑνί	μιᾷ	ἑνί	D.	δυσί(ν)

	Masc. and Fem.	Neut.		Masc. and Fem.	Neut.
N.	τρεῖς three	τρία	N.	τέσσαρες four	τέσσαρα
A.	τρεῖς	τρία	A.	τέσσαρας	τέσσαρα
G.	τριῶν	τριῶν	G.	τεσσάρων	τεσσάρων
D.	τρισί(ν)	τρισί(ν)	D.	τέσσαρσι(ν)	τέσσαρσι(ν)

Like εἷς, μία, ἕν are declined the common pronouns οὐδείς, οὐδεμία, οὐδέν and μηδείς, μηδεμία, μηδέν *nobody, nothing.* οὐδείς is used with a verb in the indicative mood, μηδείς when the verb is in other moods (cf. οὐ and μή; see p. 239). They are also used as adjectives; οὐδεὶς προφήτης *no prophet.* Like τέσσαρες is also declined δεκατέσσαρες *fourteen.* Cardinals from διακόσιοι *two hundred* onward are declined as the plural of ἅγιος, α, ον, until we reach those which employ the 3rd decl. nouns, χιλιάς, -άδος *a thousand,* and μυριάς, -άδος *ten thousand.* The other cardinals, 5 to 12, and 15 to 199 are indeclinable:

> ἑκατὸν βάτοι ἐλαίου (Luke 16.6) *a hundred baths* (measures) *of oil.*
> ἑπτὰ ἕτερα πνεύματα πονηρότερα ἑαυτοῦ (Matt. 12.45) *seven other spirits worse than himself.*

The ordinals are declined regularly, δεύτερος (α pure) like ἅγιος, α, ον, the rest like ἀγαθός, ή, όν.

FURTHER USES OF THE ARTICLE

There are several uses of the article which have not yet all received attention:

1. With an adverb it has the effect of making the adverb into a noun; e.g.

> τῷ πλησίον (Rom. 13.10) *to the neighbour.*
> ἡ αὔριον (Matt. 6.34) *the morrow* (the article is fem. as agreeing with an understood ἡμέρα).
> τὸ ἐντός (Matt. 23.26) *the inside.*

2. A prepositional phrase may be made the equivalent of a noun by having the article in front of it:

> οἱ παρὰ τὴν ὁδόν (Mark 4.15) *they* (or those) *by the way side.*

Perhaps, however, a participle is to be understood here, σπειρόμενοι: compare Mark 4.16: οἱ ἐπὶ τὰ πετρώδη σπειρόμενοι *they that are sown upon the rocky* (places);

> τὰ περὶ τοῦ Ἰησοῦ (Acts 18.25) *the things concerning Jesus.*

3. In similar fashion a prepositional phrase can be made the equivalent of an adjective by the article:

> Πάτερ ἡμῶν ὁ ἐν τοῖς οὐρανοῖς (Matt. 6.9) *our Father, who art in heaven.*
> τὴν δόξαν τὴν παρὰ τοῦ μόνου Θεοῦ οὐ ζητεῖτε (John 5.44) *the glory that* (cometh) *from the only God ye seek not.*

4. The neuter τό may be followed by a whole clause which is thus treated as a noun:

τό Εἰ δύνῃ (Mark 9.23) *the* (idea of saying) *If thou canst!*

Here Jesus repeats the words of the epileptic's father.

This τό is used in introducing a quotation, and thus practically has the force of quotation marks:

τὸ δέ 'Ανέβη τί ἐστιν εἰ μὴ . . . (Eph. 4.9) *now this* (word) *"He ascended", what is it but . . .?*

This use of τό is closely akin to its use with the infinitive (articular infinitive, see p. 101).

THE POSITION OF THE ARTICLE

The importance of the position of the article has already been briefly discussed (p. 39). The following is a summary of the points to be noted:

1. An adjective coming between an article and its noun is an epithet attached to the noun: τὸ ἀκάθαρτον πνεῦμα *the unclean spirit.*

2. Alternatively the epithet may have an article of its own and follow the noun: τὸ πνεῦμα τὸ ἀκάθαρτον.

3. An adjective without an article, and coming outside a noun and its article is a predicate: τὸ πνεῦμα ἀκάθαρτον or ἀκάθαρτον τὸ πνεῦμα *the spirit* (is) *unclean.* The verb "to be", therefore, may be left out without ambiguity, for there are few exceptions to this rule. But John 12.9 appears to be one, where ὁ ὄχλος πολύς means "the great crowd", and not "the crowd is great".

4. When a noun in the genitive case is dependent on another noun, it may, like an adjective, of which it is an equivalent, either be inserted between the article and the governing noun, or it may follow:

ὁ τοῦ Θεοῦ ἄνθρωπος (2 Tim. 3.17) *the man of God.*
ὁ τῆς δικαιοσύνης στέφανος (2 Tim. 4.8) *the crown of righteousness.*

These might equally well have been written ὁ ἄνθρωπος τοῦ Θεοῦ and ὁ στέφανος τῆς δικαιοσύνης. This is the order we have in τὸ ὄνομα τοῦ Θεοῦ (1 Tim. 6.1) *the name of God.*

The article of the governing noun may be repeated before the dependent genitive:

τὴν διδασκαλίαν τὴν τοῦ Σωτῆρος ἡμῶν (Tit. 2.10) *the doctrine of our Saviour* (literally, "the doctrine, the (one) of our Saviour").

MIDDLE PARTICIPLES

Present Participle Middle

λυόμενος, λυομένη, λυόμενον *loosing for oneself.* Declined as an adjective in -ος, -η, -ον (e.g. ἀγαθός). The pres. part. mid. of other verbs in -ω will easily be found by adding -ομενος to the present stem.

Future Participle Middle

λυσόμενος, λυσομένη, λυσόμενον *being about to loose for oneself.* Declined as ἀγαθός.

Contraction takes place in the fut. part. mid. of those verbs which have a contracted future:

The fut. part. mid. of στέλλω is στελούμενος, arrived at through the following stages: στελσόμενος, στελεσόμενος, στελεόμενος, στελούμενος (cf. p. 153).

The fut. part. mid. of ἐλπίζω is ἐλπιούμενος, arrived at through the following stages: ἐλπιδσόμενος, ἐλπισόμενος, ἐλπιεόμενος, ἐλπιούμενος (cf. p. 134).

Weak Aorist Participle Middle

λυσάμενος, λυσαμένη, λυσάμενον *having loosed for oneself.* Declined as ἀγαθός.

The wk. aor. part. mid. of a verb may be found by substituting -μενος for the final -μην of the 1st pers. sing. wk. aor. indic. mid., and omitting the augment:

ἀγγέλλω: wk. aor. indic. mid., ἠγγειλάμην; part. ἀγγειλάμενος.
φέρω: wk. aor. indic. mid., ἠνεγκάμην; part. ἐνεγκάμενος.

Strong Aorist Participle Middle

βαλόμενος, βαλομένη, βαλόμενον *having thrown for oneself.* Declined as ἀγαθός. -όμενος is added to the verb stem; be careful to distinguish this from the pres. part. mid., βαλλόμενος.

Perfect Participle Middle

λελυμένος, λελυμένη, λελυμένον *having loosed for oneself, being in a state of having loosed for oneself.* Declined as ἀγαθός, but notice that the accent is paroxytone (acute accent on the last syllable but one). The perf. part. mid. of a verb may be found by substituting -μένος for the final -μαι of the 1st pers. sing. perf. indic. mid. The reduplication is retained.

ἀγγέλλω: perf. indic. mid. ἤγγελμαι; part. ἠγγελμένος.
λαμβάνω: perf. indic. mid. εἴλημμαι; part. εἰλημμένος.

PASSIVE PARTICIPLES

Present Participle Passive

λυόμενος, λυομένη, λυόμενον *being loosed*, as for the middle.

Future Participle Passive

λυθησόμενος, λυθησομένη, λυθησόμενον *being about to be loosed*. Declined as ἀγαθός. The fut. part. pass. may be found by substituting -σομενος for the final -ν of the 1st pers. sing. of the aor. indic. pass., and omitting the augment. Note the omission of θ in the str. fut. pass. Thus στέλλω *send*, has str. aor. indic. pass. ἐστάλην and consequently fut. part. pass. σταλησόμενος.

Aorist Participle Passive

λυθείς, λυθεῖσα, λυθέν *loosed*, *being loosed*, or *having been loosed*. The accent is oxytone. -είς is substituted for the final -ην of the 1st pers. sing. of the aor. indic. pass., and the augment is omitted. Note the omission of θ in the str. aor. pass. The aor. part. pass. of στέλλω (aor. indic. pass. ἐστάλην) is σταλείς.

The declension of λυθείς (stem λυθεντ-) is as follows:

		Masculine	Feminine	Neuter
Sing.	N.	λυθείς	λυθεῖσα	λυθέν
	A.	λυθέντα	λυθεῖσαν	λυθέν
	G.	λυθέντος	λυθείσης	λυθέντος
	D.	λυθέντι	λυθείσῃ	λυθέντι
Plur.	N.	λυθέντες	λυθεῖσαι	λυθέντα
	A.	λυθέντας	λυθείσας	λυθέντα
	G.	λυθέντων	λυθεισῶν	λυθέντων
	D.	λυθεῖσι(ν)	λυθείσαις	λυθεῖσι(ν)

Other aor. participles pass. may easily be found similarly:

ἄγω: aor. indic. pass. ἤχθην;   part. ἀχθείς
βάλλω: „   „   „ ἐβλήθην;   „ βληθείς
ξηραίνω: „   „   „ ἐξηράνθην; „ ξηρανθείς
κρύπτω: „   „   „ ἐκρύβην;   „ κρυβείς

Perfect Participle Passive

λελυμένος, λελυμένη, λελυμένον *having been loosed, loosed,* as for the middle.

Vocabulary

Ἀθῆναι, ὧν, αἱ (1) (plur.)
  Athens
Ἀκύλας, ου, ὁ (1) *Aquila*
ἀσφάλεια (1) *safety*
Ἰταλία (1) *Italy*
πέτρα (1) *rock*
Πρίσκιλλα (1) *Priscilla*
προσευχή (1) *prayer*
σαγήνη (1) *net, dragnet*
τιμή (1) *price, honour*
φυλή (1) *tribe*
ἀργύριον (2) *piece of silver*
εὐαγγέλιον (2) *gospel*
θησαυρός (2) *treasure*
Κόρινθος ἡ (2) *Corinth*
Παῦλος (2) *Paul*

κλῆμα, ατος, τό (3), *branch (of vine)*
Λευεί, ὁ (indecl.) *Levi*
παραλυτικός, ή, όν *paralytic*
θεμελιόω *lay foundations, found*
ξηραίνω (for parts see p. 185) *dry up, wither*
σφραγίζω *seal*
τίκτω (for parts see p. 321) *bring forth, bear*
τιμάω *value, honour*
χωρίζω *separate (pass., depart)*
ἐπαύριον (adv.) *tomorrow*. With article, ἡ, *the next day*
ἔσω (adv.) *within, inside*
ὄπισθεν (adv.) *behind*
προσφάτως (adv.) *recently*

EXERCISE 58

Translate into English:

1. ἡ δὲ γυνὴ ἀκούσασα τὰ περὶ Ἰησοῦ, ἐλθοῦσα ἐν τῷ ὄχλῳ ὄπισθεν ἥψατο τοῦ ἱματίου αὐτοῦ. 2. καὶ προσέφερον αὐτῷ παραλυτικὸν ἐπὶ κλίνης βεβλημένον, ᾧ εἶπεν ὁ Ἰησοῦς, Ἔγειρε· ὁ δὲ ἐγερθεὶς ἀπῆλθεν εἰς τὸν οἶκον αὐτοῦ. 3. μετὰ ταῦτα χωρισθεὶς ἐκ τῶν Ἀθηνῶν, ὁ Παῦλος ἦλθεν εἰς Κόρινθον, καὶ εὑρών τινα Ἰουδαῖον ὀνόματι Ἀκύλαν, προσφάτως ἐληλυθότα ἀπὸ τῆς Ἰταλίας καὶ Πρίσκιλλαν γυναῖκα αὐτοῦ, ἔμενεν παρ' αὐτοῖς. 4. ὁμοία ἐστὶν ἡ βασιλεία τῶν οὐρανῶν θησαυρῷ τε κεκρυμμένῳ ἐν ἀγρῷ, καὶ σαγήνῃ βληθείσῃ εἰς τὴν θάλασσαν. 5. τὴν φυλακὴν εὕρομεν κεκλεισμένην

ἐν πάσῃ ἀσφαλείᾳ, ἀνοίξαντες δὲ ἔσω οὐδένα εὕρομεν. 6. πορευθέντες
εἰς τὸν κόσμον ἅπαντα κηρύξατε τὸ εὐαγγέλιον πάσῃ τῇ κτίσει. ὁ
πιστεύσας καὶ βαπτισθεὶς σωθήσεται.

EXERCISE 59

Translate into Greek, using the participle as much as possible:

1. Of the tribe of Levi were sealed twelve thousand. 2. This word,
when translated, is "God with us". 3. Where is he that is born king of
the Jews? 4. He answered and said to the man who spoke, Who is my
mother? 5. In the twelfth year they went up to Jerusalem. 6. He was
in the wilderness forty days, tempted by the devil. 7. Be like men who
obey their Lord. 8. For ten days we stayed, waiting for him. 9. The
branch which is withered is cast away. 10. This bread is good; but this
is the best bread.

EXERCISE 60

Translate into Greek, using the participle as much as possible:

1. The good shepherd will leave the ninety nine sheep in the wilderness
and will seek that which strays. 2. Seven spirits worse than the first enter
in and dwell there. 3. Thirty pieces of silver was the price of him that was
valued. 4. The house which is founded on a rock does not fall. 5. The
love which is from God shall enter your hearts and remain there. 6. Tell
my affairs to nobody for fourteen days. 7. On the next day a certain man
saw Peter while he was passing through the city, and he cried out and
besought him to come into his house. 8. About the ninth hour he saw
an angel which came to him and said, God has heard your prayers. 9. The
apostles answered and said, All the things which have been written will
be fulfilled. 10. Many of the Jews who had come followed him, hoping
to see a sign.

Chapter 22

THE ATTRACTION OF RELATIVE PRONOUNS

We find frequently in the N.T. that a relative pronoun (which agrees with its antecedent in gender and number) has been attracted also to the *case* of its antecedent, e.g.:

ὑπέστρεψαν οἱ ποιμένες δοξάζοντες καὶ αἰνοῦντες τὸν Θεὸν ἐπὶ πᾶσιν οἷς ἤκουσαν καὶ εἶδον (Luke 2.20) *the shepherds returned, glorifying and praising God for all the things that they had heard and seen.*

Here ἐπὶ πᾶσιν ἃ ἤκουσαν would have been expected, but the accusative ἃ has been attracted to the dative of its antecedent πᾶσιν.

ἤγγιζεν ὁ χρόνος τῆς ἐπαγγελίας ἧς ὡμολόγησεν ὁ Θεὸς τῷ Ἀβραάμ (Acts 7.17) *the time of the promise drew nigh, which God vouchsafed unto Abraham.*

Here ἣν ὡμολόγησεν would have been expected, but the accusative ἣν has been attracted to the genitive of its antecedent ἐπαγγελίας.

There is an extension of this in certain instances where the relative is attracted to the case of its antecedent, but the antecedent drops out, e.g.:

μηδὲν . . . ὧν εἰρήκατε (Acts 8.24) *none of the things which ye have spoken.*

This in full would be μηδὲν ἐκείνων ἃ εἰρήκατε. The genitive ἐκείνων has, as it were, attracted ἃ to its own case, and has then dropped out.

οὐδὲν ἐκτὸς λέγων ὧν οἱ προφῆται ἐλάλησαν (Acts 26.22) *saying nothing but what the prophets said.*

This in full would be οὐδὲν ἐκτὸς λέγων ἐκείνων ἃ. . . . But the genitive ἐκείνων has again attracted the ἃ to its own case, and has then dropped out.

This attraction of the relative with the omission of the antecedent is found frequently in the following phrases:

ἄχρι οὗ, μέχρις οὗ, ἕως οὗ *until.* These are the equivalent of ἄχρι (μέχρις, ἕως) τοῦ χρόνου ᾧ (or ἐν ᾧ), *until the time at which.*

ἀφ' οὗ *since*, for ἀπὸ τοῦ χρόνου ᾧ (or ἐν ᾧ) *from the time at which*.
ἐν ᾧ *while*, for ἐν τῷ χρόνῳ ἐν ᾧ.

For sentences in which some of these phrases are used, see p. 189.

THE VERBAL ADJECTIVE

The verbal adjective is passive in meaning and is part of the verb infinite. There are two forms:

1. -τέος is added to the stem, λυ-τέος, -τέα, -τέον. This conveys the idea of necessity, *requiring to be loosed*. It only occurs once in the N.T., in Luke 5.38.

2. -τός is added to the stem, λυ-τός, -τή, -τόν. There are two shades of meaning. One is hardly distinguishable from the perfect passive participle, *loosed*. Thus, ἀγαπητός *loved*, *beloved*. The other conveys the idea of capability, *able to be loosed*. The suffix -τός thus has the same effect as the English suffix *-ible* or *-able*. Thus ὁρατός *visible*.

THE GENITIVE ABSOLUTE

An absolute clause is one which, while it adds to the meaning of a sentence, usually by defining the circumstances in which the action of the main clause takes place, is nevertheless separate from (Latin, *absolutus*, "loosed from") the main clause grammatically.

English has its absolute clauses, e.g. *Dinner being over, the guests rose*. The absolute clause is "Dinner being over". It defines the circumstances in which the action of the main clause takes place. It is, however, grammatically separate from the main clause, which could stand as a complete sentence without it. Cf. "*The hour of departure having come*, they took their leave".

It will be noticed that the participle is used in the absolute clause. Note also that the absolute clause may represent a temporal clause, "when dinner was over . . .", or a causal clause, "since dinner was over . . .". It may also be the equivalent of a conditional clause, as in "this being so, I will not stay", where "this being so" represents "if this is so".

Latin also has absolute clauses. The participle is in the ablative, and its subject is a noun or pronoun also in the ablative. The construction is known as the ablative absolute. In Latin the rule quite strictly applies that there may be no grammatical connection between the absolute and the main clauses. Neither the subject nor the object of the absolute clause may appear as subject or object of the main clause; i.e. the Latin ablative absolute must be literally "absolute".

In the Greek absolute clause the participle is in the genitive, and its subject, whether noun or pronoun, is also in the genitive. The construction is known as the genitive absolute. It is fairly common, and, like the English absolute clause, may be the equivalent of a temporal, causal, or conditional clause.

καὶ ἐσθιόντων αὐτῶν λαβὼν ἄρτον εὐλογήσας ἔκλασεν (Mark 14.22) *and as they were eating, he took bread, and when he had blessed, he brake it.*

Here the present participle, ἐσθιόντων, signifies an action going on at the same time as that of the main verb.

ἐλθόντων δὲ αὐτῶν εἰς Καφαρναούμ προσῆλθον οἱ τὰ δίδραχμα λαμβάνοντες (Matt. 17.24) *and when they were come to Capernaum, they that received the half-shekel came.*

Here the aorist participle, ἐλθόντων, signifies an action prior to that of the main verb.

καὶ πορευομένων αὐτῶν ἐν τῇ ὁδῷ εἶπέν τις ... (Luke 9.57) *and as they went in the way, a certain man said ...*

Sometimes a pronoun subject is omitted, e.g.

καὶ ἐλθόντων πρὸς τὸν ὄχλον προσῆλθεν αὐτῷ ἄνθρωπος (Matt. 17.14) *and when they were come to the multitude, there came to him a man.*

Here αὐτῶν is omitted as the subject of ἐλθόντων. Cf. the lack of a subject αὐτοῦ to the participle εἰπόντος in Matt. 17.26.

The Greek genitive absolute is used more loosely than the Latin ablative absolute, and sometimes there is a grammatical connection between it and the main clause, e.g.

μὴ δυναμένου δὲ αὐτοῦ γνῶναι τὸ ἀσφαλὲς ... ἐκέλευσεν ... (Acts 21.34) *and when he could not know the certainty ... he commanded ...*

Here αὐτοῦ, the subject of the genitive participle, is the same person as the subject of the main verb ἐκέλευσεν. The clause is not strictly absolute.

FURTHER USES OF THE CASES

We have already dealt with the main uses of the cases, the accusative to express the direct object, the genitive denoting possession, the dative of indirect object, etc. (pp. 21, 30, 62, 72), and we have also noted different cases used with prepositions (pp. 36, 62ff., 137ff.) and the use of the cases to express time (p. 141). There are certain other uses of the cases which must now be treated.

THE NOMINATIVE CASE

The Nominative as Vocative

The nominative, usually with the article, is sometimes used for the vocative:

> τὸ κοράσιον, σοὶ λέγω, ἔγειρε (Mark 5.41) *damsel, I say unto thee, Arise.*

See also Matt. 11.26 and Mark 14.36.

THE ACCUSATIVE CASE

Double Accusative

In English certain verbs take two accusatives, e.g. *I teach the boys Latin; I call you my friend.* The same is true of some Greek verbs.

Greek verbs which are commonly found with two accusatives are verbs of teaching (διδάσκω), asking (αἰτέω, ἐρωτάω), making (ποιέω), putting on, clothing and unclothing (ἐνδύω, ἐκδύω), naming (ὀνομάζω), giving to drink (ποτίζω).

> ἐδίδασκεν αὐτοὺς ἐν παραβολαῖς πολλά (Mark 4.2) *he taught them many things in parables.*
>
> ἠρώτων αὐτὸν ... τὰς παραβολάς (Mark 4.10) *they asked of him the parables.*
>
> φορτίζετε τοὺς ἀνθρώπους φορτία (Luke 11.46) *ye lade men with burdens.*

Cognate Accusative

A verb may sometimes take the accusative of a noun which is cognate either in root or meaning. Such accusatives are often found with intransitive verbs (see p. 40, note 2) which cannot have an accusative of direct object. English provides examples such as "die the death" (cognate in root), "run a race" (cognate in meaning), or the playful "he sighed a great sigh". φορτίζετε τοὺς ἀνθρώπους φορτία (see under *Double Accusative* above) is an example of a cognate accusative in Greek. Compare also:

θησαυρίζετε θησαυρούς (Matt. 6.20) *lay up treasure.*

ἐχάρησαν χαρὰν μεγάλην (Matt. 2.10) *they rejoiced with great joy.*

The Greek cognate accusative, as here, must often be translated in English with the help of a preposition.

Retained Accusative

A verb which has an accusative (either direct object or cognate) in the active voice sometimes unexpectedly retains the accusative when in the passive:

ἐνδεδυμένος τρίχας (Mark 1.6) *clothed with hair.*

πεπίστευμαι τὸ εὐαγγέλιον (Gal. 2.7) *I have been entrusted with the gospel.* (Cf. also 1 Thess. 2.4.)

οἰκονομίαν πεπίστευμαι (1 Cor. 9.17) *I have been entrusted with a stewardship.*

βάπτισμα ὃ ἐγὼ βαπτίζομαι (Mark 10.38) *the baptism that I am baptized with.*

Such accusatives retained in the passive are often found with verbs which may have the double accusative, as in the first example above. They are also frequently in reality cognate accusatives. This is clear in the last of the examples above, where the accusative of the relative, ὅ, refers to βάπτισμα. The two examples with πιστεύω are also accusatives cognate in meaning, "the gospel" and "a stewardship" being substituted for the word "trust" in the phrase "I entrust a trust".

Notice how English employs the preposition *with* in translating the retained accusative.

Accusative of Respect

The accusative case may be used to express that with reference to which, or in respect of which, the action of the verb is performed or a statement is made:

πάντα ἐγκρατεύεται (1 Cor. 9.25) *he is temperate in all things* (i.e. in respect of all things).

τὸν ἀριθμὸν ὡς πεντακισχίλιοι (John 6.10) *about five thousand in number* (i.e. in respect of number).

An accusative neuter of a pronoun or adjective is frequently used in this way, and such accusatives are practically adverbs. For this reason the accusative of respect is sometimes called the adverbial accusative:

> Ἰουδαίους οὐδὲν ἠδίκηκα (Acts 25.10) *I have done no wrong to the Jews* (i.e. wronged the Jews in respect of nothing; note also the double accusative).
>
> λοιπὸν ἀπόκειταί μοι ὁ τῆς δικαιοσύνης στέφανος (2 Tim. 4.8) *henceforth there is laid up for me the crown of righteousness* (i.e. in respect of the future).

THE GENITIVE CASE

Subjective and Objective Genitives

We have seen that the genitive roughly represents our use of the preposition *of* with a noun or pronoun. Hence it commonly denotes possession. A distinction should be noted between what is called a "subjective genitive" and an "objective genitive". In a subjective genitive, e.g. ἡ ἀγάπη ἑνὸς ἑκάστου (2 Thess. 1.3) *the love of each one*, the genitive is the subject from which the ἀγάπη springs. In an objective genitive, e.g. τὸν φόβον τῶν Ἰουδαίων (John 7.13) *fear of the Jews*, the genitive is the object of the fear. It is sometimes difficult to decide between a subjective and an objective genitive: e.g. ἡ ἀγάπη τοῦ Θεοῦ: is it *our love for God*, or *God's love for us*? The context does not always make it clear.

Genitive of Separation

The genitive is also the case which denotes separation, having taken over the functions of a primitive ablative case (Latin retains an ablative case). Hence it is appropriately used with prepositions like ἐκ and ἀπό, and with verbs which express separating, hindering from, falling short of, lacking, depriving of, etc.

> μή τινος ὑστερήσατε; (Luke 22.35) *did ye lack anything?*
>
> χρῄζετε τούτων ἁπάντων (Matt. 6.32) *ye have need of all these things.*

Possibly to be considered under this heading is the genitive of comparison (see p. 158). This genitive may be used with a verb denoting comparison: πολλῶν στρουθίων διαφέρετε (Luke 12.7) *ye are of more value than* (different from) *many sparrows.*

Partitive Genitive

The genitive may denote the whole of which the dependent word is a part: e.g.

οἱ λοιποὶ τῶν ἀνθρώπων (Rev. 9.20) *the rest of mankind.*

τοὺς πτωχοὺς τῶν ἁγίων (Rom. 15.26) *the poor among (of) the saints.*

This partitive genitive is found with verbs denoting sharing:

τραπέζης Κυρίου μετέχειν (1 Cor. 10.21) *to partake of the table of the Lord.*

Genitive of Price

The genitive is used to denote price.

οὐχὶ δύο στρουθία ἀσσαρίου πωλεῖται; (Matt. 10.29) *are not two sparrows sold for a farthing?*

τοσούτου τὸ χωρίον ἀπέδοσθε; (Acts 5.8) *did ye sell the land for so much?*

Genitive after certain Verbs

The genitive is commonly found with verbs of:

(*a*) perception: ἀκούω *hear* (the strict rule, not always kept in the N.T., is genitive of person heard, accusative of thing heard); γεύομαι *taste*; ἅπτομαι *touch*; θιγγάνω *touch.*

(*b*) laying hold of: τυγχάνω *obtain*; λαγχάνω *obtain by lot*; ἀντέχομαι *hold firmly to*; κρατέω *hold.*

(*c*) remembering and forgetting: μιμνήσκομαι *remember*; μνημονεύω *remember*; ἐπιλανθάνομαι *forget.*

(*d*) ruling over: ἄρχω *rule*; κατισχύω *prevail against*; κυριεύω *be master of.*

(*e*) certain verbs expressing emotion and concern, and their opposites: σπλαγχνίζομαι *pity*; ἐπιθυμέω *desire*; ἐπιμελέομαι *care for*; φείδομαι *spare*; καταφρονέω *despise*; ἀμελέω *neglect*; ὀλιγωρέω *belittle.*

THE DATIVE CASE

Dative of Interest

The dative may be used to express the person interested in or concerned with the action of the verb or the statement made:

ἔσομαι τῷ λαλοῦντι βάρβαρος (1 Cor. 14.11) *I shall be to him that speaketh a barbarian.*

ἀστεῖος τῷ Θεῷ (Acts 7.20) *fair in God's sight* (lit. "fair to God").

13—N.T.G.

A similar dative of nouns not denoting persons may be used:

τῇ ἁμαρτίᾳ ἀπέθανεν ἐφάπαξ (Rom. 6.10) *he died once for all as far as sin is concerned* (R.V. "unto sin").

τὸν ἀσθενοῦντα τῇ πίστει (Rom. 14.1) *him that is weak as far as faith is concerned* (R.V. "in faith").

Dative of Difference

The dative is used to express difference:

πολλῷ μᾶλλον (Matt. 6.30) *much more* (lit. "more by much").

τοσούτῳ κρείττων (Heb. 1.4) *by so much better.*

Dative of Agent

A dative may be used to express agent, instead of the expected ὑπό with genitive, with a perfect passive participle:

οὐδὲν ἄξιον θανάτου ἐστὶν πεπραγμένον αὐτῷ (Luke 23.15) *nothing worthy of death hath been done by him.*

Dative after certain Verbs

The dative of indirect object is naturally found with verbs denoting speaking and giving (see p. 21). The dative is also commonly found with the following verbs:

ἀκολουθέω *follow*	παραγγέλλω *command*
ἀρέσκω *please*	πιστεύω *believe* (but εἰς or ἐν with
διακονέω *serve*	appropriate case is also used)
δουλεύω *serve*	προσκυνέω *worship*
ἐγγίζω *draw near*	ὑπακούω *obey*
ἐπιτιμάω *rebuke*	χράομαι *use*
ἐπιτρέπω *permit, allow*	

VERBS WITH STEMS ENDING IN CONSONANTS

(e) NASALS

In verbs with stems ending in nasals (μ and ν) the σ of the future and weak aorist is dropped after the nasal (cf. verbs with stems ending in liquids, p. 153). The fut. indic. act. is formed as in the case of verbs with liquid stems: κρίνω *I judge* has fut. κρινῶ, arrived at through the following stages: κρίνσω, κρινέσω, κρινέω, and so κρινῶ. In the weak aorist active a lengthening of the stem takes place in compensation for the loss of the σ: thus μένω *I remain* has aor. ἔμεινα from ἔμενσα. This lengthening in the aorist is sometimes obscured by the fact that the pres. indic. also shows a lengthening of the verb stem (though for a different reason; see below).

ν followed by θ, μ, τ, and κ usually disappears. The same is true when it is followed by σ elsewhere than in the fut. or wk. aor. Thus the 2nd pers. sing. perf. indic. pass. of κρίνω would be κέκρισαι.

Exceptions to this are the wk. aor. passives of -κτείνω *kill* (-ἐκτάνθην), and of the verbs in -ύνω and -αίνω (ἠσχύνθην, ηὐφράνθην, etc.).

	Compounds	English	Fut.	Aor. act.	Perf. act.	Perf. pass.	Aor. pass.
ἰσχύνω	καταισχύνομαι (pass.)	shame be ashamed of		ἤσχυνα			ἠσχύνθην
βαρύνω	καταβαρύνω	weigh down weigh down					ἐβαρύνθην
εὐθύνω		straighten		εὔθυνα			
εὐφραίνω		gladden					ηὐφράνθην
ἑρμαίνω		warm					
κερδαίνω		gain	κερδανῶ κερδήσω	ἐκέρδανα ἐκέρδησα			
κλίνω	ἀνακλίνω	lean make to re-cline	κλινῶ	ἔκλινα	κέκλικα		ἐκλίθην
κρίνω	ἀποκρίνομαι (dep.) κατακρίνω	judge answer condemn	κρινῶ	ἔκρινα	κέκρικα	κέκριμαι	ἐκρίθην
κτείνω	ἀποκτείνω	kill kill	-κτενῶ	-ἔκτεινα			-ἐκτάνθην
λευκαίνω		whiten		ἐλεύκανα			
μένω	προσμένω ὑπομένω	remain remain with, abide in endure	μενῶ	ἔμεινα	μεμένηκα[1]		
μηκύνω		lengthen (pass. *grow*)					
μιαίνω		defile				μεμίαμμαι	ἐμιάνθην
μωραίνω		make foolish		ἐμώρανα			ἐμωράνθην
ξηραίνω		dry up		ἐξήρανα		ἐξήραμμαι	ἐξηράνθην
ποιμαίνω		shepherd	ποιμανῶ	ἐποίμανα			
σημαίνω		signify		ἐσήμανα			
τείνω	ἐκτείνω	stretch stretch out	-τενῶ	-ἔτεινα			
τρέμω		tremble					
φαίνω		shine (pass. *appear*)	φανοῦμαι (dep.)				ἐφάνην (strong)
χύννω	ἐκχύννω	pour pour out, shed					-ἐχύθην

[1] μεμένηκα: Note that this perfect extends the stem by η, as does γίνομαι in the perf. pass., γεγένημαι.

This class includes a number of verbs whose stems are lengthened in the present by the addition of consonantal ι (the y sound), which, coming after the nasal ν, has the effect of lengthening the previous vowel sound: thus σημαν-ι-ω becomes σημαίνω.

The list on page 185 includes all the verbs of this class found in St Mark's Gospel, together with some others which, while being fairly frequent in the N.T., are not found in St Mark.

The deponent γίνομαι, having a stem ending in ν (γεν-) may be considered here, but it is irregular. The classical pres. indic. of this verb was γίγνομαι, formed by reduplication of the stem γι-γεν-ομαι (cf. πίπτω from πι-πετ-ω, p. 132). The principal parts are:

> fut., γενήσομαι; str. aor., ἐγενόμην; perf., γέγονα (N.B. varied form of stem), perf. pass., γεγένημαι; wk. aor. pass., ἐγενήθην.

It is frequent in the N.T. and has the following meanings: *become, be, come to pass, happen, be made, be done.*

Compounds are διαγίνομαι *pass, elapse*; παραγίνομαι *come, arrive.*
For verbs whose stem is lengthened in the pres. indic. by the addition of ν, see pp. 190ff.

Vocabulary

στολή (1) *robe*
ὑπερηφανία (1) *pride*
μνημεῖον (2) *tomb*
ναός (2) *sanctuary, temple*
νεανίσκος (2) *young man*
λαμπρός, ά, όν *shining*
λευκός, ή, όν *white*
πλούσιος, α, ον *rich*

διαφέρω *differ, excel, be better than*
κατασείω *wave*
μιμνήσκομαι (aor. ἐμνήσθην; perf. μέμνημαι) *remember*
νικάω *conquer*
ὑστερέω (with gen.), *come late, come short of, be inferior to*
ἔτι (adv.) *still, yet.*

EXERCISE 61

Translate into English:

1. καὶ σπλαγχνισθεὶς ἐκτείνας τὴν χεῖρα αὐτοῦ ἥψατο καὶ λέγει αὐτῷ Θέλω, καθαρίσθητι. 2. πάντων δὲ θαυμαζόντων ἐπὶ πᾶσιν οἷς ἐποίει εἶπεν πρὸς τοὺς μαθητὰς αὐτοῦ Ἀκούσατε τοὺς λόγους τούτους. 3. εἶπεν δὲ ὁ Παῦλος Δέομαί σου ἐπίτρεψόν μοι λαλῆσαι πρὸς τὸν λαόν. ἐπιτρέψαντος δὲ αὐτοῦ ὁ Παῦλος κατέσεισεν τῇ χειρὶ τῷ λαῷ. 4. αἱ δὲ γυναῖκες εἰσελθοῦσαι εἰς τὸ μνημεῖον εἶδον

νεανίσκον περιβεβλημένον στολὴν λευκήν. 5. τὸ σῶμα ὑμῶν ναὸς τοῦ ἐν ὑμῖν ʽΑγίου Πνεύματός ἐστιν, οὗ ἔχετε ἀπὸ Θεοῦ, καὶ οὐκ ἐστὲ ἑαυτῶν· ἠγοράσθητε γὰρ τιμῆς. 6. πεσοῦνται οἱ εἴκοσι τέσσαρες πρεσβύτεροι ἐνώπιον τοῦ Θεοῦ, καὶ προσκυνήσουσιν τῷ ζῶντι εἰς τοὺς αἰῶνας τῶν αἰώνων, καὶ βαλοῦσιν τοὺς στεφάνους αὐτῶν ἐνώπιον τοῦ θρόνου.

EXERCISE 62

Translate into Greek, using the genitive absolute where possible:

1. How much better is a man than a sheep? 2. As they were eating Jesus took bread. 3. When they had eaten he led them out. 4. In no respect was I inferior to the apostles. 5. They remembered the words which Paul had spoken. 6. The apostles will teach us the truth. 7. They put his own garment upon him. 8. I was entrusted by God with the gospel. 9. In all things we are conquerors. 10. Such words do not please me at all.

EXERCISE 63

Translate into Greek, using the genitive absolute where possible:

1. They who worship God must worship him in spirit and in truth. 2. As the women drew near, the Lord commanded the apostles to receive them. 3. While Peter was still speaking the Holy Spirit fell upon all who heard the word. 4. Those who use the sword shall die by the sword. 5. To forget injustice is much better than to remember pride. 6. This house was bought for a hundred pieces of silver. 7. What shall they who serve this world gain in the kingdom of heaven? 8. Our souls being weighed down by sin we cannot please God. 9. When it became day a certain rich man came and asked Pilate for the body of Jesus. 10. There appeared to them two angels clothed in shining garments.

Chapter 23

TEMPORAL CLAUSES

A temporal clause is a relative clause which defines the time at which the action of the main clause takes place; e.g. "We shall go when we have had tea". Here the temporal clause is "when we have had tea". It is introduced by the temporal conjunction *when*, which relates, or refers, to the main clause, "we shall go".

If the temporal clause refers to a definite time the verb is in the indicative mood. In this temporal clauses are like other relative clauses. But if the time referred to is indefinite, as it often is when the reference is to the future, the verb is in the subjunctive mood. In this also temporal clauses are like other relative clauses, which employ the subjunctive when the relative pronoun is indefinite (*whoever* rather than *who*). For indefinite clauses, see pp. 212f.

The usual conjunctions introducing temporal clauses in English are *when, after, since, while, until, till, before.* All these have their equivalents in Greek, of which the most common are:

ὅτε *when*:

> καὶ ὅτε ἐγγίζουσιν εἰς Ἱεροσόλυμα... ἀποστέλλει δύο τῶν μαθητῶν αὐτοῦ (Mark 11.1) *and when they draw nigh unto Jerusalem ... he sendeth two of his disciples.*

> πολλαὶ χῆραι ἦσαν... ἐν τῷ Ἰσραήλ, ὅτε ἐκλείσθη ὁ οὐρανός (Luke 4.25) *there were many widows in Israel ... when the heaven was shut up.*

> ἔρχεται ὥρα καὶ νῦν ἐστιν, ὅτε οἱ ἀληθινοὶ προσκυνηταὶ προσκυνήσουσιν τῷ Πατρὶ ἐν πνεύματι καὶ ἀληθείᾳ (John 4.23) *the hour cometh, and now is, when the true worshippers shall worship the Father in spirit and truth.*

ὁπότε, *when* (not so common as ὅτε or ὡς, below):

> οὐδὲ τοῦτο ἀνέγνωτε ὃ ἐποίησεν Δαυεὶδ ὁπότε ἐπείνασεν; (Luke 6.3) *have ye not read even this, what David did, when he was hungry?*

ὡς, *when, since, while* (the exact shade of meaning can be determined from the context):

ὡς ἐπλήσθησαν αἱ ἡμέραι . . . ἀπῆλθεν (Luke 1.23) *when the days were fulfilled . . . he departed.*

ὡς τὸ φῶς ἔχετε, πιστεύετε εἰς τὸ φῶς (John 12.36) *while ye have the light, believe on the light.*

πόσος χρόνος ἐστὶν ὡς τοῦτο γέγονεν αὐτῷ; (Mark 9.21) *how long time is it since this hath come unto him?*

ἕως, *till, until* and (sometimes) *while*:

ὁ ἀστὴρ . . . προῆγεν αὐτοὺς ἕως ἐλθὼν ἐστάθη (Matt. 2.9) *the star . . . went before them, till it came and stood.*

ἡμᾶς δεῖ ἐργάζεσθαι . . . ἕως ἡμέρα ἐστίν (John 9.4) *we must work . . . while it is day.*

The phrases noted on p. 177, being the equivalent of temporal conjunctions, are also used in temporal clauses:

ἕως οὗ or ἕως ὅτου (= ἕως τοῦ χρόνου [ἐν] ᾧ) *until, till*:

καὶ οὐκ ἐγίνωσκεν αὐτὴν ἕως οὗ ἔτεκεν υἱόν (Matt. 1.25) *and he knew her not till she had brought forth a son.*

οὐκ ἐπίστευσαν οἱ 'Ιουδαῖοι . . . ἕως ὅτου ἐφώνησαν τοὺς γονεῖς αὐτοῦ (John 9.18) *the Jews did not believe . . . until they called his parents.*

ἄχρι οὗ (= ἄχρι τοῦ χρόνου [ἐν] ᾧ) *until, till*:

ἄχρι οὗ ἀνέστη βασιλεὺς ἕτερος (Acts 7.18) *till there arose another king.*

ἐν ᾧ (= ἐν τῷ χρόνῳ ἐν ᾧ), *while*, and (occasionally) *until*:

πραγματεύσασθε ἐν ᾧ ἔρχομαι (Luke 19.13) *trade ye till I come.*

ἀφ' οὗ (= ἀπὸ τοῦ χρόνου [ἐν] ᾧ) or ἀφ' ἧς (= ἀπὸ τῆς ὥρας [ἐν] ᾗ) *since*:

αὕτη δὲ ἀφ' ἧς εἰσῆλθον οὐ διέλειπεν καταφιλοῦσά μου τοὺς πόδας (Luke 7.45) *but she, since the time I came in, hath not ceased to kiss my feet.*

πρίν or πρὶν ἤ, which mean *until* or *before*, have a different construction. Meaning *before* they are followed by the construction known as "the accusative and infinitive", the subject of the clause being in the accusative case, and the verb in the infinitive mood instead of in the indicative:

πρὶν ἀλέκτορα δὶς φωνῆσαι τρίς με ἀπαρνήσῃ (Mark 14.72) *before the cock crow twice, thou shalt deny me thrice.*

πρὶν ἢ συνελθεῖν αὐτοὺς εὑρέθη . . . (Matt. 1.18) *before they came together she was found . . .*

But meaning *until* and following a negative sentence they are followed by the subjunctive, and the particle ἄν may be added. This is the normal indefinite construction, for which see p. 212f.

RELATIVE CLAUSES OF PLACE

Akin to the relative clauses which define time (temporal clauses) are those which define place. Such clauses are introduced by the conjunctions *where, whence, whither*. E.g. in the following sentences the relative clauses are in italics: *where I am*, there ye shall be also; go to the place *whence you came*; *whither thou goest* I will go. It is to be carefully noted that the *where, whence*, and *whither* in these sentences do not ask questions. They are not interrogative adverbs, but relatives.

οὗ, *where, whither*

ἐγένετο πάροικος ἐν γῇ Μαδιάμ, οὗ ἐγέννησεν υἱοὺς δύο (Acts 7.29)
he became a sojourner in the land of Midian, where he begat two sons.

εἰς πᾶσαν πόλιν καὶ τόπον οὗ ἤμελλεν αὐτὸς ἔρχεσθαι (Luke 10.1)
into every city and place whither he himself was about to come.

ὅπου also means *where* or *whither*

ἀπεστέγασαν τὴν στέγην ὅπου ἦν (Mark 2.4) *they uncovered the roof where he was.*

ὅπου ἐγὼ ὑπάγω ὑμεῖς οὐ δύνασθε ἐλθεῖν (John 8.21) *whither I go, ye cannot come.*

ὅθεν *whence*

εἰς τὸν οἶκόν μου ἐπιστρέψω ὅθεν ἐξῆλθον (Matt. 12.44) *I will return into my house whence I came out.*

Phrases employing a relative and which are equivalent to conjunctions of place may be used to introduce relative clauses; e.g. εἰς ὅν (where the ὅν refers to a previous word like τόπος or οἶκος), and εἰς ἥν (where the ἥν refers to a previous word like γῆ or πόλις).

εὐθέως ἐγένετο τὸ πλοῖον ἐπὶ τῆς γῆς εἰς ἣν ὑπῆγον (John 6.21)
straightway the boat was at the land whither they were going.

If the place referred to is indefinite (*wherever, whithersoever*) the indefinite construction is used. See pp. 212f.

VERBS WITH STEMS LENGTHENED IN THE PRESENT

We have already had occasion to notice that in some verbs the verb stem is lengthened in the present (and consequently in the imperf. indic.) by the addition of the consonantal ι or γ sound (e.g. φυλακ-ι-ω becoming φυλάσσω), by a reduplication of the initial consonant of the verb stem (e.g. πι-πετ-ω becoming πίπτω), or by the addition of τ after a labial (e.g. κοπ-τ-ω).

Another frequent method of lengthening found in the present is by the addition of the nasal ν.   There are five variations of this:

(a) A simple addition of -ν-.
(b) The addition of -ν- and a change in the vowel of the stem.
(c) A simple addition of -αν-.
(d) The addition of -αν- and the insertion of another nasal in the stem.
(e) The addition of -νε-.

The following lists include all the verbs of this class in St Mark's Gospel together with some others found in other parts of the N.T.

(a) *Stems lengthened in the present by the addition of -ν-:*

	Stem	Compound	English	Fut.	Aor. act.	Perf. act.	Perf. pass.	Aor. pass.
δύνω¹	δυ-	ἐνδύνω	*sink* *put on*		$\left\{\begin{array}{l}\text{ἔδυν²}\\\text{ἔδυσα}\end{array}\right.$		-δέδυμαι	
τίνω	πι- and πο-		*drink*	πίομαι (dep.)	ἔπιον (strong)	πέπωκα		ἐπόθην
τέμνω	τεμ-	περιτέμνω	*cut* *cut round, circumcize*		-ἔτεμον (strong)		-τέτμημαι³	-ἐτμήθην³
φθάνω	φθα-		*anticipate, come quickly*		ἔφθασα	ἔφθακα		

(b) *Stems lengthened in the present by the addition of -ν- together with a change in the vowel of the stem:*

	Stem	Compound	English	Future	Aor. act.	Perf. act.	Perf. pass.	Aor. pass.
βαίνω	βη-		*go, come*	-βήσομαι (dep.)	-ἔβην²	-βέβηκα		
		ἀναβαίνω	*go up*					
		ἐμβαίνω	*embark*					
		καταβαίνω	*go down*					
		προβαίνω	*go forward*					
		συμβαίνω	*come to pass, happen*					
		συναναβαίνω	*go up with*					
Λαύνω	ἐλα-		*drive*		-ἤλασα	ἐλήλακα⁴		

1 δύνω is another form of δύω.
2 For these strong unthematic aorists, see p. 192.
3 -τέμνω. In the perf. pass. and aor. pass. the verb stem τεμ- is extended by η, the ε dropping out: thus τεμ-η, τμη-. This verb is not found in the N.T. as a simplex.
4 See next page for this "Attic" reduplication.

-ἔβην and ἔδυν are strong unthematic aorists of βαίνω and δύνω (cf. ἔγνων, p. 194). The thematic vowels ο and ε are not employed in their flexion:

Sing.	1.	-ἔβην	ἔδυν	Aor. imperative		
	2.	-ἔβης	ἔδυς	Sing.	2.	-βηθι or -βα
	3.	-ἔβη	ἔδυ		3.	-βάτω
Plur.	1.	-ἔβημεν	ἔδυμεν	Plur.	2.	-βατε
	2.	-ἔβητε	ἔδυτε		3.	-βάτωσαν
	3.	-ἔβησαν	ἔδυσαν	Infinitive: -βῆναι		
				Participle: -βάς, -βᾶσα,		
				-βάν		

ἐλήλακα (the perfect of ἐλαύνω) is an instance of the so-called Attic reduplication. The first syllable, ἐλ, of the stem is prefixed, ἐλ-ελα-, and the first vowel of the stem is lengthened, ἐλ-ήλα-κα. See p. 55.

(c) *Stems lengthened in the present by the addition of -αν-:*

	Stem	Compounds	English	Future	Aor. act.	Perf. act.	Perf. pass.	Aor. pass.
αἰσθάνομαι (dep.)	αἰσθ-		perceive		ᾐσθόμην (strong)			
ἁμαρτάνω	ἁμαρτ-		sin	ἁμαρτήσω	ἥμαρτον (strong) or ἡμάρτησα	ἡμάρτηκα		
αὐξάνω	αὐξ-		increase	αὐξήσω	ηὔξησα			ηὐξήθην
βλαστάνω[1]	βλαστ-		sprout		ἐβλάστησα			

[1] Lengthened form of βλαστάω.

(d) *Stems lengthened in the present by the addition of -αν- and the insertion of another nasal in the stem:*

	Stem	Compounds	English	Future	Aor. act.	Perf. act.	Perf. pass.	Aor. pass.
λαγχάνω	λαχ-		draw by lot		ἔλαχον (strong)			
λαμβάνω	λαβ- and ληβ-		take	λήμψομαι (dep.)	ἔλαβον (strong)	εἴληφα	εἴλημμαι	ἐλήμφθην
		ἀναλαμβάνω	take up					
		ἀπολαμβάνω	take aside, receive back					
		ἐπιλαμβάνομαι (mid.)	lay hold of					
		καταλαμβάνω	seize					
		παραλαμβάνω	take to oneself					

	Stem	Compounds	English	Future	Aor. act.	Perf. act.	Perf. pass.	Aor. pass.
		προλαμβάνω προσλαμβάνομαι (mid.) συλλαμβάνω	*anticipate take to oneself seize, conceive (of woman)*					
ανθάνω	λαθ- and ληθ-		*be hidden*		ἔλαθον (strong)			-λέλησμαι
		ἐπιλανθάνομαι (mid.)	*forget*					
ανθάνω	μαθ-		*learn*		ἔμαθον (strong)	μεμάθηκα		
τυνθάνομαι (dep.)	πυθ-		*ascertain*		ἐπυθόμην (strong)			
υγχάνω	τυχ-		*happen*		ἔτυχον (strong)	τέτυχα[1] τέτευχα τετύχηκα		

[1] Different manuscripts provide these three variants in Heb. 8.6.

(e) Stems lengthened in the present by the addition of -νε-:

This is not so common. One N.T. example is -ἱκνέομαι (stem ἱκ-). It has a str. aor. -ἱκόμην. The simplex is not found in N.T. Compounds are: ἀφικνέομαι *arrive at, reach*; διικνέομαι *go through*; ἐφικνέομαι *arrive at, reach*.

VERBS WITH STEMS LENGTHENED IN THE PRESENT BY THE ADDITION OF -σκ- OR -ισκ-

In verbs of this class a reduplication of the initial consonant of the stem is also sometimes seen; e.g. in μιμνήσκω, βιβρώσκω, γινώσκω (classical Greek γιγνώσκω: stem, γνω-).

	Stem	Compounds	English	Future	Aor. act.	Perf. act.	Perf. pass.	Aor. pass.
ρέσκω	ἀρε-		*please*	ἀρέσω	ἥρεσα			
ιβρώσκω	βρω-		*eat*			βέβρωκα		
όσκω	βο-		*feed*					
ηράσκω	γηρα-		*grow old*		ἐγήρασα			
ινώσκω	γνω-		*perceive, know*	γνώσομαι (dep.)	ἔγνων[1]	ἔγνωκα		ἐγνώσθην
		ἀναγινώσκω ἐπιγινώσκω	*read recognize*					
ιδάσκω[2]	διδαχ-		*teach*	διδάξω	ἐδίδαξα			ἐδιδάχθην
διδύσκω	δυ- (simplex not found)		*put on*					

	Stem	Compounds	English	Future	Aor. act.	Perf. act.	Perf. pass.	Aor. pass.
εὑρίσκω	εὑρε-		find	εὑρήσω	εὗρον (strong)	εὕρηκα		εὑρέθην
-θνῄσκω	θαν-		die	-θανοῦμαι (dep.)	-ἔθανον (strong)	τέθνηκα³		
		ἀποθνῄσκω συναποθνῄσκω	die die with					
μιμνῄσκω	μνη-		remind (pass. remember)	-μνήσω	-ἔμνησα		μέμνημαι	ἐμνήσθην
		ἀναμιμνῄσκω	remind					
πάσχω⁴	παθ- πονθ- }		suffer		ἔπαθον (strong)	πέπονθα		
πιπράσκω	πρα-		sell			πέπρακα		ἐπράθην

¹ ἔγνων is a strong unthematic aorist; cf. ἔβην and ἔδυν, p. 192.

Aorist imperative

Sing.	1.	ἔγνων	Sing.	2.	γνῶθι
	2.	ἔγνως		3.	γνώτω
	3.	ἔγνω	Plur.	2.	γνῶτε
Plur.	1.	ἔγνωμεν		3.	γνώτωσαν
	2.	ἔγνωτε	Infinitive:		γνῶναι
	3.	ἔγνωσαν	Participle:		γνούς, γνοῦσα, γνόν.

² διδάσκω is not an example of the reduplicated present. The verb stem is διδαχ-.
³ τέθνηκα is not compounded. The other parts of -θνῄσκω, however, are not found in the simplex but are compounded with ἀπο-; ἀποθνῄσκω *I die*. The meaning of the perfect, τέθνηκα, is "I have died", and so "I am dead".
⁴ The stem παθ- is extended to παθ-σκ- and so to πασχ-. The verb has two stems, παθ- and πονθ-.

Vocabulary

ὀψία (1) *evening*
Σατανᾶς, ᾶ, ὁ (1) *Satan* (cf. βορρᾶς, p. 280)
ἔλαιον (2) *oil*
παρθένος, ου, ἡ (2) *maiden, virgin*
ἀποκάλυψις, εως, ἡ (3) *revelation*
δῶμα, ατος, τό (3) *housetop*
βασιλικός, ή, όν *royal* (as noun, courtier, nobleman)

ἴδιος, α, ον *one's own*
κομψός, ή, όν *elegant, fine*
μωρός, ά, όν *foolish*
κακῶς (adv.) *badly, ill*
κομψότερον (comparative adv.) *more finely, better*
οὐδέ (neg. particle) *and not, neither* (as conjunction) *not even* (as adv.)

ἔχειν with an adverb may mean *to be*; thus ἑτοίμως ἔχειν *to be ready*.

EXERCISE 64

Translate into English:

1. ὀψίας δὲ γενομένης, ὅτε ἔδυσεν ὁ ἥλιος, ἔφερον πρὸς αὐτὸν πάντας τοὺς κακῶς ἔχοντας καὶ τοὺς δαιμονιζομένους. 2. λέγει πρὸς

αὐτὸν ὁ βασιλικός Κύριε, κατάβηθι πρὶν ἀποθανεῖν τὸ παιδίον μου. λέγει αὐτῷ ὁ Ἰησοῦς Πορεύου· ὁ υἱός σου ζῇ. 3. ἤδη δὲ αὐτοῦ καταβαίνοντος οἱ δοῦλοι ὑπήντησαν αὐτῷ. ἐπύθετο οὖν τὴν ὥραν παρ' αὐτῶν ἐν ᾗ κομψότερον ἔσχεν ὁ παῖς. 4. οὗτοι δέ εἰσιν οἱ παρὰ τὴν ὁδὸν ὅπου σπείρεται ὁ λόγος, καὶ εὐθὺς ἔρχεται ὁ Σατανᾶς καὶ αἴρει τὸν λόγον τὸν ἐσπαρμένον εἰς αὐτούς. 5. καὶ ἀποκριθεὶς πρὸς αὐτοὺς εἶπεν ὁ Ἰησοῦς Οὐδὲ τοῦτο ἀνέγνωτε ὃ ἐποίησεν Δαυεὶδ ὁπότε ἐπείνασεν αὐτὸς καὶ οἱ μετ' αὐτοῦ ὄντες; 6. οὐδὲ γὰρ ἐγὼ παρὰ ἀνθρώπου παρέλαβον τὸ εὐαγγέλιον οὔτε ἐδιδάχθην, ἀλλὰ δι' ἀποκαλύψεως Ἰησοῦ Χριστοῦ.

EXERCISE 65

Translate into Greek:

1. These things the Lord did until he was taken up from us. 2. The seed fell where it had not much earth. 3. The second brother took her and died, leaving no seed. 4. Let not him that is on the housetop come down. 5. His master ordered him to be sold. 6. I sinned when I forgot my promise. 7. And Jesus, when he was baptized, immediately went up out of the water. 8. The foolish virgins took their lamps and took no oil with them. 9. They did what they were taught. 10. I never knew you; depart from me.

EXERCISE 66

Translate into Greek:

1. Those who did not reach the river suffered and died in the wilderness. 2. While you are drinking this wine I will go down and buy bread. 3. They remained in Egypt until an angel commanded Joseph to take the child and return to the land of Israel. 4. Where your treasure is there will your heart be also. 5. Not even Titus who was with me was compelled to be circumcised. 6. Ye will receive him who comes in his own name. 7. These things happened to him at the hands of the Jews when he went up to Jerusalem. 8. When evening came the people from the villages brought the dying to Jesus. 9. He who sins has neither seen him nor known him. 10. Know the things of the Spirit, for the things of the flesh are not pleasing to God.

Chapter 24

REPORTED SPEECH

"I have lost my hat", said the boy.

The boy said that he had lost his hat.

The first sentence gives the actual words of the speaker. In English this is indicated by the use of inverted commas. This is called Direct Speech, or in Latin *oratio recta*.

The second sentence has exactly the same meaning, but it expresses the speaker's words indirectly. This construction is known as Reported or Indirect Speech, or in Latin *oratio obliqua*.

It will be noticed that reported speech necessitates certain alterations. In the above example the pronouns *I* and *my* of the *oratio recta* are changed to *he* and *his*, and the tense of the verb is changed from perfect, *have lost*, to pluperfect, *had lost*.

The term "reported speech" also includes reported or indirect thought:

"We shall go soon", they thought (*oratio recta*).

They thought that they would go soon (*oratio obliqua*).

Note the change of *we* into *they*, and of *shall go* into *would go*.

Oratio obliqua may express indirectly statements (as in the examples above), questions or commands:

"Who is it?" he asked (*oratio recta*).

He asked who it was (*oratio obliqua*).

"Pick up the book", said the master (*oratio recta*).

The master told him to pick up the book (*oratio obliqua*).

INDIRECT STATEMENT

We shall deal with indirect *statement* first. Two constructions are possible in Greek: (*a*) a noun clause introduced by ὅτι; (*b*) the accusative and infinitive construction.

(*a*) INDIRECT STATEMENT IN A NOUN CLAUSE

We have seen that in English an indirect statement is usually expressed in a clause introduced by the conjunction *that* after the verb of saying or thinking. The Greek construction is very similar, the conjunction ὅτι being used after the verb of saying or thinking to introduce the clause which contains the words spoken or thought:

> ἐσθίομεν ἄρτον *we are eating bread* (oratio recta).
> λέγουσιν ὅτι ἐσθίουσιν ἄρτον *they say that they are eating bread* (oratio obliqua).

The clause ὅτι ἐσθίουσιν ἄρτον is what is called a noun clause, supplying the object of the verb λέγουσιν.

> εἶπον ὅτι ἐσθίουσιν ἄρτον *they said that they were eating bread.*

Here the same words are reported, but this time the introductory word of saying is in past time. Notice that this does not involve any change of tense in the reported speech in Greek. In English *are eating* must be altered to *were eating*, but in Greek the tense remains unchanged. Greek retains in *oratio obliqua* after ὅτι the tense actually used by the speaker or thinker. A change must, of course, be made in person; the ἐσθίομεν of *oratio recta* must become ἐσθίουσιν.

It will sometimes be noticed, however, that not only the tense but also the person of the direct speech is retained after ὅτι. When this is so the ὅτι is to be omitted in translation, for its function is simply that of the English inverted commas, and what follows is in reality *oratio recta*. Mark 7.20 is an example of this: ἔλεγεν δὲ ὅτι Τὸ ἐκ τοῦ ἀνθρώπου ἐκπορευό-μενον, ἐκεῖνο κοινοῖ τὸν ἄνθρωπον *and he said, That which proceedeth out of the man, that defileth the man.*

But examples of genuine reported speech are:

> εἶπεν ὅτι θεραπεύσει τὸν τυφλόν *he said that he would heal the blind man.* The actual words were "I will heal". The future tense is retained in *oratio obliqua*.
> ἐνόμισαν ὅτι πλεῖον λήμψονται (Matt. 20.10) *they thought that they would receive more.* Their thought was "we shall receive more". The future tense is retained in *oratio obliqua*.
> εἶπεν ὅτι ἔλαβε τὸ ἀργύριον *he said that he had taken the money.* His actual words were "I took the money". The aorist tense is, therefore, used in Greek *oratio obliqua*. Notice that in English the tense is changed to the pluperfect.

(b) Indirect Statement in the Accusative and Infinitive

An alternative method of expressing an indirect statement in English is by the use of the accusative and infinitive after the verb of saying or thinking. E.g. *oratio recta*: "He is insane"; *oratio obliqua*: "They certified him to be insane", where the accusative *him* replaces the nominative *he* of *oratio recta*, and the infinitive *to be* replaces the indicative *is*. This construction is not usual in English with verbs of saying, but is more common with verbs of thinking: "We said her to be late" is not usual English, but "We thought her to be late" is not unusual.

In Greek the accusative and infinitive may be used after a verb of saying or thinking, but the construction is by no means so frequent in the N.T. as the object clause introduced by ὅτι.

πῶς λέγουσιν τὸν Χριστὸν εἶναι Δαυεὶδ υἱόν; (Luke 20.41) *how say they that the Christ is David's son?* Notice that υἱόν must be accusative being the predicate of τὸν Χριστόν and in agreement with it.

The negative μή is used with the infinitive.

καὶ ἔρχονται Σαδδουκαῖοι πρὸς αὐτόν, οἵτινες λέγουσιν ἀνάστα-σιν μὴ εἶναι (Mark 12.18) *there come unto him Sadducees which say that there is no resurrection* (lit. "say a resurrection not to be").

If, however, the subject of the infinitive is the same as the subject of the main verb (as in "we say that we are good") it is not written, and any word which is in agreement with it is put in the nominative:

λέγομεν εἶναι ἀγαθοί *we say that we are good.*
θέλεις τέλειος εἶναι (Matt. 19.21) *thou wouldest be perfect* (lit. "thou willest (thyself) to be perfect").

Occasionally what appears to be a mixture of the accusative and infinitive and the noun clause constructions is found:

ἀποδεικνύντα ἑαυτὸν ὅτι ἐστὶν Θεός (2 Thes. 2.4) *proclaiming himself to be God.* This looks like a mixture of the two constructions ἀποδεικνύντα ἑαυτὸν εἶναι Θεόν and ἀποδεικνύντα ὅτι ἐστὶν Θεός.

THE SUBJUNCTIVE

The subjunctive comprises the primary tenses of the conjunctive mood. The conjunctive mood is so called because it is mostly employed in clauses which are "joined with" (Latin, *coniungo*) and subordinate to the main clause of a sentence. The conjunctive has two divisions, the optative (the historic tenses; see pp. 217ff.) and the subjunctive. In some grammar books the term "conjunctive" is not used, and the subjunctive and optative are treated as separate moods rather than as the primary and historic tenses of a single mood.

The subjunctive takes its name from the Latin *subiungo*, "subjoin", because it is chiefly found in subordinate clauses.

THE SUBJUNCTIVE OF εἰμί *I am*

Sing.	1.	ὦ
	2.	ᾖς
	3.	ᾖ
Plur.	1.	ὦμεν
	2.	ἦτε
	3.	ὦσι(ν)

There is no aor. or perf. subj. of εἰμί.

THE SUBJUNCTIVE OF λύω *I loose*

PRESENT	*Active*	*Middle*	*Passive*
Sing. 1.	λύω	λύωμαι	λύωμαι
2.	λύῃς	λύῃ	λύῃ
3.	λύῃ	λύηται	λύηται
Plur. 1.	λύωμεν	λυώμεθα	λυώμεθα
2.	λύητε	λύησθε	λύησθε
3.	λύωσι(ν)	λύωνται	λύωνται

WEAK AORIST	*Active*	*Middle*	*Passive*
Sing. 1.	λύσω	λύσωμαι	λυθῶ
2.	λύσῃς	λύσῃ	λυθῇς
3.	λύσῃ	λύσηται	λυθῇ
Plur. 1.	λύσωμεν	λυσώμεθα	λυθῶμεν
2.	λύσητε	λύσησθε	λυθῆτε
3.	λύσωσι(ν)	λύσωνται	λυθῶσι(ν)

WEAK PERFECT	*Active* (Formed by perf. part. act., and subj. of εἰμί.)	*Middle* (Formed by perf. part. mid. and subj. of εἰμί.)	*Passive* (Formed by perf. part. pass. and subj. of εἰμί.)
Sing. 1.	λελυκὼς ὦ	λελυμένος ὦ	
2.	λελυκὼς ᾖς	λελυμένος ᾖς	
3.	λελυκὼς ᾖ	λελυμένος ᾖ	
Plur. 1.	λελυκότες ὦμεν	λελυμένοι ὦμεν	
2.	λελυκότες ἦτε	λελυμένοι ἦτε	
3.	λελυκότες ὦσι(ν)	λελυμένοι ὦσι(ν)	

N.B. the participle will be fem. or neut. if the subject of the perf. subj. is fem. or neut.

Refer to pp. 294ff. for the relation of the subjunctive to the rest of the verb.

SUBJUNCTIVES OF STRONG TENSES

βάλλω (str. aor. indic. act., ἔβαλον), -ἀλλάσσω (str. aor. indic. pass. -ἠλλάγην) and γράφω (str. perf. indic. act. γέγραφα) provide the examples.

STRONG AORIST	*Active*	*Middle*	*Passive*
Sing. 1.	βάλω	βάλωμαι	-ἀλλαγῶ
2.	βάλῃς	βάλῃ	-ἀλλαγῇς
3.	βάλῃ	βάληται	-ἀλλαγῇ
Plur. 1.	βάλωμεν	βαλώμεθα	-ἀλλαγῶμεν
2.	βάλητε	βάλησθε	-ἀλλαγῆτε
3.	βάλωσι(ν)	βάλωνται	-ἀλλαγῶσι(ν)

STRONG PERFECT	*Active*	*Middle* and *Passive*
Sing. 1.	γεγραφὼς ὦ	γεγραμμένος ὦ
2.	γεγραφὼς ᾖς	γεγραμμένος ᾖς
3.	γεγραφὼς ᾖ	γεγραμμένος ᾖ
Plur. 1.	γεγραφότες ὦμεν	γεγραμμένοι ὦμεν
2.	γεγραφότες ἦτε	γεγραμμένοι ἦτε
3.	γεγραφότες ὦσι(ν)	γεγραμμένοι ὦσι(ν)

Refer to pp. 304ff. for the relation of the subjunctive to the other strong tenses.

Subjunctives of Contracted Verbs

The rules of contraction (see p. 109) are observed in the formation of the pres. subj. of verbs whose stems end in α, ε, and ο.

Present Subjunctive of τιμῶ (τιμά-ω):

	Active	Mid. and Pass.
Sing. 1.	τιμῶ (ά-ω)	τιμῶμαι (ά-ωμαι)
2.	τιμᾷς (ά-ης)	τιμᾷ (ά-η)
3.	τιμᾷ (ά-η)	τιμᾶται (ά-ηται)
Plur. 1.	τιμῶμεν (ά-ωμεν)	τιμώμεθα (α-ώμεθα)
2.	τιμᾶτε (ά-ητε)	τιμᾶσθε (ά-ησθε)
3.	τιμῶσι(ν) (ά-ωσι)	τιμῶνται (ά-ωνται)

Present Subjunctive of φιλῶ (φιλέ-ω).

	Active	Mid. and Pass.
Sing. 1.	φιλῶ (έ-ω)	φιλῶμαι (έ-ωμαι)
2.	φιλῇς (έ-ης)	φιλῇ (έ-η)
3.	φιλῇ (έ-η)	φιλῆται (έ-ηται)
Plur. 1.	φιλῶμεν (έ-ωμεν)	φιλώμεθα (ε-ώμεθα)
2.	φιλῆτε (έ-ητε)	φιλῆσθε (έ-ησθε)
3.	φιλῶσι(ν) (έ-ωσι)	φιλῶνται (έ-ωνται)

Present Subjunctive of δηλῶ (δηλό-ω):

	Active	Mid. and Pass.
Sing. 1.	δηλῶ (ό-ω)	δηλῶμαι (ό-ωμαι)
2.	δηλοῖς (ό-ης)	δηλοῖ (ό-η)
3.	δηλοῖ (ό-η)	δηλῶται (ό-ηται)
Plur. 1.	δηλῶμεν (ό-ωμεν)	δηλώμεθα (ο-ώμεθα)
2.	δηλῶτε (ό-ητε)	δηλῶσθε (ό-ησθε)
3.	δηλῶσι(ν) (ό-ωσι)	δηλῶνται (ό-ωνται)

Refer to pp. 299–303 for the relation of the subjunctive to the rest of the verb.

On the above models the subjunctives of other verbs in -ω may easily be found:

Present Subjunctive Active

To the present stem (and it must be remembered that this may be a lengthening of the verb stem) -ω, -ῃς, ῃ, etc., are added. Care must be taken to follow the rules of contraction where necessary. Thus:

The pres. subj. act. of λέγω is λέγω, λέγῃς, λέγῃ, etc.

„ „ „ „ „ βάλλω is βάλλω, βάλλῃς, βάλλῃ, etc.

„ „ „ „ „ νικῶ (νικά-ω) is νικῶ, νικᾷς, νικᾷ, etc.

Present Subjunctive Middle and Passive

In the same way -ωμαι, -ῃ, -ηται, etc., are added to the present stem, the rules of contraction being observed where necessary.

Weak Aorist Subjunctive Active

The final -α, -ας, -ε(ν), etc., of the indicative are replaced by -ω, ῃς, ῃ, etc., and the augment is omitted. Thus:

The aor. subj. act. of στέλλω (aor. indic. ἔστειλα) is στείλω, -ῃς, -ῃ, etc.

The aor. subj. act. of μένω (aor. indic. ἔμεινα) is μείνω, -ῃς, -ῃ, etc.

Weak Aorist Subjunctive Middle

In the same way -ωμαι, -ῃ, -ηται, replace the -άμην, -ω, -ατο of the indicative, and the augment is omitted. Thus:

The aor. subj. of the middle deponent verb ἐργάζομαι *work* (aor. indic. ἠργασάμην) is ἐργάσωμαι, ἐργάσῃ, ἐργάσηται, etc.

Strong or Weak Aorist Subjunctive Passive

-ῶ, -ῇς, -ῇ, etc., replace the -ην, -ης, -η, etc., of the indicative, and the augment is omitted. Thus:

The wk. aor. subj. pass. of τιμῶ (aor. indic. pass. ἐτιμήθην) is τιμηθῶ, -ῇς, -ῇ, etc.

The wk. aor. subj. pass. of λαμβάνω (aor. indic. pass. ἐλήμφθην) is λημφθῶ, -ῇς, -ῇ, etc.

The str. aor. subj. pass. of σπείρω *soiv* (aor. indic. pass. ἐσπάρην) is σπαρῶ, -ῃς, -ῃ.

The Perfect Subjunctive

In the N.T. this is invariably formed by using the perfect participle together with the subjunctive of εἰμί *I am*. For the perf. subj. act. the perf. part. act. is, of course, used (e.g. βεβληκὼς ὤ), and for the perf. subj. mid. or pass. the perf. part. mid. or pass. is used (βεβλημένος ὤ).

Note the subjunctives of the strong unthematic aorists -ἔβην and ἔγνων (βαίνω and γινώσκω).

Sing.	1.	-βῶ	γνῶ
	2.	-βῇς	γνῷς
	3.	-βῇ	γνῷ or γνοῖ
Plur.	1.	-βῶμεν	γνῶμεν
	2.	-βῆτε	γνῶτε
	3.	-βῶσι(ν)	γνῶσι(ν)

THE USE OF THE SUBJUNCTIVE

(1) Final Clauses

Final clauses are clauses subordinate to the main verb of the sentence; they express purpose. In the sentence "We eat in order that we may live" the final clause is "in order that we may live". Note that in English there are several ways of expressing purpose: e.g. in order to live; so that we may live; so as to live; to live. In Greek also there are several ways of expressing purpose. We have already noted the use of the future participle (page 167). For the moment we are only concerned with the use of the subjunctive to express purpose. For other constructions see p. 236.

When the subjunctive is used to express purpose the final clause is introduced by the particle ἵνα or (more rarely) ὅπως; if the final clause is negative it is introduced by ἵνα μή, ὅπως μή, or simply μή. Thus the nearest English equivalent of ἵνα and ὅπως is "in order that", and of ἵνα μή and μή "in order that not" or "lest".

> δόξασόν σου τὸν Υἱόν, ἵνα ὁ Υἱὸς δοξάσῃ σέ (John 17.1) *glorify thy Son, that the Son may glorify thee.* (δοξάσῃ is 3rd pers. sing. wk. aor. subj. act. of δοξάζω.)

> οὐ γὰρ ἀπέστειλεν ὁ Θεὸς τὸν Υἱὸν εἰς τὸν κόσμον ἵνα κρίνη τὸν κόσμον, ἀλλ' ἵνα σωθῇ ὁ κόσμος δι' αὐτοῦ (John 3.17) *for God sent not the Son into the world to judge* (lit. "in order that he might judge") *the world; but that the world should be saved through him.* (κρίνη is 3rd pers. sing. wk. aor. subj. act. of κρίνω; σωθῇ is 3rd pers. sing. wk. aor. subj. pass. of σώζω.)

> γρηγορεῖτε καὶ προσεύχεσθε, ἵνα μὴ εἰσέλθητε εἰς πειρασμόν (Matt. 26.41) *watch and pray, that ye enter not* (or *lest ye enter*) *into temptation.*

[Those who have some knowledge of classical Greek should note carefully a difference here between classical and N.T. Greek. In the N.T. the subjunctive is used in the final clause whether the verb in the main clause is in a primary or in an historic tense. In classical Greek, if the main verb was in an historic tense, the optative could be used in the final clause. Thus a main verb in a primary tense was followed by the subjunctive (one of the primary tenses of the conjunctive mood), whilst a main verb in an historic tense was often followed by the optative (one of the historic tenses of the conjunctive mood). This construction is known as the Sequence of Tenses. But in the N.T. the optative is not used in this way, and its use is practically confined to the expression of wishes. (See p. 219.)]

In a final clause the present subjunctive is used if the action of the verb is prolonged or repeated, and the aorist subjunctive if a single action is described, or if no particular stress is laid on its being prolonged or repeated: e.g. in ἵνα ὁ κόσμος πιστεύῃ (John 17.21) *that the world may believe*, the pres. subj. is used since the belief is not to be a momentary action, but is to be prolonged; but in προσηύξαντο περὶ αὐτῶν ὅπως λάβωσιν Πνεῦμα Ἅγιον (Acts 8.15) *they prayed for them that they might receive the Holy Ghost*, the aor. subj. λάβωσιν is used, since the action of receiving is a single and momentary one. Thus the distinction in their use between the pres. and the aor. subj. is much the same as that between the pres. and aor. imperative (see p. 89) and between the pres. and the aor. infinitive (see p. 101).

The perfect subjunctive is used in a final clause when it is desired to draw special attention to the completed state of the action purposed. Thus, αἰτεῖτε, καὶ λήμψεσθε, ἵνα ἡ χαρὰ ὑμῶν ᾖ πεπληρωμένη (John 16.24) *ask, and ye shall receive, that your joy may be fulfilled*. The use of the perf. subj. pass. here presents the idea of a completely realized and continuing joy. The pres. subj. pass., πληρῶται, would have meant that the joy was to be in process of fulfilment; the aor. subj. pass., πληρωθῇ, would have referred simply to the moment of fulfilment, and would have placed no emphasis on the fact of its continuance. The perfect subjunctive is exactly right in this context.

The use of the subjunctive in the following final clauses should be studied:

καὶ παρετήρουν αὐτὸν ... ἵνα κατηγορήσωσιν αὐτοῦ (Mark 3.2) *and they watched him ... that they might accuse him.*

καὶ εἶπεν τοῖς μαθηταῖς αὐτοῦ ἵνα πλοιάριον προσκαρτερῇ αὐτῷ διὰ τὸν ὄχλον, ἵνα μὴ θλίβωσιν αὐτόν (Mark 3.9) *and he spake to his disciples, that a little boat should wait on him because of the crowd, lest they should throng him.*

καὶ διεστείλατο αὐτοῖς πολλὰ ἵνα μηδεὶς γνοῖ τοῦτο (Mark 5.43) *and he charged them much that no man should know this.*

καλῶς ἀθετεῖτε τὴν ἐντολὴν τοῦ Θεοῦ, ἵνα τὴν παράδοσιν ὑμῶν τηρήσητε (Mark 7.9) *well do ye reject the commandment of God, that ye may keep your tradition.*

Vocabulary

ὀπτασία (1) *vision*
σταυρός (2) *cross*
Βηθλεέμ, ἡ (indecl.) *Bethlehem*
διαμαρτύρομαι (dep.) *affirm solemnly*

κατηγορέω (with gen.) *accuse*
μαίνομαι (dep.) *be mad*
νομίζω *think*
περισσεύω *abound*

EXERCISE 67

Translate into English:

1. οὐχ ἡ γραφὴ εἶπεν ὅτι ἐκ τοῦ σπέρματος Δαυείδ, καὶ ἀπὸ Βηθλεὲμ τῆς κώμης ὅπου ἦν Δαυείδ, ἔρχεται ὁ Χριστός; 2. τὸ Πνεῦμα τὸ Ἅγιον κατὰ πόλιν διαμαρτύρεταί μοι λέγον ὅτι δεσμὰ καὶ θλίψεις με μένουσιν. 3. αἱ δὲ γυναῖκες μὴ εὑροῦσαι τὸ σῶμα αὐτοῦ ἦλθον λέγουσαι ὀπτασίαν ἀγγέλων ἑωρακέναι, οἳ λέγουσιν αὐτὸν ζῆν. 4. καὶ ἐποίησεν δώδεκα ἵνα ὦσιν μετ᾽ αὐτοῦ, καὶ ἵνα ἀποστέλλῃ αὐτοὺς κηρύσσειν καὶ ἔχειν ἐξουσίαν ἐκβάλλειν τὰ δαιμόνια. 5. οὗτός ἐστιν ὁ ἄρτος ὁ ἐκ τοῦ οὐρανοῦ καταβαίνων, ἵνα τις ἐξ αὐτοῦ φάγῃ καὶ μὴ ἀποθάνῃ. 6. ὁ Θεὸς ἠλέησεν αὐτόν, οὐκ αὐτὸν δὲ μόνον ἀλλὰ καὶ ἐμέ, ἵνα μὴ λύπην ἐπὶ λύπην σχῶ.

EXERCISE 68

Translate into Greek:

1. Do you think that the workmen will build this house in three months? 2. Do you think that your father saw the thief? 3. Do you think that these things cause scandal to the people? 4. Did you think that the shepherds were guarding the sheep? 5. They watched him in order to accuse him. 6. This I pray, that your love may abound. 7. The Son did not come to condemn the world. 8. The apostles prayed that they might receive the Holy Spirit. 9. The Spirit said to Peter that two men were seeking him. 10. They said that Paul was a god.

EXERCISE 69

Translate into Greek:

1. I say unto you that many prophets and kings wished to see the things which you see, and did not see them.   2. I have sent a beloved brother to you that you may know our affairs, and that he may comfort your hearts. 3. Even so must the Son of man be lifted up, that every one who believes in him may have eternal life.   4. These things are written, that ye might believe that Jesus is the Christ.   5. The prophets said that the Messiah would come in the last days.   6. I come again, and will receive you to myself, that where I am you may be also.   7. We prayed that the blind man might receive his sight.   8. We say that Christ died upon the cross and was raised on the third day.   9. Some said that the apostles were mad; others said that they spoke the words of God.   10. John said that they had seen a certain man casting out evil spirits, and had forbidden him because he did not follow the apostles.

Chapter 25

ADVERBS AND THEIR COMPARISON

Some adverbs are formed from adjectives in one of the two following ways:

1. The accusative neuter of the adjective may be used as an adverb: e.g.

πρῶτον *first*   ταχύ *quickly*
δεύτερον *the second time*   πρότερον *before*
τρίτον *the third time*   ὕστερον *later*
μικρόν *for a little while, a little way*   πολύ *greatly*
μόνον *alone, only*

2. The suffix -ως may be employed (i.e. ς is substituted for ν at the end of the gen. plur. of the adjective): e.g.

ἀληθῶς *truly*   ὁμοίως *likewise*
κακῶς *badly, ill*   περισσῶς *abundantly*
καλῶς *rightly, well*   ταχέως *quickly*
ὅλως *altogether*   φανερῶς *openly*

But there are many adverbs which are not formed from any adjective. Among the common adverbs in the N.T. not already noted above are:

ADVERBS OF TIME

ἀεί *always, unceasingly*   *οὐδέποτε *never*
ἄρτι *now, just now*   *οὐδέπω *not yet*
αὔριον *tomorrow*   *οὐκέτι *no longer*
εἶτα *then, next*   *οὔπω *not yet*
ἐπαύριον *on the morrow*   πάλιν *again*
ἔπειτα *then, thereupon*   πάντοτε *always*
εὐθέως, εὐθύς *straightway*   πολλάκις *often*
ἐχθές, χθές *yesterday*   πρωΐ *in the morning, early*
ἤδη *now, already*   πώποτε *ever, yet*
νῦν, νυνί *now*   τότε *then*

Those marked with an asterisk have corresponding forms with the negative μή, μηδέποτε, μηδέπω, etc., which are used when μή would be used instead of οὐ, i.e. usually with verbs in a mood other than the indicative.

The adverbial numerals (see p. 169) are also adverbs of time.

ADVERBS OF PLACE

ἀλλαχοῦ *elsewhere*	ἔξωθεν *from without*
ἀλλαχόθεν *from another place*	ἔσω *within*
ἄνω *up, upwards, above*	κάτω *down, beneath*
ἄνωθεν *from above* (also of time, *again*)	μακράν *far*
	μακρόθεν *from afar*
δεῦρο *hither* (often used as imperative, *come hither*)	μεταξύ *between*
	ὀπίσω *back, behind, after*
δεῦτε *hither*	πανταχοῦ *everywhere*
ἐγγύς *near*	πάντοθεν *from all sides*
ἐκεῖ *there*	πέραν *on the other side*
ἐκεῖθεν *thence*	πλησίον *near*
ἔμπροσθεν *before, in front*	πόρρω *far off*
ἐνθάδε *here, hither*	πόρρωθεν *from afar*
ἔνθεν *hence*	χαμαί *on the ground*
ἔξω *outside*	ὧδε *hither, here*

Note that the suffix -θεν denotes place from.

Other adverbs are sometimes called Adverbs of Manner, as describing the manner in which an action is performed. The division of adverbs into classes, however, is of no importance, except, perhaps, to provide students with useful categories in which to commit them to memory.

ADVERBS OF MANNER

ἀμήν *verily*	ὁμοῦ *together*
δωρεάν *freely, in vain*	ὅμως *nevertheless, yet*
Ἑβραϊστί *in Hebrew*	ὄντως *really*
εἰκῆ, εἰκῆ *in vain, without reason*	οὕτως *so, thus*
Ἑλληνιστί *in Greek*	Ῥωμαϊστί *in Latin*
εὖ *well*	σπουδαίως *earnestly*
ἡδέως *sweetly*	σφόδρα *exceedingly*
ἰδίᾳ *separately*	τάχα *perhaps*
κρυφῇ, κρυφῆ *secretly*	χωρίς *separately*
λάθρᾳ, λάθρα *secretly*	

ἀμήν is a Greek transliteration of a Hebrew word.

καθ' εἷς *one by one*, and κατ' ἰδίαν *privately*, are adverbial phrases rather than adverbs, employing the preposition κατά in an idiomatic way (see p. 64).

Note the idiomatic use of ἔχω with an adverb where we should expect εἰμί with an adjective:

> ἐσχάτως ἔχει (Mark 5.23) *she is at the point of death* (where we might expect ἐσχάτη ἐστίν).
> οἱ κακῶς ἔχοντες (Mark 2.17) *they that are sick* (for οἱ κακοὶ ὄντες).
> ἑτοίμως ἔχω (Acts 21.13) *I am ready* (for ἕτοιμός εἰμι).

THE COMPARISON OF ADVERBS

The comparative of an adverb is provided by the accusative neuter singular of the comparative of the adjective, and the superlative by the accusative neuter plural of the superlative of the adjective.   Thus:

> adjective, σοφός *wise*; σοφώτερος *wiser*; σοφώτατος *wisest*.
> adverb σοφῶς *wisely*; σοφώτερον *more wisely*; σοφώτατα *most wisely*.

But there are frequent irregularities.   The student will find the following table of the adverbs with irregular comparison useful:

Positive adverb	English	Comparative	Superlative
ἄγχι	near	ἆσσον	
ἄνω	up, above	ἀνώτερον	
ἐγγύς	near	ἐγγύτερον	ἔγγιστα
εὖ	well	κρεῖσσον ⎫ βέλτιον ⎭	κράτιστα βέλτιστα
ἡδέως	sweetly		ἥδιστα
κακῶς	badly, ill	ἧσσον ⎫ ἧττον ⎬ χεῖρον ⎭	
καλῶς	rightly, well	κάλλιον	
κάτω	down, beneath	κατωτέρω	
[μάλα] (positive not found in N.T.)	very, exceedingly	μᾶλλον *more*	μάλιστα *most*
μικρόν	for a little way, a little while	ἔλασσον ⎫ ἔλαττον ⎬ *less*	ἐλάχιστα *least*
πέραν	across	περαιτέρω *further*	

Positive adverb	English	Comparative	Superlative
περισσῶς	abundantly	περισσότερον ⎱ περισσοτέρως ⎰	
πολύ	greatly	πλεῖον ⎱ πλέον ⎰	
πόρρω	far off	πορρώτερον	
ταχύ	quickly	τάχειον ⎱ τάχιον ⎰	τάχιστα

It will be noticed that the comparative occasionally employs the ending -τέρως instead of -τερον. It employs -τέρω in the case of κατωτέρω and περαιτέρω.

The superlative adverb, like the superlative adjective, is rare in the N.T.: the comparative is capable of doing duty for it.

The superlative may denote either the highest degree, or a very high degree. Thus ἥδιστα may mean "most sweetly" or "very sweetly".

There is an idiomatic use of the superlative adverb after ὡς. Thus, ὡς ἥδιστα as sweetly as possible; ὡς τάχιστα (Acts 17.15) as quickly as possible.

INDIRECT QUESTIONS

Indirect questions are introduced by the same interrogative pronouns, e.g. τίς, τί who, what, or adverbs, e.g. ποῦ where, πῶς how, as are used to introduce direct questions. The rule of classical Greek, that ὅστις rather than τίς and ὅπως rather than πῶς, etc., introduced indirect questions, is very rarely observed in the N.T. Εἰ if, whether, is also used to introduce indirect questions.

The usual verbs of asking questions are ἐρωτῶ (ἐρωτά-ω) ask, and πυνθάνομαι inquire. (Αἰτῶ(έ-ω) ask, is used of requests.) But it must be noted that an indirect question is not necessarily introduced by a verb of asking. In the sentence "I am not sure what I shall say", the clause "what I shall say" is an indirect question, representing "What shall I say?" in oratio recta. In the sentence "Tell me where I can go", the clause "where I can go" is an indirect question, representing "Where can I go?" in oratio recta. In neither of these sentences is the indirect question introduced by a verb of asking.

The verb in the indirect question clause retains the same tense as that of the direct question, and is normally in the indicative:

ἐπυνθάνετο παρ' αὐτῶν ποῦ ὁ Χριστὸς γεννᾶται (Matt. 2.4) he inquired of them where the Christ should be born. The direct question is ποῦ ὁ Χριστὸς γεννᾶται; the pres. indic. γεννᾶται has here a quasi-future sense "is to be born".

εἰπὲ τίς ἐστιν (John 13.24) *tell us who it is.*

πῶς δὲ νῦν βλέπει οὐκ οἴδαμεν (John 9.21) *but how he now seeth, we know not.*

When the verb of the direct question would be in the subjunctive, the indirect question has the verb in the subjunctive also:

εἰπὲ τί ποιήσωμεν, *tell us what we are to do.* (The *oratio recta* is τί ποιήσωμεν; a deliberative subjunctive, see p. 214.)

When the introductory verb is in past time, the verb of the indirect question clause is sometimes in the optative (see p. 220.) But in the N.T. this occurs only in St Luke's writings (Gospel and Acts):

ἐπυνθάνετο τίς εἴη (Acts 21.33) *he inquired who he was.*

ἐδέξαντο τὸν λόγον ... ἀνακρίνοντες τὰς γραφὰς εἰ ἔχοι ταῦτα οὕτως (Acts 17.11) *they received the word ... examining the scriptures, whether these things were so.*

INDIRECT COMMANDS

Direct commands are in the imperative mood, and direct prohibitions may either have μή with the present imperative or μή with the aorist subjunctive (see pp. 94 and 215). In indirect speech after an introductory verb of command the infinitive may be used, or an object clause with verb in the subjunctive and introduced by ἵνα, or in the case of an indirect prohibition by ἵνα μή or μή.

κελεύω is a frequent verb of command. It takes an accusative of the person commanded, and the infinitive is used to express what is commanded. The infinitive, it is to be remembered, is a verbal noun (see p. 99) representing here a second object. κελεύω is not followed by the ἵνα construction:

ἐκέλευσεν αὐτὸν ὁ κύριος πραθῆναι (Matt. 18.25) *his* (the) *lord commanded him to be sold.*

ὁ Ἰησοῦς ἐκέλευσεν αὐτὸν ἀχθῆναι πρὸς αὐτόν (Luke 18.40) *Jesus commanded him to be brought unto him.*

There is not infrequently an ellipse of the accusative (i.e. the acc. is omitted, although it must be understood):

ἐκέλευσεν ἀπελθεῖν εἰς τὸ πέραν (Matt. 8.18) *he commanded* (them) *to depart unto the other side.* The ellipse can be reproduced in English by translating with the R.V. "He gave commandment to depart. ..."

ἐντέλλομαι has a dative of the person commanded, but there is frequently an ellipse of this dative. The infinitive and the object clause are both used after ἐντέλλομαι:

> τί οὖν Μωϋσῆς ἐνετείλατο δοῦναι ...; (Matt. 19.7) *why then did Moses command* (us) *to give ...?*
> καὶ τῷ θυρωρῷ ἐνετείλατο ἵνα γρηγορῇ (Mark 13.34) *he commanded also the porter to watch.*

παραγγέλλω has either acc. or dat. of person. There may be an ellipse of the acc. or dat. The infinitive and the object clause introduced by ἵνα are both found after παραγγέλλω:

> παραγγείλας καὶ τοῖς κατηγόροις λέγειν (Acts 23.30) *charging his accusers also to speak.*
> ... παραγγέλλειν τε τηρεῖν τὸν νόμον (Acts 15.5) *and to charge* (them) *to keep the law.*
> καὶ παρήγγειλεν αὐτοῖς ἵνα μηδὲν αἴρωσιν εἰς ὁδόν (Mark 6.8) *and he charged them that they should take nothing for their journey.*

THE USE OF THE SUBJUNCTIVE

(2) INDEFINITE CLAUSES

The subjunctive is used in indefinite clauses, i.e. in clauses introduced by relative pronouns or temporal or local conjunctions which do not refer to definite persons, times, or places. Indefiniteness may be given to a pronoun or adverb by placing after it the particle ἄν or ἐάν; ὅς means *who*, ὃς ἄν means *whoever* or *whosoever*; ὅτε means *when*, ὅταν (for ὅτε ἄν) means *whenever*; ὅπου *where*, ὅπου ἄν *wherever*.

> ὃς πιστεύει *he who believes.*
> ὃς ἂν πιστεύσῃ *whosoever believes.*
> ὅτε ἐξεχύννετο τὸ αἷμα Στεφάνου (Acts 22.20) *when the blood of Stephen was shed.*
> ὅταν ἄγωσιν ὑμᾶς (Mark 13.11) *when* (indefinite "whensoever") *they lead you.*
> ὅπου γάρ ἐστιν ὁ θησαυρός σου, ἐκεῖ ἔσται καὶ ἡ καρδία σου (Matt. 6.21) *for where thy treasure is, there will thy heart be also.*
> ὅπου ἐὰν εἰσέλθητε εἰς οἰκίαν, ἐκεῖ μένετε (Mark 6.10) *wheresoever ye enter into a house, there abide.*

Notice that ὅταν *whenever*, and ὅπου ἄν *wherever*, may be used with reference to the past, in which case the verb is in the indicative:

ὅταν αὐτὸν ἐθεώρουν (Mark 3.11) *whensoever they beheld him.*
καὶ ὅταν ὀψὲ ἐγένετο (Mark 11.19) *and whenever evening came.*
καὶ ὅπου ἄν εἰσεπορεύετο εἰς κώμας (Mark 6.56) *and wheresoever he entered into villages.*

ἕως and ἕως ἄν

A clause in past time introduced by ἕως *until*, is definite, and the verb is indicative (see p. 189):

ὁ ἀστὴρ ... προῆγεν αὐτοὺς ἕως ... ἐστάθη (Matt. 2.9) *the star ... went before them till ... it stood.*

But when ἕως refers to the future it is nearly always indefinite, and the verb is subjunctive. ἄν may be added, but it is often omitted, as ἕως in a future context is felt to be sufficiently indefinite and not to need the indefinite particle:

ἴσθι ἐκεῖ ἕως ἄν εἴπω σοι (Matt. 2.13) *be thou there until I tell thee.*
ἕως καὶ τὸ ἔσχατον λεπτὸν ἀποδῷς (aor. subj. of ἀποδίδωμι, see p. 242) (Luke 12.59) *till thou have paid the very last mite.*

ἕως οὖ, ἕως ὅτου (see p. 189) and, after a negative sentence, πρὶν ἤ with or without ἄν may also introduce indefinite clauses with the verb in the subjunctive:

ἕως οὖ ἀπολύσῃ τοὺς ὄχλους (Matt. 14.22) *till he should send the multitudes away.*
ἕως ὅτου σκάψω περὶ αὐτήν (Luke 13.8) *till I dig about it.*
... μὴ ἰδεῖν θάνατον πρὶν ἤ ἄν ἴδῃ τὸν Χριστὸν Κυρίου (Luke 2.26) *... not see death, before he had seen the Lord's Christ.*

When the aor. subj. is used in an indefinite clause the sense is often future perfect, although it would usually be cumbrous to bring this out in English translation.

The future perfect as a tense is obsolete in N.T. Greek, but students of Latin, in which it has survived, will be familiar with it: e.g. amavero, the English equivalent of which is "I shall have loved". It refers to an action in the future, but an action which is regarded as completed in relation to some other action. The retention of the fut. perf. in Latin enables that language to express the sequence of future events more accurately than English normally does; e.g. "I shall go when *I have finished* my tea" is adequate English. But a moment's thought will reveal that *I have finished* is not really a simple perfect. It refers in fact to a future action, but an action which is to be completed before that of the main verb *I shall go*. Latin here would use the future perfect tense, the literal English equivalent of which is the cumbrous "I shall have finished".

The future perfect sense can be detected in some of the examples above in which the aor. subj. is used in indefinite clauses: ἕως ἂν εἴπω *until I shall have told*; ἕως ... ἀποδῷς *until thou shalt have paid*. In the last example given above, πρὶν ἢ ἂν ἴδῃ, the R.V. translation "before he *had* seen" is necessitated by the rules of sequence of tenses in English, as the subordinate clauses are governed by the main clause "it had been revealed unto him".

(3) The Hortatory Subjunctive

The subjunctive is not only used in subordinate clauses. It may be used in a main clause to express an exhortation, thus supplying the lack of a first person in the imperative mood.

δι' ὑπομονῆς τρέχωμεν (Heb. 12.1) *let us run with patience.*

φάγωμεν καὶ πίωμεν, αὔριον γὰρ ἀποθνήσκομεν (1 Cor. 15.32) *let us eat and drink, for tomorrow we die.*

(4) The Deliberative Subjunctive

The subjunctive may also be used to express a deliberative question, i.e. a question which deliberates or weighs two courses of action:

δῶμεν ἢ μὴ δῶμεν; (Mark 12.14) *shall we give* ("are we to give") *or shall we not give?*

τί ποιήσωμεν; (Acts 2.37) *what shall we do* ("what are we to do")?

(5) Prohibitions

The imperative mood is used to express commands (see pp. 89ff). The present imperative with μή prohibits an action which is thought of as already in process:

μὴ κλαίετε (Luke 8.52) *weep not* (the context makes clear that they were already weeping).

The present imperative with μή is also used in prohibitions when the verb used is such as to describe a state or a necessarily prolonged action:

μὴ κωλύετε αὐτά (Mark 10.14) *forbid them not.*

μὴ θησαυρίζετε ὑμῖν θησαυροὺς ἐπὶ τῆς γῆς (Matt. 6.19) *lay not up for yourselves treasures upon the earth.*

μὴ κρίνετε, ἵνα μὴ κριθῆτε (Matt. 7.1) *judge not, that ye be not judged.*

Where the aorist imperative would be used in a positive command (i.e. where the action is thought of as instantaneous, or where no special emphasis is placed on the continuance or repetition of the action), in a prohibition μή is used with the aorist subjunctive (not aorist imperative):

μὴ νομίσητε ὅτι ἦλθον καταλῦσαι τὸν νόμον (Matt. 5.17) *think not that I came to destroy the law.*

μὴ εἰσενέγκῃς ἡμᾶς εἰς πειρασμόν (Luke 11.4) *bring us not into temptation.*

μηδένα διασείσητε (Luke 3.14) *do violence to no man.*

(6) STRONG DENIAL

The aorist subjunctive may be used after the double negative οὐ μή to express denial. In classical Greek the construction always conveyed a very firm negative, but there are N.T. instances where there seems to be no particular emphasis:

θάνατον οὐ μὴ θεωρήσῃ (John 8.51) *he shall not see death.*

οὐ μὴ παρέλθῃ ἡ γενεὰ αὕτη (Matt. 24.34) *this generation shall not pass away.*

Occasionally οὐ μή is followed by the future indicative with the same meaning:

οὐ μὴ ἔσται σοι τοῦτο (Matt. 16.22) *this shall not be unto thee.* (In this example the emphasis is certainly present).

Vocabulary

ἐλεημοσύνη (1) *almsgiving, alms*
δέσμιος (2) *prisoner*
᾽Ιάκωβος (2) *James*
νομικός (2) *lawyer*
νυμφίος (2) *bridegroom*
πάσχα, τό (indecl.) *Passover*

εὔκοπος, ον *easy*
ἐπαισχύνομαι (dep.) *be ashamed of*
καταλύω *destroy*
προσπορεύομαι (dep.) *come near*
σαλπίζω *sound a trumpet*
ὥσπερ (adv.) *even as, as*
ἤ (conjunction) *or*

EXERCISE 70

Translate into English:

1. ὃς γὰρ ἐὰν ἐπαισχυνθῇ με καὶ τοὺς ἐμοὺς λόγους, καὶ ὁ Υἱὸς τοῦ ἀνθρώπου ἐπαισχυνθήσεται αὐτόν, ὅταν ἔλθῃ ἐν τῇ δόξῃ τοῦ Πατρὸς αὐτοῦ. 2. οὐ μὴ γεύσωνται θανάτου ἕως ἂν ἴδωσιν τὴν βασιλείαν τοῦ Θεοῦ ἐληλυθυῖαν ἐν δυνάμει. 3. ὅταν οὖν ποιῇς

15—N.T.G.

ἐλεημοσύνην, μὴ σαλπίσῃς ἔμπροσθέν σου, ὥσπερ οἱ ὑποκριταὶ ποιοῦσιν, ὅπως δοξασθῶσιν ὑπὸ τῶν ἀνθρώπων. 4. εἶπον οὖν πρὸς αὐτόν Τί ποιῶμεν ἵνα ἐργαζώμεθα τὰ ἔργα τοῦ Θεοῦ; 5. καὶ προσπορεύονται αὐτῷ Ἰάκωβος καὶ Ἰωάνης λέγοντες αὐτῷ Διδάσκαλε, θέλομεν ἵνα ὃ ἐὰν αἰτήσωμέν σε ποιήσῃς ἡμῖν. 6. ἐπυνθάνοντο δὲ παρὰ τοῦ πατρὸς αὐτοῦ πῶς θέλει καλεῖσθαι τὸν υἱόν.

EXERCISE 71

Translate into Greek:

1. The lawyer asked him what was the great commandment. 2. Jesus asks them whose son the Christ is. 3. The disciples asked when these things would happen. 4. The woman asked him to cast the evil spirit out of her daughter. 5. For the third time the king commanded the soldiers to kill the prisoners. 6. What are we to eat, or what are we to drink? 7. Do not answer until you have heard what I say. 8. Let us seek the wandering sheep elsewhere. 9. I will certainly not listen to such words. 10. When we came nearer we saw much better.

EXERCISE 72

Translate into Greek:

1. We shall build the temple as well as possible. 2. Where are we to prepare for you to eat the passover? 3. The twelve asked where they were to make preparations for (εἰς) the feast. 4. And in the morning he went out, and remained outside a little while. 5. They said that they had never seen so great deeds. 6. Wherever you enter a house, stay there until you depart thence. 7. Whosoever causes one of these little ones to stumble shall be condemned. 8. Whensoever the bridegroom is taken away from them, then shall they fast in those days. 9. He speaks more easily in Greek than in Hebrew. 10. Do not think that I came to destroy the law or the prophets.

Chapter 26

THE OPTATIVE

The Optative consists of the historic tenses of the conjunctive mood (see the paradigms set out in full at the end, especially that of λύω, pp. 294–8). It takes its name from the Latin *opto*, "I wish", because one of its chief uses is to express wishes. It is not widely used in the N.T., but is found more frequently in St Luke's writings (Gospel and Acts) than elsewhere.

The optative tenses of λύω are as follows:

		Pres. opt. act.	*Pres. opt. mid.*	*Pres. opt. pass.*
Sing.	1.	λύοιμι	λυοίμην	λυοίμην
	2.	λύοις	λύοιο	λύοιο
	3.	λύοι	λύοιτο	λύοιτο
Plur.	1.	λύοιμεν	λυοίμεθα	λυοίμεθα
	2.	λύοιτε	λύοισθε	λύοισθε
	3.	λύοιεν	λύοιντο	λύοιντο

		Wk. aor. opt. act.	*Wk. aor. opt. mid.*	*Wk. aor. opt. pass.*
Sing.	1.	λύσαιμι	λυσαίμην	λυθείην
	2.	λύσαις	λύσαιο	λυθείης
	3.	λύσαι	λύσαιτο	λυθείη
		or (rarely) λύσειε(ν)		
Plur.	1.	λύσαιμεν	λυσαίμεθα	λυθείημεν
	2.	λύσαιτε	λύσαισθε	λυθείητε
	3.	λύσαιεν	λύσαιντο	λυθείησαν
		or λύσειαν		

The perfect optative active is not found in the N.T. The perfect optative middle and passive is formed by the perfect participle middle and passive and the present optative of εἰμί *to be* (for which see p. 316).

PERFECT OPTATIVE MIDDLE AND PASSIVE

Sing. 1.	λελυμένος	εἴην
2.	„	εἴης
3.	„	εἴη
Plur. 1.	λελυμένοι	εἴημεν
2.	„	εἴητε
3.	„	εἴησαν

THE OPTATIVE OF STRONG TENSES

Verbs which have a strong aorist form the aorist optative from the verb stem with the same endings as for the present optative. The strong aorist middle is only found in the N.T. in γένοιτο, the 3rd pers. sing. aor. opt. of γίνομαι, but the tense is given here in full for the sake of completeness. The strong aorist optative passive is not found in the N.T.

	Str. aor. opt. act. (of βάλλω)	*Str. aor. opt. mid.* (of βάλλω)
Sing. 1.	βάλοιμι	βαλοίμην
2.	βάλοις	βάλοιο
3.	βάλοι	βάλοιτο
Plur. 1.	βάλοιμεν	βαλοίμεθα
2.	βάλοιτε	βάλοισθε
3.	βάλοιεν	βάλοιντο

No optatives of contracted forms are found in the N.T. Nor are optatives of the common -μι verbs ἵστημι and τίθημι, or their compounds. Of δίδωμι *give*, only the 3rd pers. sing. str. aor. opt. act. is found, δῴη.

The present optative of δύναμαι *I am able, I can* (see p. 225) occurs:

Sing. 1.		δυναίμην
2.		δύναιο
3.		δύναιτο
Plur. 1.		δυναίμεθα
2.		δύναισθε
3.		δύναιντο

THE USE OF THE OPTATIVE

1. THE OPTATIVE IN WISHES

The optative may be used to express a wish (hence its name, from the Latin *opto*, "I wish"). The negative is μή. Nearly half the examples of this use in the N.T. are μὴ γένοιτο *may it not happen* (translated in A.V. "God forbid"):

αὐτὸς δὲ ὁ Θεὸς τῆς εἰρήνης ἁγιάσαι ὑμᾶς ... καὶ ... ὑμῶν τὸ πνεῦμα ... τηρηθείη (1 Thess. 5.23) *and may the God of peace himself sanctify you ... and ... may your spirit ... be preserved.*

2. THE OPTATIVE IN CONDITIONS

The optative may be used in a conditional clause when the condition is in future time and of relative uncertainty: the negative is μή.

εἰ πάσχοιτε (1 Peter 3.14) *if you should suffer.*
ἔσπευδεν γάρ, εἰ δυνατὸν εἴη αὐτῷ ... γενέσθαι (Acts 20.16) *for he was hastening, if it were possible for him ... to be....* (Here the "if clause" is future in relation to the action of ἔσπευδεν).

For further treatment of conditional sentences involving the use of the optative, see pp. 299f.

3. POTENTIAL OPTATIVE

The optative may be used with or without ἄν in potential clauses, i.e. clauses which state that a thing *could* or *might* be. Such a clause is in reality the main clause (*apodosis*: see p. 227) of a conditional sentence, the if-clause (*protasis*) being unexpressed. Examples in English are:

I could eat a hearty meal (if one were put in front of me).
I might be prepared to do it (if you asked me nicely).

This type of potential clause, therefore, is closely linked with that type of conditional sentence which employs the optative (see pp. 229f.). Examples in the N.T. of the potential optative are:

εὐξαίμην ἂν τῷ Θεῷ (Acts 26.29) *I could pray to God.*
τί ἂν θέλοι ὁ σπερμολόγος οὗτος λέγειν; (Acts 17.18) *what might this babbler be wanting to say?*

The N.T. has no examples of this use of the optative with a negative. In classical Greek the negative was οὐ.

4. Optative in Indirect Questions

In indirect questions the optative may (but need not) be used when the introductory verb is in past time (see p. 211).

5. Optative in Subordinate Clauses in Indirect Speech

In classical Greek the verb of a subordinate clause in indirect speech introduced by a verb in past time might be put in the optative if in the *oratio recta* it would have been a primary tense. Thus, to take a very simplified example:

> Direct speech: πιστεύω ἃ λέγω *I believe what I say.*
> Indirect speech (introductory verb in past time): εἶπεν ὅτι πιστεύοι ἃ λέγοι *he said that he believed what he said.*

Here the verb of the subordinate clause ("what he said") retains in Greek the same *tense* (present) as in *oratio recta*, but the *mood* is changed from indicative to optative. Classical Greek, however, often retained in indirect speech the same mood as well as tense, and the above sentence could equally well be εἶπεν ὅτι πιστεύει ἃ λέγει.

This use of the optative appears in the temporal clause introduced by πρὶν ἤ in Acts 25.16[1]: πρὸς οὓς ἀπεκρίθην ὅτι οὐκ ἔστιν ἔθος Ῥωμαίοις χαρίζεσθαί τινα ἄνθρωπον πρὶν ἢ ὁ κατηγορούμενος κατὰ πρόσωπον ἔχοι τοὺς κατηγόρους *to whom I answered, that it is not the custom of the Romans to give up any man, before that the accused have the accusers face to face.*

Oratio recta here would be οὐκ ἔστιν ἔθος ... χαρίζεσθαι ... πρὶν ἢ ἄν ... ἔχῃ.

VERBS IN -μι (1) ἵστημι

Hitherto we have dealt only with the various classes of verbs whose 1st pers. sing. pres. indic. act. ends in -ω. There are also verbs whose 1st pers. sing. pres. indic. act. ends in -μι. Verbs of this class are not numerous, but several of them have many compounds, and are of very frequent occurrence.

Their inflection differs from that of the -ω verbs in the present and imperfect (active, middle and passive) and in the strong aorist (active and middle). In these tenses the -μι verbs are non-thematic, i.e. they do not

---

[1] J. H. Moulton, however (*An Introduction to the Study of N.T. Greek*, 5th ed., p. 140), explains this optative as an example of the classical rule by which, when the principal clause is negative, πρίν is followed by the subjunctive with ἄν in primary time, and in past time by the optative without ἄν.
Probably both rules apply, and, so to speak, ἔχοι is doubly optative.

employ the thematic vowels ο and ε before the personal endings. The other tenses have the same characteristics as the corresponding tenses of the -ω verbs.

The most common of these verbs are ἵστημι *make to stand, set up, place*; τίθημι *place, lay, set*; and δίδωμι *give*.

We first deal with ἵστημι *I make to stand*: the verb stem varies between στα- and στη-, being lengthened in the present by reduplication with the aid of the vowel ι: ἵστημι (for σίστημι). (Not all the forms here given are actually found in the N.T. For example there are no instances of the imperf. indic. act. But they are all found in, or to be deduced from, *Koine* Greek, and are given here for the sake of completeness.)

PRESENT AND IMPERFECT ACTIVE OF ἵστημι

		Pres. indic.	Imperf. act.	Pres. imperative act.	Pres. subjunctive act.
Sing.	1.	ἵστημι	ἵστην		ἱστῶ
	2.	ἵστης	ἵστης	ἵστη	ἱστῇς
	3.	ἵστησι(ν)	ἵστη	ἱστάτω	ἱστῇ
Plur.	1.	ἵσταμεν	ἵσταμεν		ἱστῶμεν
	2.	ἵστατε	ἵστατε	ἵστατε	ἱστῆτε
	3.	ἵστασι(ν)	ἵστασαν	ἱστάτωσαν	ἱστῶσι(ν)

Present infinitive active: ἱστάναι.

Present participle active: ἱστάς (gen. ἱστάντος), ἱστᾶσα, ἱστάν.

See pp. 308–9 for the verb in full.

Transitive and Intransitive Uses of ἵστημι

Transitive verbs are those which require a direct object to complete the idea they present. The verb *hit* is a transitive verb. "I hit" does not present a complete idea, for one needs to know what is hit. An object must be expressed to complete the idea of the verb. The term "transitive" is from the Latin *transitivus*, derived from the verb *transeo*, "pass over", and it implies that the action of a verb "passes over" into an object.

Intransitive verbs, on the other hand, are those which can have no direct object, e.g. *run, walk, hang, stay*, and other verbs which describe movement or position. The student must not be misled by the fact that such expressions are possible as "run a race" and "walk a mile". *Race* and *mile* here are not direct objects of the verbs; they represent adverbial phrases, "run in a race", "walk for a mile".

Certain verbs may have both transitive and intransitive force; e.g. "I stopped the horse". Here the verb *stop* is used transitively. "When I saw him, I stopped". Here it is used intransitively. So also *hang*, mentioned above as intransitive, may be used transitively; e.g. "to hang a picture"; and *stay* can also be used transitively: e.g. "he stayed his hand".

So in Greek certain verbs have both transitive and intransitive uses, and ἵστημι is one of them:

The pres. and imperf. indic. act. (and consequently the pres. imperative, subjunctive, infinitive, and participle) are transitive: ἵστημι *I stand* in the sense of "I make to stand", "set up", "place".

The weak aorist indic. act. ἔστησα (inflected as ἔλυσα) is also transitive, *I stood* in the sense of "I made to stand".

The strong aorist indic. act. ἔστην (and consequently the str. aor. imperative, subjunctive, infinitive, and participle) is intransitive, *I stood*, as for example in "I stood amazed" (no direct object possible).

STRONG AORIST ACTIVE OF ἵστημι

	Strong aor. indic. act.	*Strong aor. imperative act*	*Strong aor. subjunctive act.*
Sing. 1.	ἔστην		στῶ
2.	ἔστης	στῆθι or -στα	στῇς
3.	ἔστη	στήτω	στῇ
Plur. 1.	ἔστημεν		στῶμεν
2.	ἔστητε	στῆτε	στῆτε
3.	ἔστησαν	στήτωσαν	στῶσι(ν)

Strong aorist infinitive active, στῆναι.

Strong aorist participle active, στάς (gen. στάντος), στᾶσα, στάν.

ἔστησαν is 3rd pers. plur. both of the strong and of the weak aor. indic. act. The context will declare which is employed.

PERFECT AND PLUPERFECT OF ἵστημι

The perfect active is intransitive and has a present meaning, *I stand*. ἕστηκα thus supplies the lack of a pres. indic. act. of ἵστημι with intransitive meaning. The pluperfect consequently refers simply to past time, *I stood* (intransitive).

	Perfect indic. act.	Pluperfect indic. act.	(No perfect imperative in N.T.)	Perfect subjunctive active
Sing. 1	ἕστηκα	εἱστήκειν or ἑστήκειν, etc.		ἑστηκὼς ὦ or ἑστώς (str.) ὦ, etc.
2.	ἕστηκας	εἱστήκεις		
3.	ἕστηκε(ν)	εἱστήκει		
Plur. 1.	ἑστήκαμεν	εἱστήκειμεν		ἑστηκότες ὦμεν or ἑστῶτες (str.) ὦμεν, etc.
2.	ἑστήκατε	εἱστήκειτε		
3.	ἑστήκασι(ν)	εἱστήκεισαν		

Both strong and weak forms of the perfect infinitive and participle occur (but the N.T. has no instance of the weak perfect infinitive):

Strong perfect infinitive active: ἑστάναι.
(Weak perfect infinitive active: ἑστηκέναι: not found in N.T.)
Strong perfect participle active: ἑστώς (gen. ἑστῶτος), ἑστῶσα, ἑστός.
Weak perfect participle active: ἑστηκώς (gen. ἑστηκότος), ἑστηκυῖα, ἑστηκός.

PRESENT MIDDLE AND PASSIVE OF ἵστημι

The middle voice has the meaning "I place for myself", and the passive "I am being placed" or "I am placed", and so "I stand" (intransitive). Thus the pres. indic. pass., like the perf. indic. act., ἕστηκα, supplies the lack of a pres. indic. act. with intransitive meaning.

	Pres. indic. mid. and pass.	Imperfect indic. mid. and pass.	Pres. imperative mid. and pass.	Pres. subjunctive mid. and pass.
Sing. 1.	ἵσταμαι	ἱστάμην		ἱστῶμαι
2.	ἵστασαι	ἵστασο	ἵστασο	ἱστῇ
3.	ἵσταται	ἵστατο	ἱστάσθω	ἱστῆται
Plur. 1.	ἱστάμεθα	ἱστάμεθα		ἱστώμεθα
2.	ἵστασθε	ἵστασθε	ἵστασθε	ἱστῆσθε
3.	ἵστανται	ἵσταντο	ἱστάσθωσαν	ἱστῶνται

Present infinitive mid. and pass.: ἵστασθαι.
Present participle mid. and pass.: ἱστάμενος, ἱσταμένη, ἱστάμενον.

OTHER TENSES OF ἵστημι

The other tenses of ἵστημι are inflected like the corresponding tenses of λύω.

> Future active: στήσω *I shall cause to stand* (transitive).
> 1st or weak aorist: ἔστησα *I caused to stand* (transitive).
> Future middle: στήσομαι *I shall place for myself* (transitive), *I shall stand* (intransitive).
> Future passive: σταθήσομαι *I shall be placed, I shall stand* (intransitive).
> 1st or weak aorist passive: ἐστάθην *I was placed, I stood* (intransitive).

Compounds of ἵστημι

The compounds of ἵστημι found in the N.T. are as follows, those occurring in St Mark being marked with an asterisk:

	Meaning of transitive tenses	Meaning of intransitive tenses
ἀνθίστημι	*set against*	*withstand, resist*
*ἀνίστημι	*raise up*	*rise*
ἀντικαθίστημι	*replace*	*withstand*
*ἀποκαθίστημι	*restore*	
ἀφίστημι	*put away, move to revolt*	*depart from, fall away*
διίστημι	*set apart, intervene*	*part, withdraw*
ἐνίστημι	*place in*	*be at hand, be present, threaten*
ἐξανίστημι	*raise up*	*rise*
*ἐξίστημι	*put out, confound, amaze*	*retire from, be mad, be amazed*
*ἐπανίστημι	*raise up against*	*rise up against*
ἐφίστημι	*set upon, set by*	*stand upon, be set over, be at hand*
καθίστημι	*set in order, appoint, establish*	
κατεφίστημι		*rise up against*
μεθίστημι	*change, remove, pervert*	
*παρίστημι	*place beside, provide, prove, present*	*stand by, be present*
περιίστημι	*place round*	*stand round*
προΐστημι	*put before, set over*	*rule, preside*
συνεφίστημι	*place over*	*rise together*
συνίστημι	*commend, prove*	*stand with, consist of*

New Verbs formed from ἵστημι

In *Koine* Greek there is a tendency to assimilate all verbs to -ω forms. Consequently we find a new thematic verb στήκω *I stand* (intransitive), formed from the perfect stem of ἵστημι. ἱστάνω also and στάνω *I cause to stand* (transitive), are formed from the present. But these by no means oust ἵστημι itself.

Other Verbs like ἵστημι

δύναμαι *I am able*, is conjugated like ἵσταμαι.

> Imperfect indicative: ἐδυνάμην or ἠδυνάμην.
> Future indicative: δυνήσομαι.
> Aorist indicative: ἠδυνήθην.

Like ἵσταμαι also are ἐπίσταμαι *I understand*; and κρέμαμαι *I hang* (intransitive): 1st aor act. indic. ἐκρέμασα; aor. pass. indic. ἐκρεμάσθην.

φημί, whose stem varies between φα- and φη-, is found only in a few forms:

> 1st pers. sing. pres. indic.: φημί.
> 3rd pers. sing.   ,,   ,,   φησί.
> 3rd pers. plur.   ,,   ,,   φασί(ν).
> 2nd pers. sing. imperative: φαθί
> 3rd pers. sing. imperf. indic.: ἔφη.

The above forms of the pres. indic. are enclitic; i.e. the accent is thrown back on to the preceding word where it is able to receive it:

> ὁ δέ φησιν Οὔ (Matt. 13.29) *but he saith, Nay.*

φημί is sometimes used in quoting the actual words used by a speaker.

Vocabulary

ἀπογραφή (1) *census, enrolment*
ἀπώλεια (1) *destruction*
δωρεά, ᾶς (1) *gift*
κριτής (1) *judge*
πτερύγιον (2) *turret, pinnacle*
λεγιών, ῶνος, ἡ (3) *legion*
ὅραμα, ατος, τό (3) *vision*
πυλών, ῶνος, ὁ (3) *porch, gateway*
χρῆμα, ατος, τό (3) *thing*; plur., *wealth*

Γαλιλαῖος, αία, αῖον *Galilaean*
ἑδραῖος, ον *steadfast*
μέσος, η, ον *middle*
ἐν μέσῳ (with gen.) *in the midst*
διαπορέω *be in perplexity*
ἐπικαλέω *name, call*
κτάομαι *acquire, get*
ξενίζω *entertain*; pass., *lodge*

EXERCISE 73

Translate into English:

1. καὶ λαβὼν παιδίον ἔστησεν αὐτὸ ἐν μέσῳ αὐτῶν, καὶ ἐναγκαλισάμενος αὐτὸ εἶπεν αὐτοῖς Ὃς ἂν ἓν τῶν τοιούτων παιδίων δέξηται ἐπὶ τῷ ὀνόματί μου, ἐμὲ δέχεται. 2. Πέτρος δὲ εἶπεν πρὸς αὐτόν Τὸ ἀργύριόν σου σὺν σοὶ εἴη εἰς ἀπώλειαν, ὅτι τὴν δωρεὰν τοῦ Θεοῦ ἐνόμισας διὰ χρημάτων κτᾶσθαι. 3. οὔτε θάνατος οὔτε ζωὴ οὔτε τις κτίσις ἑτέρα δυνήσεται ἡμᾶς χωρίσαι ἀπὸ τῆς ἀγάπης τοῦ Θεοῦ τῆς ἐν Χριστῷ Ἰησοῦ. 4. δοκεῖς ὅτι οὐ δύναμαι παρακαλέσαι τὸν Πατέρα μου, καὶ παραστήσει μοι ἄρτι πλείους ἢ δώδεκα λεγιῶνας ἀγγέλων; 5. ἔφη αὐτῷ ὁ κύριος αὐτοῦ Εὖ, δοῦλε ἀγαθὲ καὶ πιστέ, ἐπὶ ὀλίγα ἦς πιστός, ἐπὶ πολλῶν σε καταστήσω· εἴσελθε εἰς τὴν χαρὰν τοῦ κυρίου σου. 6. ὡς δὲ ἐν ἑαυτῷ διηπόρει ὁ Πέτρος τί ἂν εἴη τὸ ὅραμα ὃ εἶδεν, ἰδοὺ οἱ ἄνδρες οἱ ἀπεσταλμένοι ὑπὸ τοῦ ἑκατοντάρχου ἐπέστησαν ἐπὶ τὸν πυλῶνα, καὶ ἐπυνθάνοντο εἰ Σίμων ὁ ἐπικαλούμενος Πέτρος ἐνθάδε ξενίζεται.

EXERCISE 74

Translate into Greek:

1. May I enter into the house of the Lord. 2. His kinsmen asked what he wished his son to be called. 3. Who appointed me a judge over you? 4. Two men stood with Jesus on the mountain. 5. He spoke these things because of the crowd which stood around. 6. God will raise the faithful up at the last day. 7. Jesus says to her, Your brother shall rise. 8. May nobody eat fruit of thee. 9. The devil set him on the pinnacle of the temple. 10. And Peter arose and ran to the tomb.

EXERCISE 75

Translate into Greek:

1. May the Lord cleanse your heart and purify your soul. 2. I could say that you are not a true disciple. 3. Jesus said that the scribes were not willing themselves to do what they taught. 4. On these two commandments hangs the whole law. 5. They went up to Jerusalem to present him to the Lord, as it is written in the law. 6. When they saw that the lame man could walk the multitudes were amazed. 7. You were not able to resist the Spirit of God. 8. After him arose Judas the Galilaean in the days of the enrolment and moved the people to revolt. 9. Stand steadfast when men rise up against you, and you will be able to confound them. 10. And Peter denied, and said to the maid, I do not understand what you say.

Chapter 27

CONDITIONAL SENTENCES

Conditional sentences speak of something as happening conditionally upon something else happening. A complete conditional sentence has two clauses, the "if-clause", known as the *protasis* (πρότασις *premiss*) and the main clause, known as the *apodosis* (ἀπόδοσις *answering clause*).

Conditional sentences may broadly be divided into three classes:

A. Those which express a plain fact, although it is a conditional fact. They may refer to past, present, or future time: "if he did this, he did wrong"; "if he does this, he does wrong"; "if he does this (i.e. shall do this), he will do wrong".

B. Those which imply that in fact the condition was or is unfulfilled: "if he had done this, he would have done wrong"; "if he were doing this, he would be doing wrong".

C. Those which express a future contingency with tentativeness: "if he were to do this, he would do wrong".

We shall consider these three main classes in turn.

A. CONDITIONAL SENTENCES OF SIMPLE FACT

In these the apodosis states what was, is, or will be the fact if the condition was or is fulfilled. The protasis makes no assumption about whether the condition was or is fulfilled or not. The protasis is introduced by εἰ *if*, and the verb in both protasis and apodosis is in the indicative mood. The negative is usually οὐ, but a few instances with μή are found.

1. εἰ ταῦτα ἐποίησεν, ἡμάρτησεν *if he did this, he did wrong*.

Alternatively, if the sense required it, the imperfect indicative could be used:

εἰ ταῦτα ἐποίει, ἡμάρτανεν *if he was doing this, he was doing wrong*.
2. εἰ ταῦτα ποιεῖ, ἁμαρτάνει *if he does (or is doing) this, he does wrong*.
3. εἰ ταῦτα ποιήσει, ἁμαρτήσει *if he does (i.e. shall do) this, he will do wrong*.

But when the reference is to the future as in this last type, A.3, εἰ with the fut. indic. in the protasis is less frequently used than ἐάν with the pres. or aor. subjunctive (negative μή):

4. ἐὰν ταῦτα ποιῇ (or ποιήσῃ), ἁμαρτήσει.

This same construction, A.4, is also commonly used when there is no particular reference to the future, but rather a general reference:

ἐὰν τὰς ἐντολάς μου τηρήσητε, μενεῖτε ἐν τῇ ἀγάπῃ μου (John 15.10) *if ye keep my commandments, ye shall abide in my love.*

Here τηρήσητε has a general reference. So with:

ἐὰν τις φάγῃ ἐκ τούτου τοῦ ἄρτου, ζήσει (John 6.51) *if any man eat of this bread, he shall live.*
ἐὰν θέλῃς δύνασαί με καθαρίσαι (Mark 1.40) *if thou wilt* (i.e. at any time) *thou canst make me clean.*

The types given above are instances in which both protasis and apodosis refer to the same time.   But the apodosis need not refer to the same time as the protasis; e.g. If he started yesterday, he is there now (*or* he will be there tomorrow).   Two more types in this class can therefore be added:

5. εἰ Ἀβραάμ . . . ἐδικαιώθη, ἔχει . . . (Rom. 4.2) *if Abraham was justified, he hath. . . .*

The apodosis in fact need not be a tense of the indicative mood.   An imperative is frequent:

6. εἰ θέλεις εἰς τὴν ζωὴν εἰσελθεῖν, τήρει τὰς ἐντολάς (Matt. 19.17) *if thou wouldest enter into life, keep the commandments.*

B. UNFULFILLED CONDITIONAL SENTENCES

Of these there are two types, referring to past and to present time.   The protasis implies that the condition was not, or is not, fulfilled, and the apodosis states what would have been, or what would be, the result if the condition had been, or were, fulfilled.

The protasis is introduced by εἰ (negative μή)[1], and the particle ἄν is used in the apodosis to emphasize the contingent nature of the statement which is being made.   The verb in both clauses is in the aorist indicative

[1] Note: Mark 14.21, where we have the negative οὐ, is an exception.

when the reference is to past time, and in the imperfect indicative when the reference is to the present.

1. *Past time*: εἰ ταῦτα ἐποίησεν, ἡμάρτησεν ἄν *if he had done this, he would have done wrong.*
2. *Present time*: εἰ ταῦτα ἐποίει, ἡμάρτανεν ἄν *if he were doing this, he would be doing wrong.*

A mixture of these two types is possible:

3. The protasis referring to past time, and the apodosis to the present: εἰ ἐδιδάχθη, ἀπεκρίνετο ἄν *if he had been taught, he would answer.*
4. The protasis referring to present time, and the apodosis to the past: εἰ ἠγαπᾶτέ με, ἐχάρητε ἄν (John 14.28) *if ye loved me* (i.e. now), *ye would have rejoiced.*

C. INDISTINCT FUTURE CONDITIONS

In classical Greek future conditions of a more indefinite type than A.3 and A.4 above were expressed by εἰ with the optative mood (negative μή) in the protasis, and the optative with ἄν (negative οὐ) in the apodosis:

εἰ ταῦτα ποιοίη, ἁμαρτάνοι ἄν *if he were to (or should) do this he would do wrong.*

But the N.T. does not fully employ this type of conditional sentence. We have, however, the protasis of this type in:

ἀλλ' εἰ καὶ πάσχοιτε διὰ δικαιοσύνην, μακάριοι (1 Pet. 3.14) *but and if ye should suffer for righteousness' sake, blessed* (would ye be).

Here the verb of the apodosis has to be supplied.

κρεῖττον γὰρ... εἰ θέλοι τὸ θέλημα τοῦ Θεοῦ, πάσχειν (1 Pet. 3.17) *for* (it is) *better, if the will of God should so will, to suffer.* . . .

Here again the verb of the apodosis has to be supplied, "it is", or more strictly, "it would be".

And we have an apodosis of this type in:

Πῶς γὰρ ἂν δυναίμην ἐὰν μή τις ὁδηγήσει με; (Acts 8.31) *for how could I, except some one should* (literally, "shall") *guide me?*

Here the protasis, which in classical Greek would have been εἰ μή τις ὁδηγήσαι με, assumes a more definite form. But note that where we should expect ἐὰν μή with the subjunctive ὁδηγήσῃ (as in type A.4 above), we actually have the still more vivid future indicative ὁδηγήσει.

THE PARTICLE ἄν

We have now illustrated the most common uses of the particle ἄν, which we here summarize:

1. It is used after relative pronouns (e.g. ὃς ἄν) or conjunctions (e.g. ὅταν, πρὶν ἄν, ἕως ἄν), and is the equivalent of the English -ever. For examples of this use, see pp. 212f.

2. It is used, attached closely to the verb, in the apodosis of certain types of conditional sentences, those which state unfulfilled conditions (see p. 229), and those which state indistinct future conditions (see p. 229–30). Here the effect of the ἄν is to heighten the contingency.

3. Occasionally ἄν is used in a purpose clause after ὅπως; e.g.

ὅπως ἂν ἔλθωσιν καιροὶ ἀναψύξεως (Acts 3.20) *that so there may come seasons of refreshing.*

Here, too, the effect of the ἄν is to heighten the contingency, "if so be that".

κἄν (for καὶ ἄν) is sometimes found and seems practically to be the equivalent of καί *even.* (This is to be distinguished from κἄν as the equivalent of καὶ ἐάν *and if,* or sometimes, *even if.*)

ἵνα κἂν τοῦ κρασπέδου τοῦ ἱματίου αὐτοῦ ἅψωνται (Mark 6.56) *that they might touch if it were but the border of his garment.*

ἵνα . . . κἂν ἡ σκιὰ ἐπισκιάσῃ (Acts 5.15) *that . . . at the least his shadow might overshadow. . . .*

But here also the ἄν possibly serves to emphasize contingency, the R.V. representing this by the phrases "if it were" and "at the least".

C. F. D. Moule (*An Idiom-Book of New Testament Greek*) remarks "if a plain καί had been used, it is difficult to imagine that the ἄν would have been missed". The archaic "haply", or the modern "perhaps" almost exactly represent the force of this ἄν or κἄν.

VERBS IN -μι (2) τίθημι

Τίθημι *I place, lay,* is another common -μι verb. The verb stem varies between θε- and θη-, lengthened in the present by reduplication with the aid of the vowel ι, τί-θη-μι.

It may be noted as an indication of the way in which the -μι verbs were being assimilated (even in classical times) to the -ω verbs that ἐτίθεις, ἐτίθει in the imperf. indic. act., and τίθει in the pres. imperative act. are forms borrowed from verbs of the type of φιλέω.

See pp. 310–11 for the verb in full.

PRESENT AND IMPERFECT ACTIVE OF τίθημι

	Pres. indic. act.	Imperfect indic. act.	Pres. imperative act.	Pres. subjunctive act.
Sing. 1.	τίθημι	ἐτίθην		τιθῶ
2.	τίθης	ἐτίθεις	τίθει	τιθῇς
3.	τίθησι(ν)	ἐτίθει	τιθέτω	τιθῇ
Plur. 1.	τίθεμεν	ἐτίθεμεν		τιθῶμεν
2.	τίθετε	ἐτίθετε	τίθετε	τιθῆτε
3.	τιθέασι(ν) or τιθιᾶσι(ν)	ἐτίθεσαν or ἐτίθουν	τιθέτωσαν	τιθῶσι(ν)

Present infinitive active: τιθέναι.
Present participle active: τιθείς (gen. τιθέντος), τιθεῖσα, τιθέν.

AORIST ACTIVE OF τίθημι

The singular of the aor. indic. act. is weak and has a perfect form, showing the characteristic κ of the perfect before the ending, whilst the plural is strong.

	Aor. indic. act.	Strong aor. imperative act.	Strong aor. subjunctive act.
Sing. 1.	ἔθηκα		θῶ
2.	ἔθηκας	θές	θῇς
3.	ἔθηκε(ν)	θέτω	θῇ
Plur. 1.	ἔθεμεν		θῶμεν
2.	ἔθετε	θέτε	θῆτε
3.	ἔθεσαν	θέτωσαν	θῶσι(ν)

Strong aorist infinitive active: θεῖναι.
Strong aorist participle active: θείς (gen. θέντος), θεῖσα, θέν.

16—N.T.G.

THE MIDDLE VOICE OF τίθημι

The passive of the present and imperfect tenses of τίθημι does not occur, but is supplied by the verb κεῖμαι (see p. 234).

The middle voice is as follows:

		Present indic. mid.	Imperfect indic. mid.	Pres. imperative mid.	Pres. subjunctive mid.
Sing.	1.	τίθεμαι	ἐτιθέμην		τιθῶμαι
	2.	τίθεσαι	ἐτίθεσο	τίθεσο	τιθῇ
	3.	τίθεται	ἐτίθετο	τιθέσθω	τιθῆται
Plur.	1.	τιθέμεθα	ἐτιθέμεθα		τιθώμεθα
	2.	τίθεσθε	ἐτίθεσθε	τίθεσθε	τιθῆσθε
	3.	τίθενται	ἐτίθεντο	τιθέσθωσαν	τιθῶνται

Present infinitive middle: τίθεσθαι.
Present participle middle: τιθέμενος, τιθεμένη, τιθέμενον.

		Strong aor. indic. mid.	Strong aor. imperative mid.	Strong aor. Subjunctive mid.
Sing.	1.	ἐθέμην		θῶμαι
	2.	ἔθου	θοῦ	θῇ
	3.	ἔθετο	θέσθω	θῆται
Plur.	1.	ἐθέμεθα		θώμεθα
	2.	ἔθεσθε	θέσθε	θῆσθε
	3.	ἔθεντο	θέσθωσαν	θῶνται

Strong aorist infinitive middle: θέσθαι.
Strong aorist participle middle: θέμενος, θεμένη, θέμενον.

OTHER TENSES OF τίθημι

The other tenses of τίθημι are inflected like the corresponding tenses of λύω.

Future Active: θήσω.
Future Middle: -θήσομαι.
Future Passive: -τεθήσομαι.

Perfect Active: τέθεικα.
Perfect Middle and Passive: τέθειμαι.
Pluperfect Middle and Passive: -ἐτεθείμην.
Weak Aorist Passive: ἐτέθην.

Compounds of τίθημι

The compounds of τίθημι found in the N.T. are as follows, those occurring in St Mark being marked with an asterisk:

ἀνατίθημι *set up*; in mid., *set forth, declare*
ἀντιδιατίθημι in mid. *oppose*
ἀποτίθημι in mid. *put off*
διατίθημι in mid., *dispose of, make a testament*
ἐκτίθημι *expose, set forth, expound*
*ἐπιτίθημι *place upon*; in mid., *attack*
*κατατίθημι *lay down*; in mid., *lay up for oneself*
μετατίθημι *transfer, change*
*παρατίθημι *place beside, set before*
*περιτίθημι *place around*
προσανατίθημι *lay on besides*, in mid., *undertake besides, consult*
*προστίθημι *add*
προτίθημι *set before, set forth*; in mid., *propose, purpose*
συγκατατίθημι *deposit together*; in mid., *agree with, assent to*
συνεπιτίθημι *help in putting on*; in mid., *join in attacking*
συντίθημι *put together*; in mid., *agree*
ὑποτίθημι *place under, lay down*; in mid., *suggest*

Other Verbs like τίθημι

Like τίθημι are conjugated the present and imperfect of:

πίμπλημι *fill*[1] (aor. indic. act., ἔπλησα; perf. indic. pass., πέπλησμαι; aor. indic. pass., ἐπλήσθην).
πίμπρημι *burn, cause to swell*; mid., *to swell* (aor. indic. act., -ἔπρησα).

Κεῖμαι

Κεῖμαι *lie* (verb stem κει-) also supplies the passive of τίθημι in the present and imperfect. It is conjugated very much like a perfect mid. or pass.

---

[1] πίμπλημι, and other verbs (and adjectives) implying *filling*, or *being full* are used with the genitive of that which fills.

		Pres. indic.	Imperfect indic.
Sing.	1.	κεῖμαι	ἐκείμην
	2.	κεῖσαι	ἔκεισο
	3.	κεῖται	ἔκειτο
Plur.	1.	κείμεθα	ἐκείμεθα
	2.	κεῖσθε	ἔκεισθε
	3.	κεῖνται	ἔκειντο

Present infinitive: κεῖσθαι.
Present participle: κείμενος, κειμένη, κείμενον.

The imperative and conjunctive moods are not found.

Compounds of κεῖμαι

Those occurring in St Mark are marked with an asterisk.

*ἀνάκειμαι *be laid up, recline at table, sit at meat*
ἀντίκειμαι *resist*
ἀπόκειμαι *be laid up, be in store*
ἐπίκειμαι *lie on, threaten, press upon*
*κατάκειμαι *lie down, recline at table*
παράκειμαι *lie beside, be present*
*περίκειμαι *be placed round*
πρόκειμαι *be set before*
*συνανάκειμαι *recline with at table*

Vocabulary

ἡλικία (1) *age, stature*
ὑδρία (1) *water-pot*
θεμέλιος, ὁ *or* θεμέλιον, τό (2) *foundation*
συκάμινος, ου, ἡ (2) *sycamine, mulberry tree*
νυμφών, ῶνος, ὁ (3) *bridechamber*
ἀποσυνάγωγος, ον *expelled from the synagogue, excommunicated*

φωτεινός, ή, όν *light, bright*
ἐκριζόω *root out*
παρατίθημι (with dat. of indirect obj.) *place beside, set before*
βεβαίως (adv.) *firmly*
δωρεάν (adv.) *freely, in vain*
ἐπιμελῶς (adv.) *carefully*
σήμερον (adv.) *today*

EXERCISE 76

Translate into English:

1. εἰ ὁ κόσμος ὑμᾶς μισεῖ, γινώσκετε ὅτι ἐμὲ πρῶτον ὑμῶν μεμίσηκεν. εἰ ἐκ κόσμου ἦτε, ὁ κόσμος ἂν τὸ ἴδιον ἐφίλει. 2. εἰ εἴχετε πίστιν ὡς κόκκον σινάπεως, ἐλέγετε ἂν τῇ συκαμίνῳ ταύτῃ Ἐκριζώθητι καὶ φυτεύθητι ἐν τῇ θαλάσσῃ· καὶ ὑπήκουσεν ἂν ὑμῖν. 3. ἐὰν οὖν ᾖ ὁ ὀφθαλμός σου ἁπλοῦς, ὅλον τὸ σῶμά σου φωτεινὸν ἔσται. 4. καὶ ἐὰν βασιλεία ἐφ᾽ ἑαυτὴν μερισθῇ, οὐ δύναται σταθῆναι ἡ βασιλεία ἐκείνη. 5. ἤδη γὰρ συνετέθειντο οἱ Ἰουδαῖοι ἵνα ἐάν τις αὐτὸν ὁμολογήσῃ Χριστόν, ἀποσυνάγωγος γένηται. 6. καὶ ἐξελθόντες οἱ δοῦλοι ἐκεῖνοι εἰς τὰς ὁδοὺς συνήγαγον πάντας οὓς εὗρον, πονηρούς τε καὶ ἀγαθούς· καὶ ἐπλήσθη ὁ νυμφὼν ἀνακειμένων.

EXERCISE 77

Translate into Greek:

1. If they persecuted me, they will persecute you. 2. If you laid the foundation firmly, the house will stand. 3. If you sat at table with him, you would know him better. 4. If the Lord had not sent the apostles, the gospel would not have been heard there. 5. Which of you can add one cubit to his stature? 6. You ordered the servants to place the loaves on the table. 7. The water-pots which were placed there were filled with wine. 8. We persuaded the teacher to set this parable before the people. 9. They wished to see where the body had been laid. 10. If you seek diligently, you will find the money.

EXERCISE 78

Translate into Greek:

1. While the twelve sat at meat, Judas purposed to go to the chief priests. 2. The good shepherd will lay down (use τίθημι) his life for the sheep. 3. If the disciples had asked, they would have received. 4. If this should happen, we should wonder at it. 5. If you seek first the kingdom, God will add all these things for you. 6. The father besought Jesus to place his hands upon his daughter, that she might be saved. 7. If any man wants to come after me, let him deny himself, and let him take up his cross daily. 8. If I had not come and spoken unto them, they would have no sin. 9. If the Lord had not placed our enemies under our feet, we should be slaves today. 10. If righteousness comes by the law, Christ died in vain.

Chapter 28

PURPOSE CLAUSES

Purpose may be expressed in several ways, some of which have already been described. We here summarize five ways in which N.T. Greek expresses purpose.

1. By a final clause introduced by ἵνα or ὅπως, the verb being in the subjunctive. See pp. 203f.
2. By the simple infinitive. See p. 103.
3. By the future participle. See p. 167.
4. By the infinitive with the article (articular infinitive) after εἰς or πρός.

> πρὸς τὸ θεαθῆναι αὐτοῖς (Matt. 6.1) *to be seen by them.* This clause is parallel to ὅπως δοξασθῶσιν ὑπὸ τῶν ἀνθρώπων *that they may have glory of men,* in the next verse.
> εἰς τὸ εἶναι αὐτὸν δίκαιον (Rom. 3.26) *that he might himself be just.*

Notice that where a subject of the infinitive is expressed it is in the accusative case.

5. By the infinitive with the article in the genitive case.

> μέλλει γὰρ Ἡρῴδης ζητεῖν τὸ παιδίον τοῦ ἀπολέσαι αὐτό (Matt. 2.13) *for Herod intends to seek the young child to destroy him.*
> προπορεύσῃ ... ἑτοιμάσαι ... τοῦ δοῦναι γνῶσιν (Luke 1.76–7) *thou shalt go before ... to make ready ... to give knowledge.*

Here the simple infinitive and the infinitive with article in the genitive are both used to express purpose.

THE USE OF ἵνα

1. ἵνα *expressing Purpose*

The normal use of ἵνα is to introduce final or purpose clauses. The verb of the final clause is in the subjunctive. The negative is ἵνα μή or simply μή, *in order that not, lest.* See p. 203f.

Occasionally ἵνα is followed by the future indicative to express purpose. Thus:

ἵνα καὶ οἱ μαθηταί σου θεωρήσουσιν τὰ ἔργα σου (John 7.3) *that thy disciples also may behold thy works.*

2. ἵνα *expressing Consequence*

The student of Latin will know how very difficult it sometimes is, when dealing with clauses introduced by *ut*, to decide whether they are final clauses denoting purpose, or consecutive clauses denoting result. The same difficulty sometimes appears in English with the use of the various phrases which may be used to introduce final and consecutive clauses. Thus: "he clutched the rail so as to save himself". In this sentence it is not easy to say whether "so as to save himself" explains the purpose of clutching the rail, or merely describes the result of so doing. The probability is that it is a final clause. But the clutching of the rail might have been an involuntary action, in which case "so as to save himself" would simply state the consequence. There is, at any rate, enough ambiguity to make one hesitate.

On the other hand "he clutched the rail in order to save himself" is clearly final; and "he clutched the rail so that he saved himself" is clearly consecutive.

Classical Greek practically avoided this difficulty by introducing purpose or final clauses by ἵνα, ὅπως, and, more rarely, ὡς; and consecutive or result clauses by ὥστε or ὡς. There is no example in classical Greek literature of ἵνα used of result.

But in the New Testament there are some grounds for believing that the uncertain distinction between the concepts of purpose and result has produced some examples of ἵνα or its negatives ἵνα μή and μή used to introduce consecutive rather than final clauses. A well-nigh intolerable sense is removed if this is so in Mark 4.11–12:

ἐκείνοις δὲ τοῖς ἔξω ἐν παραβολαῖς τὰ πάντα γίνεται, ἵνα βλέποντες βλέπωσιν καὶ μὴ ἴδωσιν *but unto them that are without, all things are done in parables: so that they look and look and do not see* (rather than "that seeing they may not see" of R.V.).

In Matt. 1.22 "so that" gives much better sense than "in order that":

τοῦτο δὲ ὅλον γέγονεν ἵνα πληρωθῇ τὸ ῥηθὲν ... *now all this is come to pass, so that it was fulfilled which was spoken* ... (rather than "that it might be fulfilled" of R.V.).

Other possible examples may be studied in Luke 9.45; John 9.2; Gal. 5.17; 1 Thess. 5.4; and 1 John 1.9.

The question, however, is hotly debated, and there are scholars who affirm that ἵνα never introduces a consecutive clause, however attractive it is on occasion so to translate it.

3. ἵνα introducing Commands

Occasionally ἵνα followed by the subjunctive expresses a command:

τὸ θυγάτριόν μου ἐσχάτως ἔχει, ἵνα ἐλθὼν ἐπιθῇς τὰς χεῖρας αὐτῇ (Mark 5.23) my little daughter is at the point of death; come and lay thy hands on her (R.V. translates "(I pray thee) that thou come . . .").

ὥσπερ ἐν παντὶ περισσεύετε . . . ἵνα καὶ ἐν ταύτῃ τῇ χάριτι περισσεύητε (II Cor. 8.7) as ye abound in everything . . . abound in this grace also (R.V. translates "(see) that ye abound . . .").

As the R.V. insertions imply, the origin of this use of ἵνα may be in dependence on an omitted verb, e.g. δέομαι or θέλω in Mark 5.23, and βλέπετε in II Cor. 8.7.

4. ἵνα expressing Content

A clause with its verb in the subjunctive may be introduced by ἵνα to explain or fill out the content of a word previously used:

ἴδετε ποταπὴν ἀγάπην δέδωκεν ἡμῖν ὁ Πατὴρ ἵνα τέκνα Θεοῦ κληθῶμεν (1 John 3.1) behold what manner of love the Father hath bestowed upon us, that we should be called children of God. Here the ἵνα clause expresses the content of the word ἀγάπην.

αὕτη δέ ἐστιν ἡ αἰώνιος ζωή, ἵνα γινώσκωσιν σέ (John 17.3) and this is life eternal, that they should know thee. Here the ἵνα clause fills out the content of the noun ζωή.

Ἀβραὰμ ὁ πατὴρ ὑμῶν ἠγαλλιάσατο ἵνα ἴδῃ τὴν ἡμέραν τὴν ἐμήν (John 8.56) your father Abraham rejoiced to see my day. Here the ἵνα clause defines more fully the verb ἠγαλλιάσατο.

This kind of clause is in some grammars called epexegetic, or explanatory. It is most common in the Johannine writings; but note also:

καὶ πόθεν μοι τοῦτο ἵνα ἔλθῃ ἡ μήτηρ τοῦ Κυρίου μου πρὸς ἐμέ; (Luke 1.43) and whence is this to me, that the mother of my Lord should come unto me? Here the ἵνα clause gives the content of the pronoun τοῦτο.

οὐ AND μή

1. The broad distinction between the negative particles οὐ and μή is that οὐ is used in connection with statements of fact, and μή with movements of the mind and will, and in hypothetical statements. Hence, generally speaking, οὐ is used with the indicative mood, and μή with the imperative, subjunctive and optative.

In classical Greek οὐ was used with infinitives and participles which represented statements (e.g. with the infinitive representing a statement in indirect speech, and with a participle used in a temporal sense). But in the N.T. μή is regularly used with the infinitive and the participle. To this there are very few exceptions; e.g. ὁ μισθωτὸς καὶ οὐκ ὢν ποιμήν (John 10.12) *he that is a hireling, and not a shepherd.*

2. The distinction between οὐ and μή in asking questions is described on p. 77. Οὐ expects the answer *yes*, and μή expects the answer *no*. Μή can also put a very hesitant question, e.g. μήτι οὗτός ἐστιν ὁ Χριστός; (John 4.29) *can this be the Christ?* It can be used similarly in indirect questions:

σκόπει οὖν μὴ τὸ φῶς ... σκότος ἐστίν (Luke 11.35) *look therefore whether the light ... be not darkness.* Here the direct question would be μή (or μή ποτε) τὸ φῶς σκότος ἐστίν; *can it be that (or perhaps) the light is darkness?*

3. There is a closely allied use of μή after verbs denoting fear or caution:

φοβηθεὶς ὁ χιλίαρχος μὴ διασπασθῇ ὁ Παῦλος ... (Acts 23.10) *the chief captain, fearing lest Paul should be torn in pieces....* J. H. Moulton suggests that there are really two clauses here: "the chief captain feared: could it be that Paul would be torn in pieces?"

φοβοῦμαι δὲ μή πως ... φθαρῇ τὰ νοήματα ὑμῶν (2 Cor. 11.3) *but I fear lest by any means ... your minds should be corrupted* (i.e. "I fear—can it be that your minds will be corrupted?").

βλέπετε μή τις ὑμᾶς πλανήσῃ (Mark 13.5) *take heed that no man (or lest any man) lead you astray* (i.e. "take heed—can it be that any man will lead you astray?").

The above examples, in which a verb of fearing or caution is followed by μή and the subjunctive, all refer to future fears. If the fear refers to the past, μή is followed by the indicative:

φοβοῦμαι ὑμᾶς μή πως εἰκῇ κεκοπίακα εἰς ὑμᾶς (Gal. 4.11) *I am afraid of you, lest by any means I have bestowed labour on you in vain.*

In all these examples μή may accurately be translated *lest*. Fear that something *may happen*, or *has happened*, is thus to be translated by a clause introduced by μή. Fear that something *may not* happen is, therefore, naturally expressed by a clause introduced by μή followed by οὐ:

> φοβοῦμαι μή πως οὐχ οἵους θέλω εὕρω ὑμᾶς (2 Cor. 12.20) *I fear lest by any means I should not find you such as I would.*

4. The use of οὐ μή in strong denials is described on p. 215.

5. A compound negative following another negative strengthens the negative idea (i.e. they do not cancel each other):

> οὐ δύνασθε ποιεῖν οὐδέν (John 15.5) *ye can do nothing.*

There may be more than two such negatives:

> οὗ οὐκ ἦν οὐδεὶς οὔπω κείμενος (Luke 23.53) *where never man had yet lain.*

PERIPHRASTIC TENSES

The word *periphrastic* is derived from περιφράζομαι *to express in a roundabout way*. A periphrastic tense is one which employs a participle and a part of the verb "to be" or "to have". English is compelled to use periphrastic tenses since it has no other means of expressing continued action; *I am going, I shall be going, I was going*, etc. English also has no simple perfect or pluperfect tense, and therefore uses the periphrastic *I have gone, I had gone*.

Greek, although possessing a more complete list of tenses than English, employs periphrastic tenses especially to emphasize continued action. A few examples are here given of periphrasis employed in various tenses.

Present (Present participle with present tense of εἰμί). This is rare, because the simple pres. indic. adequately expresses continuing action.

> ἡ διακονία . . . οὐ μόνον ἐστὶν προσαναπληροῦσα (2 Cor. 9.12) *the ministration . . . not only filleth up. . . .*

Imperfect (Present participle with past tense of εἰμί).

> πᾶν τὸ πλῆθος ἦν τοῦ λαοῦ προσευχόμενον (Luke 1.10) *the whole multitude of the people were praying.*
> ἀκούοντες ἦσαν (Gal. 1.23) *they heard* (with the force of "kept on hearing").

Future (Present participle with future tense of εἰμί).

> ἀνθρώπους ἔσῃ ζωγρῶν (Luke 5.10) *thou shalt catch men.*
>
> οἱ ἀστέρες ἔσονται ... πίπτοντες (Mark 13.25) *the stars shall be falling* (so R.V.).

Perfect (Perfect participle with present tense of εἰμί).

> ἡ ἀγάπη αὐτοῦ τετελειωμένη ἐν ἡμῖν ἐστιν (1 John 4.12) *his love is perfected in us.*
>
> τῇ γὰρ χάριτί ἐστε σεσωσμένοι (Eph. 2.8) *for by grace have ye been saved.*

Pluperfect (Perfect participle with past tense of εἰμί).

> ἦν αὐτῷ κεχρηματισμένον (Luke 2.26) *it had been revealed unto him.*

VERBS IN -μι, (3) δίδωμι

Δίδωμι *I give*, is another common verb in -μι. The verb stem varies between δο- and δω-, lengthened in the present by reduplication with the aid of the vowel ι, δι-δω-μι.

Assimilation to the -ω verb forms is seen in the imperfect indicative active, where ἐδίδουν, ἐδίδους, etc., are on the analogy of the imperfect of δηλόω.

See pp. 312–13 for the verb in full.

PRESENT AND IMPERFECT ACTIVE OF δίδωμι

	Pres. indic. act.	Imperfect indic. act.	Pres. imperative act.	Pres. subjunctive act.
Sing. 1.	δίδωμι	ἐδίδουν		διδῶ
2.	δίδως	ἐδίδους	δίδου	διδῷς or διδοῖς
3.	δίδωσι(ν)	ἐδίδου	διδότω	διδῷ or διδοῖ
Plur. 1.	δίδομεν	ἐδίδομεν		διδῶμεν
2.	δίδοτε	ἐδίδοτε	δίδοτε	διδῶτε
3.	διδόασι(ν)	ἐδίδοσαν	διδότωσαν	διδῶσι(ν)

Present infinitive active: διδόναι.
Present participle active: διδούς (gen. διδόντος), διδοῦσα, διδόν.

AORIST ACTIVE OF δίδωμι

The singular of the aor. indic. act. is weak, and has a perfect form, showing the characteristic κ of the perfect before the ending.   The plural is strong.   In the other moods of the aor. the tense is strong.

	Aor. indic. act.	Strong aor. imperative act.	Strong aor. subjunctive act.
Sing. 1.	ἔδωκα		δῶ
2.	ἔδωκας	δός	δῷς or δοῖς
3.	ἔδωκε(ν)	δότω	δῷ, δοῖ, or δώῃ
Plur. 1.	ἔδομεν		δῶμεν
2.	ἔδοτε	δότε	δῶτε
3.	ἔδοσαν	δότωσαν	δῶσι(ν)

Strong aorist infinitive active: δοῦναι.
Strong aorist participle active: δούς (gen. δόντος), δοῦσα, δόν.

The aorist optative active is found in the 3rd pers. sing. only, δῴη (strong).

MIDDLE AND PASSIVE VOICES OF δίδωμι

	Pres. indic. mid. and pass.	Imperfect indic. mid. and pass.	Pres. imperative mid. and pass.	Pres. subjunctive mid. and pass.
Sing. 1.	δίδομαι	ἐδιδόμην		διδῶμαι
2.	δίδοσαι	ἐδίδοσο	δίδοσο	διδῷ
3.	δίδοται	ἐδίδοτο	διδόσθω	διδῶται
Plur. 1.	διδόμεθα	ἐδιδόμεθα		διδώμεθα
2.	δίδοσθε	ἐδίδοσθε	δίδοσθε	διδῶσθε
3.	δίδονται	ἐδίδοντο	διδόσθωσαν	διδῶνται

Present infinitive middle and passive: δίδοσθαι.
Present participle middle and passive: διδόμενος, διδομένη, διδόμενον.

STRONG AORIST INDICATIVE MIDDLE

Sing. 1.	ἐδόμην
2.	ἔδου
3.	ἔδοτο

Plur. 1.   ἐδόμεθα
      2.   ἔδοσθε
      3.   ἔδοντο

The other moods of the aorist middle are not found in the N.T.   The classical forms are noted in the table on p. 313.

OTHER TENSES OF δίδωμι

The other tenses of δίδωμι are inflected like the corresponding tenses of λύω:

> Future active: δώσω.
> Future middle: -δώσομαι.
> Future passive: δοθήσομαι.
> Perfect active: δέδωκα.
> Pluperfect active: (ἐ)δεδώκειν.
> Perfect mid. and pass.: δέδομαι.
> Weak aorist passive: ἐδόθην.

Compounds of δίδωμι

The compounds of δίδωμι found in the N.T. are as follows, those occurring in St Mark being marked with an asterisk.

> ἀναδίδωμι *give up, hand over*
> ἀνταποδίδωμι *give back as a recompense*
> *ἀποδίδωμι *restore, return, render*
> διαδίδωμι *distribute*
> *ἐκδίδωμι (in mid.) *let out for hire*
> ἐπιδίδωμι *give over, give in*
> μεταδίδωμι *give a share of*
> *παραδίδωμι *give, hand over, hand down, deliver*
> προδίδωμι *give first, betray*

VERBS IN -μι (4) ἵημι

Ἵημι *send*, is a verb found only in compound form in the N.T.   The verb stem varies between ἑ- and ἡ-, lengthened in the present by reduplication with the aid of the vowel ι, but with the loss of the aspirate on the stem, ῐ-η-μι.

Some tenses are entirely lacking in the N.T., and of the others only one or two parts are found.   For the sake of completeness those tenses which

are found at all in the N.T. are given here in full, the parts not found being
enclosed in brackets.

THE PRESENT ACTIVE OF ἵημι

	Pres. indicative act.	Pres. imperative act.	Pres. subjunctive act.
Sing. 1.	-ἵημι		[ἱῶ]
2.	[ἵης]	[ἵει]	[ἱῇς]
3.	-ἵησι(ν)	-ἱέτω	-ἱῇ
Plur. 1.	[ἵεμεν]		[ἱῶμεν]
2.	-ἵετε	-ἵετε	-ἱῆτε
3.	-ἱᾶσι(ν)	[ἱέτωσαν]	-ἱῶσι(ν)

Present infinitive active: -ἱέναι.
Present participle active: -ἱείς (gen. -ἱέντος) [ἱεῖσα], [ἱέν].

AORIST ACTIVE OF ἵημι

The singular of the aor. indic. act. is weak and has a perfect form, showing the characteristic κ of the perfect before the ending.   A weak 3rd
pers. plur. -ἧκαν is also found.   The plural is strong, but does not occur
in the N.T.

	Aor. indicative act.	Strong aor. imperative act.	Strong aor. subjunctive act.
Sing. 1.	-ἧκα		-ὦ
2.	-ἧκας	-ἕς	[ᾖς]
3.	-ἧκε(ν)	[ἕτω]	-ᾖ
Plur. 1.	[εἷμεν]		[ὦμεν]
2.	[εἷτε]	-ἕτε	-ἧτε
3.	[εἷσαν] or -ἧκαν	[ἕτωσαν]	-ὦσι[ν]

Aorist infinitive active: -εἷναι.
Aorist participle active: -εἵς (gen. -έντος) [εἷσα], [ἕν].

MIDDLE AND PASSIVE VOICES OF ἵημι.

Few parts are found in the N.T.   The imperfect indicative, the present
imperative, the present subjunctive, and the present infinitive are lacking.

PRESENT INDICATIVE MIDDLE AND PASSIVE

Sing. 1. [ἵεμαι]
2. [ἵεσαι]
3. -ἵεται
Plur. 1. [ἱέμεθα]
2. [ἵεσθε]
3. -ἵενται

Present participle mid. and pass.: -ἱέμενος, -ἱεμένη, -ἱέμενον.

Of the aorist middle only the participle occurs: ἕμενος, ἑμένη, ἕμενον.

OTHER TENSES OF ἵημι

The other tenses of ἵημι are inflected like the corresponding tenses of λύω.

Future active: -ἥσω.
Future passive: -ἑθήσομαι.
Perfect active: -εἶκα.
Perfect mid. and pass.: -ἕωμαι, but the participle: -εἵμενος.

Compounds of ἵημι

The compounds of ἵημι found in the N.T. are as follows, those occurring in St Mark being marked with an asterisk:

ἀνίημι *send up, send back, let go, loosen*
*ἀφίημι *send away, forgive, neglect, permit*
καθίημι *send down, let down*
παρίημι *pass by, disregard, relax*
*συνίημι *bring together, understand*

New Verbs formed from ἵημι

This verb illustrates the thematizing process which the -μι verbs undergo in *Koine* Greek. We find a number of forms which are from -ἵω instead of the non-thematic -ἵημι. Thus:

1st pers. plur. pres. indic. act.: ἀφίομεν.
3rd pers. plur. pres. indic. act.: ἀφίουσι(ν).
3rd pers sing. imperf. indic. act.: ἤφιε.[1]
3rd pers. plur. pres. indic. mid. and pass.: ἀφίονται.

[1] Notice that the prefix is augmented, as though it were part of the stem of the verb.

Vocabulary

μεθοδία (1) *deceit*
πανοπλία (1) *full armour*
γεωργός (2) *farm worker, husband-man*
Καῖσαρ, αρος, ὁ (3) *Caesar*
παράπτωμα, ατος, τό (3) *trespass*
τεῖχος, ους, τό (3) *wall*
ὑπόδειγμα, ατος, τό (3) *example*
Αἰνών, ἡ (indecl.) *Aenon*

βλέπω *see, take heed*
σκοπέω *look at, take heed*
εὐκαίρως (adv.) *in due season, at an opportune moment*
διότι (conjunction) *because*
ἰδού (aor. imper. mid. of ὁράω used as demonstrative particle) *behold, lo*

EXERCISE 79

Translate into English:

1. ἰδοὺ ἀφίεται ὑμῖν ὁ οἶκος ὑμῶν. λέγω δὲ ὑμῖν, οὐ μὴ ἴδητέ με ἕως εἴπητε Εὐλογημένος ὁ ἐρχόμενος ἐν ὀνόματι Κυρίου. 2. μὴ φοβοῦ ἀλλὰ λάλει καὶ μὴ σιωπήσῃς, διότι ἐγώ εἰμι μετὰ σοῦ καὶ οὐδεὶς ἐπιθήσεταί σοι τοῦ κακῶσαί σε, διότι λαός ἐστί μοι πολὺς ἐν τῇ πόλει ταύτῃ. 3. ἦν δὲ ᾿Ιωάνης βαπτίζων ἐν Αἰνών, ὅτι ὕδατα πολλὰ ἦν ἐκεῖ, καὶ παρεγίνοντο καὶ ἐβαπτίζοντο· οὔπω γὰρ ἦν βεβλημένος εἰς τὴν φυλακὴν ᾿Ιωάνης. 4. ἐνδύσασθε τὴν πανοπλίαν τοῦ Θεοῦ πρὸς τὸ δύνασθαι ὑμᾶς στῆναι πρὸς τὰς μεθοδίας τοῦ διαβόλου. 5. καὶ ὅταν στήκητε προσευχόμενοι, ἀφίετε εἴ τι ἔχετε κατά τινος, ἵνα καὶ ὁ Πατὴρ ὑμῶν ὁ ἐν τοῖς οὐρανοῖς ἀφῇ ὑμῖν τὰ παραπτώματα ὑμῶν. 6. μετὰ δύο ἡμέρας τὸ πάσχα γίνεται, καὶ ὁ Υἱὸς τοῦ ἀνθρώπου παραδίδοται εἰς τὸ σταυρωθῆναι.

EXERCISE 80

Translate into Greek:

1. Judas sought how he might betray Jesus. 2. Take heed, therefore, lest the light that is in thee be darkness. 3. I cannot give you the bread you ask for. 4. Much shall be required from him to whom much was given. 5. I have come to do thy will. 6. The disciples let Paul down by night through the wall. 7. Jesus washed the disciples' feet to give them an example. 8. Take heed that you do not cause the children to stumble. 9. Take heed how you walk, lest you fall. 10. Do you understand what the will of the Lord is?

EXERCISE 8 I

Translate into Greek:

1. The things that are God's must not be rendered to Caesar. 2. This I say, that all men may believe that the Son of man can forgive sins. 3. I will certainly not give up my sword, even if you kill me. 4. The Pharisees disregarded the great commandment, but kept the small points of the law. 5. A certain man had a vineyard which he let out to husbandmen. 6. My meat is to obey the Father who sent me. 7. If you steal you will be handed over to prison. 8. The betrayer was one of the twelve who sat at meat with Jesus. 9. Distribute all that you have to the poor, and you will have treasure in heaven. 10. If he had not been victorious, the crown would not have been given to him.

Chapter 29

PERFECTS WITH PRESENT MEANING: οἶδα, εἴωθα

Οἶδα, the perfect of a stem ϝειδ- which does not exist in the present (it is the same stem as in εἶδον *I saw*, infinitive, ἰδεῖν), has a present meaning, *I know*.

Its irregular conjugation in classical Greek (perfect, οἶδα, οἶσθα, οἶδε(ν), ἴσμεν, ἴστε, ἴσασιν; pluperfect, ᾔδη, ᾔδησθα, ᾔδει(ν), ᾔσμεν, ᾖστε, ᾔδεσαν or ᾖσαν) has given way to the regular perfect and pluperfect endings, although ἴσασι(ν) is sometimes found for the 3rd pers. plur. perf. indic.

	Indicative		Imperative	Subjunctive
	Primary perfect	*Historic pluperfect*		
Sing. 1.	οἶδα	ᾔδειν		εἰδῶ
2.	οἶδας	ᾔδεις	ἴσθι	εἰδῇς
3.	οἶδε(ν)	ᾔδει	ἴστω	εἰδῇ
Plur. 1.	οἴδαμεν	ᾔδειμεν		εἰδῶμεν
2.	οἴδατε	ᾔδειτε	ἴστε	εἰδῆτε
3.	οἴδασι(ν)	ᾔδεισαν	ἴστωσαν	εἰδῶσι(ν)

Infinitive: εἰδέναι.
Participle: εἰδώς (gen. εἰδότος), εἰδυῖα, εἰδός.

Εἴωθα, the perfect of ἔθω, which does not occur in the N.T., has a present meaning, *I am accustomed*. The pluperfect, *I was accustomed*, is εἰώθειν.

See also p. 255.

FURTHER NOTES ON THE TENSES

What has already been said about the meaning of the Greek tenses (pp. 15, 17, 24, 25f, 43, 53, 61) needs some amplification. For the Greek tense not only refers to the *time* of an action, but also to its *nature*. By this is meant that an action may be regarded as instantaneous, or as prolonged. J. H. Moulton has suggested the use of the words *linear* and *punctiliar* to describe these two modes of action, *linear* because a prolonged action may appropriately be represented by a line, and *punctiliar* because instantaneous action may be represented by a point.   We shall use these convenient words to describe the two kinds of action.

English largely employs auxiliary verbs to express linear action, e.g. *I am coming, I was coming, I shall be coming.* We have seen (p. 240) that Greek can employ similar periphrastic tenses, but it also uses some of the simple tenses primarily for linear action, whilst others denote punctiliar action.   It follows, therefore, that the Greek tenses are not necessarily the exact equivalents of the English tenses which most closely correspond. Usually the correspondence is close enough to cause the student little difficulty; but occasionally he will come across a Greek tense which seems to defy adequate translation.

We shall here set out a list of the Greek indicative active tenses, showing whether they are used for linear or punctiliar action.   We shall then add some notes on each of the tenses with particular reference to their less normal uses.

> Pres. indic. act., λύω *I am loosing*: usually present linear action
> Imperf. indic. act., ἔλυον *I was loosing*: past linear action
> Fut. indic. act., λύσω *I shall loose*: usually future punctiliar, but can be future linear, I shall be loosing
> Aor. indic. act., ἔλυσα *I loosed*: past punctiliar action
> Perf. indic. act., λέλυκα *I have loosed*: past punctiliar, but with effect extending into the present
> Pluperf. indic. act., ἐλελύκειν *I had loosed*: past punctiliar, but with effect extending to a subsequent time also in the past

THE PRESENT INDICATIVE

The pres. indic. normally expresses linear action in the present: λύω *I am loosing*.   But note that there are some verbs whose meaning can only be punctiliar, e.g. ῥαπίζω *I slap*. The punctiliar λέγω *I say*, is very frequent.   But no difficulties of translation are raised here.

Iterative Present

The pres. indic. may also represent repeated action:

> καθ᾽ ἡμέραν ἀποθνῄσκω (1 Cor. 15.31) *I die daily.*
> ἄλλος πρὸ ἐμοῦ καταβαίνει (John 5.7) *another* (always) *steps down before me.*

Again there is no difficulty in translation.

Conative Present

There are a few instances in the N.T. of the pres. indic. being used to represent attempted action (Conative Present), e.g. λιθάζετε (John 10.32) *ye try to stone*, or *want to stone*.

Special Uses of the Present Indicative

Historic Present

Greek, like English, especially in a piece of narrative, can use the pres. indic. instead of a past tense for the sake of vividness. This is common in St Mark. Compare, e.g., the pres. indic. ἐκβάλλει, in Mark 1.12, with the main verbs in past tense which precede and follow.

Futuristic Present

Greek, again like English (e.g. *I sail next week*), occasionally uses the pres. indic. with a future sense:

> ὁ Υἱὸς τοῦ ἀνθρώπου παραδίδοται (Mark 9.31) *the Son of Man will be betrayed.*

The use of the present tense here implies assurance.

Gnomic Present

Again as in English, the pres. indic. is used to express a general truth (γνώμη).

> πᾶν δένδρον ἀγαθὸν καρποὺς καλοὺς ποιεῖ (Matt. 7.17) *every good tree bringeth forth good fruit.*

Matt. 9.16–17 also contains examples.

The Imperfect Indicative

The imperf. indic. expresses linear action in the past. Corresponding to similar uses in the pres. indic. it may also express repeated action (Iterative Imperfect) or attempted action (Conative Imperfect). Thus ἔλυον may be *I was loosing, I kept on loosing* (or *I used to loose*), or *I tried to loose.*

> ὃν ἐτίθουν[1] καθ᾽ ἡμέραν πρὸς τὴν θύραν (Acts 3.2) *whom they used to lay each day at the door* (Iterative Imperfect).
> καὶ ἐκάλουν αὐτὸ . . . Ζαχαρίαν (Luke 1.59) *and they wanted to call him . . . Zacharias* (Conative Imperfect).

[1] ἐτίθουν is an instance of the assimilation of -μι verbs to -ω forms. It is 3rd pers. plur. imperf. indic. (on the analogy of ἐφίλουν) for ἐτίθεσαν.

SPECIAL USES OF THE IMPERFECT INDICATIVE

Inceptive Imperfect

The imperf. may be used to express the beginning of a course of action, e.g. ἐδίδασκεν (Mark 1.21) *he began to teach.*

In Conditional Sentences

The use of the imperf. indic. in conditions which are unfulfilled in present time is explained on pp. 228–9.

THE FUTURE INDICATIVE

The fut. indic. represents normally punctiliar action in the future: ἄξω *I shall bring.* But it may also represent future linear action, ἄξω *I shall lead,* though the N.T. frequently employs the periphrastic future for the linear sense (p. 241).

THE AORIST INDICATIVE

The aor. indic. usually expresses punctiliar action in past time. It is therefore usually equivalent to the English simple past tense:

καὶ εὐθὺς ἐκάλεσεν αὐτούς (Mark 1.20) *and straightway he called them.*

But, when the verb is one which by its nature describes linear rather than punctiliar action, the aor. indic. expresses a continuing action or state in past time, when there is no particular emphasis on the linear nature of the action, and it is regarded simply as a fact in past time:

ἐκάθισεν δὲ ἐνιαυτὸν καὶ μῆνας ἕξ (Acts 18.11) *and he dwelt* (there) *a year and six months.*

SPECIAL USES OF THE AORIST INDICATIVE

Ingressive or Inceptive Aorist

The aor. indic. of a verb which by its nature describes linear action may be used to denote the beginning of the action:

ἐσίγησαν (Luke 20.26) *they fell silent.*
νεκρὸς ἦν καὶ ἔζησεν (Luke 15.32) *he was dead, and came to life.* (R.V. "is alive again". But note that an English *perfect* makes a good translation here, "has come to life". Cf. Rev. 2.8.)

ἤδη ἐπλουτήσατε· χωρὶς ἡμῶν ἐβασιλεύσατε (1 Cor. 4.8) *ye have already become rich, ye have become kings apart from us* (here also an English perfect is the natural translation).

Constative or Summary Aorist

The aor. indic. of a verb which describes effort may be used to denote the successful completion of an action:

τεσσεράκοντα καὶ ἓξ ἔτεσιν οἰκοδομήθη ὁ ναὸς οὗτος (John 2.20) *this temple took forty six years to build* (C. F. D. Moule's example).

ἐποίησεν ὁ δεῖνα (not in N.T.: a craftsman's signature on a work of art) *this was made by so-and-so* (J. H. Moulton's example).

μόλις κατέπαυσαν τοὺς ὄχλους (Acts 14.18) *they scarcely restrained the multitudes.*

Aorist of Immediate Past

Occasionally the student will find an aorist for which an English present seems the only possible translation.   Instances noted by Moule and Moulton are:

ἐν σοὶ εὐδόκησα (Mark 1.11) *in thee I am well pleased* (cf. Matt. 17.5).

ἔγνων τί ποιήσω (Luke 16.4) *I know what I will do.*

ἐμνήσθημεν ὅτι ἐκεῖνος ὁ πλάνος εἶπεν (Matt. 27.63) *we remember that that deceiver said....*

In all these instances the aorist strictly refers to a moment in the immediate past: εὐδόκησα, the approval was given at the moment of baptism; ἔγνων, the speaker's mind was made up a moment before he spoke; ἐμνήσθημεν, the Pharisees remembered what Jesus had said some little time before they addressed Pilate.

Gnomic Aorist

The aor. indic. may be used instead of the pres. indic. (Gnomic Present) to describe general truths of the proverb type.   It will be translated by an English present:

ἀνέτειλεν γὰρ ὁ ἥλιος σὺν τῷ καύσωνι καὶ ἐξήρανεν τὸν χόρτον, καὶ τὸ ἄνθος αὐτοῦ ἐξέπεσεν (James 1.11) *for the sun ariseth with the scorching wind, and withereth the grass; and the flower thereof falleth.*[1]

---

[1] Moule suggests that these aorists are not gnomic, but are used "to emphasize the suddenness and completeness of the withering".   But this would not seem to explain the aorist ἀνέτειλεν.

The Epistolary Aorist

The aor. indic. is sometimes used in letters instead of a present tense when the writer, putting himself at the standpoint of the reader, refers to an event which is present to himself, but which will be past when the letter is read:

ἐξαυτῆς ἔπεμψα πρὸς σέ (Acts 23.30) *I send* (him) *to you forthwith.*

So also ἔγραψα in 1 Cor. 5.11; ἐξῆλθεν in 2 Cor. 8.17; συνεπέμψαμεν in 2 Cor. 8.18, 22; ἔπεμψα in 2 Cor. 9.3; ἡγησάμην in 2 Cor. 9.5; ἔγραψα in Gal. 6.11 (see R.V. margin); ἔπεμψα in Eph. 6.22; ἡγησάμην in Phil. 2.25; ἔπεμψα in Phil. 2.28 and Col. 4.8; ἀνέπεμψα in Philem. 12.

In Conditional Sentences

The use of the aor. indic. in conditions which are unfulfilled in past time is explained on pp. 228-9.

The Perfect Indicative

The Greek perf. indic. normally describes an action which was completed at a point of time in the past, the result of the action being thought of as continuing into the present:

ἡ πίστις σου σέσωκέν σε (Mark 5.34) *thy faith hath saved thee.*

The use of the perfect here emphasizes the present state of the woman resulting from the initial act of saving faith.

> Ἰωάνης ὁ Βαπτίζων ἐγήγερται ἐκ νεκρῶν (Mark 6.14) *John the Baptist is risen from the dead* (i.e. "he rose, and is still risen"; cf. the use of the aorist ἠγέρθη in 6.16, where the reference is simply to the past event. Herod, as though clinching the argument says "John, whom I beheaded, rose".

In these, and many other examples, the correspondence between the Greek and the English perfects is sufficiently close. But in fact the English perfect is not the equivalent of the Greek. Three differences may here be noted:

(*a*) The English perfect may be used to denote a past action when the time is not definitely stated, and in this it is much more akin to the Greek aorist than to the Greek perfect. (When the time of a past action is definite, English uses the simple past tense: cf. "I have read this book"—no definite time stated—and "I read this book last week"—definite time).

(*b*) The English perfect lays no particular stress on the continuing effects of the past action.   The Greek perfect nearly always does.

(*c*) Greek idiom allows a definite reference to time to be made in conjunction with a perfect tense.   In English this is unusual.

For these reasons some English perfects would be best rendered into Greek by the aorist, and conversely we come across some Greek aorists which demand an English perfect in translation.

We likewise find some Greek perfects which demand an English simple past tense.   We here give a few examples of both.

Greek Aorists translated by English Perfects

Instances of this have been given on pp. 251–2.   The following are also examples:

ἄλλα πολλά ἐστιν ἃ παρέλαβον κρατεῖν (Mark 7.4) *many other things there are which they have received to hold.*

Here the Greek aorist simply states a past fact with no reference to a particular time.   It is a typical Greek aorist, but English demands a perfect.

οὐκ ἀνέγνωτε ἐν τῇ βίβλῳ Μωϋσέως . . .; (Mark 12.26) *have ye not read in the book of Moses . . .?*

Here again the Greek aorist refers to past indefinite time, and the English perfect is the necessary translation.

Other examples may be studied in Matt. 11.25, 12.3, 12.28, 27.8; Mark 16.6; Luke 5.26, 7.16, 14.20; John 12.19, 13.1.

Greek Perfects translated by English Simple Past Tense

ὅτι ἐγήγερται τῇ ἡμέρᾳ τῇ τρίτῃ (1 Cor. 15.4) *that he was raised on the third day.*

Here the use of the Greek perfect emphasizes the present results—he was raised and is alive.   But a definite note of time is not usual with an English perfect, and a simple past tense is natural.   But R.V. translates "hath been raised".

τεθέαμαι τὸ Πνεῦμα καταβαῖνον (John 1.32) *I beheld the Spirit descending.*

Here also the Greek perfect emphasizes the present effect which the past vision has on the speaker.   But the vision took place at a definite moment in the past, and the English idiom therefore prefers a simple past tense.   But again R.V. translates with a perfect: "I have beheld".

οἳ καὶ πρὸ ἐμοῦ γέγοναν ἐν Χριστῷ (Rom. 16.7) *who were in Christ before me.*

The note of time implied in πρὸ ἐμοῦ necessitates an English simple past tense in translation.

Greek Perfects used as Aorists

There are one or two Greek perfects in the N.T. which appear to have no perfect force (i.e. no reference to present effects of a past act), but to be the equivalent of the aorist:

οὐκ ἔσχηκα ἄνεσιν (2 Cor. 2.13) *I had no relief.*

There is certainly no reference to present effect here. There are several other instances of this use of ἔσχηκα by St Paul (2 Cor. 1.9, 7.5; Rom. 5.2), though in some of them perhaps a perfect force can be discerned. In Mark 5.15 the perfect participle ἐσχηκότα is used with a clearly aoristic meaning.

J. H. Moulton suggests that the aorist ἔσχον appropriated to itself the meaning "got", "received", and that therefore the perfect ἔσχηκα was taken over as an aorist to mean "possessed", "had".

Γέγονα is sometimes cited as another instance of an aoristic perfect, but in the majority of its occurrences it refers to present time ("I have become", "I am"). But in Matt. 25.6 it seems to be aoristic: μέσης δὲ νυκτὸς κραυγὴ γέγονεν *at midnight there was a cry.* Again, there is no reference here to present result. Moulton, however, suggests that it may be an historic present, and so R.V. translates.

Πεποίηκα in 2 Cor. 11.25, ἀπέσταλκεν in 1 John 4.14, and πέπρακεν in Matt. 13.46 are also possible instances.

PERFECTS WITH PRESENT MEANING

A few Greek perfects have a present meaning. In most of these, however, the perfect force is perceptible:

ἔγνωκα *I know* (I have perceived and therefore know)
εἴωθα *I am accustomed* (I have grown accustomed)
ἔοικα *I am like* (I have become like, and so am like)
ἕστηκα *I stand* (I have taken my stand, and so stand)
ἤγγικα *I am near* (I have drawn near, and so am near)
κεῖμαι *I am lying down* (I have lain down, and am still lying down)
μέμνημαι *I remember* (I have called to mind, and so remember)
οἶδα *I know* (I have discovered, and so know)
πέποιθα *I trust* (I have placed my trust in, and so trust)
τέθνηκα *I am dead* (I have died, and so am dead)

The Pluperfect Indicative

The English pluperfect is commonly used to express a past action which was prior in relation to another past action, e.g. "They left on Monday; we had gone already". But the Greek pluperfect is not used in this way. The Greek aorist is used where English would use the pluperfect of relative past time:

καὶ ἐπελάθοντο λαβεῖν ἄρτους (Mark 8.14) *and they had forgotten to take bread.*

The Greek pluperfect is the historic tense of the perfect, and expresses an action completed at a point of time in the past of which the result is thought of as continuing to the point of time in the past envisaged in the sentence as a whole:

τεθεμελίωτο γὰρ ἐπὶ τὴν πέτραν (Matt. 7.25) *for it had been founded upon the rock.*

When the perfect of a verb expresses present meaning (see the list above), the pluperfect is to be translated by an English simple past tense:

ᾔδειν *I knew.*

ἑστήκειν *I stood.*

VERBS IN –μι (5) δείκνυμι

Δείκνυμι *I show,* is one of a class of -μι verbs which in the present add -νυ- to the stem before the ending; thus δείκ-νυ-μι, ῥήγ-νυ-μι, πήγ-νυ-μι.

The subjunctive endings are as for λύω.

Present and Imperfect Active of δείκνυμι

		Pres. indicative act.	Imperfect indic. act.	Pres. imperative act.	Pres. subjunctive act.
Sing.	1.	δείκνυμι	ἐδείκνυν		δεικνύω
	2.	δείκνυς	ἐδείκνυς	δείκνυ	δεικνύῃς
	3.	δείκνυσι(ν)	ἐδείκνυ	δεικνύτω	δεικνύῃ
Plur.	1.	δείκνυμεν	ἐδείκνυμεν		δεικνύωμεν
	2.	δείκνυτε	ἐδείκνυτε	δείκνυτε	δεικνύητε
	3.	δεικνύασι(ν)	ἐδείκνυσαν	δεικνύτωσαν	δεικνύωσι(ν)

Present infinitive active: δεικνύναι.

Present participle active: δεικνύς (gen. δεικνύντος) δεικνῦσα, δεικνύν.

PRESENT AND IMPERFECT MIDDLE AND PASSIVE OF δείκνυμι

	Pres. indic. mid. and pass.	Imperfect indic. mid. and pass.	Pres. imperative mid. and pass.	Pres. subjunctive mid. and pass.
Sing. 1.	δείκνυμαι	ἐδεικνύμην		δεικνύωμαι
2.	δείκνυσαι	ἐδείκνυσο	δείκνυσο	δεικνύῃ
3.	δείκνυται	ἐδείκνυτο	δεικνύσθω	δεικνύηται
Plur. 1.	δεικνύμεθα	ἐδεικνύμεθα		δεικνυώμεθα
2.	δείκνυσθε	ἐδείκνυσθε	δείκνυσθε	δεικνύησθε
3.	δείκνυνται	ἐδείκνυντο	δεικνύσθωσαν	δεικνύωνται

Present infinitive mid. and pass.: δείκνυσθαι.
Present participle mid. and pass.: δεικνύμενος, δεικνυμένη, δεικνύμενον.

OTHER TENSES OF δείκνυμι

The other tenses of δείκνυμι are inflected like the corresponding tenses of λύω.

Future active: δείξω.
Aorist active: ἔδειξα.
Aorist middle: ἐδειξάμην.
Perfect mid. and pass.: δέδειγμαι.
Weak aorist pass.: ἐδείχθην.

Compounds of δείκνυμι

The compounds of δείκνυμι found in the N.T. are as follows. None of them occurs in St Mark.

ἀναδείκνυμι *show forth, declare, consecrate*
ἀποδείκνυμι *show forth, declare*
ἐνδείκνυμι *show forth, prove*
ἐπιδείκνυμι *show, exhibit*
ὑποδείκνυμι *show secretly, teach*

New Verbs formed from δείκνύμι

The tendency in *Koine* Greek to substitute -ω verb forms for -μι verb forms is shown by the appearance in some parts of a verb δεικνύω *show*, inflected like λύω.

Other Verbs like δείκνυμι

Other verbs which, like δείκνυμι, insert the suffix -νυ- in the present before the ending are as follows. Those which occur in St Mark are marked with an asterisk.

	Compounds	English	Future	Aor. act.	Perf. act.	Perf. pass.	Aor. pass.
-ἄγνυμι		break	-ἐάξω	-ἔαξα			-ἐάγην
	κατάγνυμι	break					
-ἔννυμι		clothe					
	ἀμφιέννυμι	clothe				ἠμφίεσμαι	
ζεύγνυμι		yoke		-ἔζευξα			
	*συνζεύγνυμι	yoke to-gether					
ζώννυμι		gird	ζώσω	-ἔζωσα		-ἔζωσμαι	
	διαζώννυμι	gird round					
	ὑποζώννυμι	undergird					
κεράννυμι		mix		ἐκέρασα		{κεκέρασμαι / κέκραμαι}	
κορέννυμι		satiate				κεκόρεσμαι	ἐκορέσθην
-μίγνυμι		mix		-ἔμιξα		μέμιγμαι	
	συναναμίγνυμι	mix up together					
*-ὄλλυμι		destroy	{-ὀλέσω / -ὀλῶ}	-ὤλεσα	-ὄλωλα		
	*ἀπόλλυμι	destroy					
	συναπόλλυμι	destroy together					
*ὄμνυμι		swear		ὤμοσα			
πετάννυμι		spread		-ἐπέτασα			
	ἐκπετάννυμι	spread out					
πήγνυμι		fasten		ἔπηξα			
*ῥήγνυμι		break, rend	ῥήξω	{ἔρρηξα / ἔρηξα}			
ῥώννυμι		strengthen				ἔρρωμαι	
*σβέννυμι		quench	σβέσω	ἔσβεσα			
*στρώννυμι		strew, furnish		ἔστρωσα		ἔστρωμαι	-ἐστρώθην
	καταστρώννυμι	overthrow					
	ὑποστρώννυμι	spread under					

In this whole class thematization is frequent, and the N.T. has many thematic instead of non-thematic -μι forms. Thus there are forms from

ἀπολλύω instead of ἀπόλλυμι

ζωννύω instead of ζώννυμι

ὀμνύω instead of ὄμνυμι

στρωννύω instead of στρώννυμι

All these are conjugated as λύω.

ῥήγνυμι has a synonym ῥήσσω.

ἀπόλλυμι *I destroy*. For the principal parts see the list above. Note carefully that the active voice is transitive, *to destroy* or *to lose utterly*, while the middle voice is intransitive, *to perish* or *to be lost*. Thus:

Present middle: ἀπόλλυμαι *I perish*.
Future middle: ἀπολοῦμαι *I shall perish*.
Aorist middle: ἀπωλόμην *I perished*.

The strong perfect ἀπόλωλα, also has an intransitive meaning, *I have perished*, or *I am lost*.

Vocabulary

θυσία (1) *sacrifice*
οἰκοδεσπότης (1) *master of a house, householder*
πλεονέκτης (1) *covetous man*
ἀνάγαιον (2) *upper room*
ἀσκός (2) *wineskin*
δηνάριον (2) *denarius, penny*
Ναζαρηνός (2) *Nazarene*
πόρνος (2) *fornicator*
ποτήριον (2) *cup*
δεξιός, ά, όν *right*

ἐκ δεξιῶν *on the right*
λοιπός, ή, όν *remaining, the rest*
μαλακός, ή, όν *soft*
ὕστερος, α, ον *later, latter*
ψυχρός, ά, όν *cold*
κοιμάω *put to sleep* (mid. and pass., *fall asleep*)
ὀφείλω, *owe, be obliged, ought*
ποῖος, α, ον (interr. adj.) *what kind of? what?*
ὕστερον (adv.) *later, afterwards*

EXERCISE 82

Translate into English:

1. παρῆσαν δέ τινες ἐν αὐτῷ τῷ καιρῷ ἀπαγγέλλοντες αὐτῷ περὶ τῶν Γαλιλαίων ὧν τὸ αἷμα Πειλᾶτος ἔμιξεν μετὰ τῶν θυσιῶν αὐτῶν. 2. γρηγορεῖτε οὖν, ὅτι οὐκ οἴδατε ποίᾳ ἡμέρᾳ ὁ Κύριος ὑμῶν ἔρχεται. ἐκεῖνο δὲ γινώσκετε ὅτι εἰ ᾔδει ὁ οἰκοδεσπότης ποίᾳ φυλακῇ ὁ κλέπτης ἔρχεται, ἐγρηγόρησεν ἄν. 3. ὕστερον δὲ ἔρχονται καὶ αἱ λοιπαὶ παρθένοι λέγουσαι Κύριε Κύριε, ἄνοιξον ἡμῖν. ὁ δὲ ἀποκριθεὶς εἶπεν Ἀμὴν λέγω ὑμῖν, οὐκ οἶδα ὑμᾶς. 4. νῦν δὲ ἔγραψα ὑμῖν μὴ συναναμίγνυσθαι ἐάν τις ἀδελφὸς ὀνομαζόμενος ᾖ πόρνος ἢ πλεονέκτης. 5. ὁ δὲ λέγει αὐταῖς Μὴ ἐκθαμβεῖσθε· Ἰησοῦν ζητεῖτε τὸν Ναζαρηνὸν τὸν ἐσταυρωμένον· ἠγέρθη, οὐκ ἔστιν ὧδε· ἴδε ὁ τόπος ὅπου ἔθηκαν αὐτόν. 6. καὶ ὃς ἐὰν ποτίσῃ ἕνα τῶν μικρῶν τούτων ποτήριον ψυχροῦ (sc. ὕδατος) μόνον εἰς ὄνομα μαθητοῦ, ἀμὴν λέγω ὑμῖν, οὐ μὴ ἀπολέσῃ τὸν μισθὸν αὐτοῦ.

EXERCISE 83

Translate into Greek:

1. They tried to prevent him.   2. Have you come to destroy us? 3. He that believes in Christ shall not perish.   4. We know who you are. 5. Repent; for the kingdom of God is at hand.   6. If you had sat on his right you would have been honoured.   7. You are doing what you ought not.   8. This servant owed five hundred pence, the other fifty.   9. If your son is dead, bring him to the disciples.   10. Ask the master of the house to show you the furnished upper room.

EXERCISE 84

Translate into Greek:

1. When he had cast out the husbandmen, he destroyed the vineyard. 2. When the crowds came to him, Jesus began to teach them as his custom was.   3. If they put new wine into old wineskins, the wineskins are rent. 4. One of them, seeing that he had been healed, turned back, glorifying God.   5. Peter swore that he did not know Jesus.   6. They did not go out to see a man clothed in soft clothing.   7. Many remain until this day, but some have fallen asleep.   8. We have seen his mighty deeds, and are witnesses of them.   9. Unless you repent, you will all perish likewise. 10. Nobody dared to tell what he had seen.

Chapter 30

COMPOUND VERBS

THE SIGNIFICANCE OF PREPOSITIONS IN COMPOUND VERBS

A preposition when used as a prefix in a compound verb often retains the meaning it has as a preposition, but it often has a different meaning.

1. *Prefix retaining prepositional meaning*

Instances are many: e.g. ἐκβάλλω *cast out*; προβαίνω *go forward*; κατατίθημι *lay down*; παρίστημι *place beside*; συνάγω *bring together*.

2. *Prefix bearing original adverbial meaning of preposition*

Some prepositions used as prefixes bear a meaning which they may once have had when used separately as adverbs, but have since lost. Thus:

> διά sometimes introduces the idea of *dividing*, suggesting that it once possessed the meaning *between*: διαδίδωμι *distribute*; διακρίνω *distinguish*; διασκορπίζω *scatter*; διασπάω *tear asunder*.
>
> ἀνά as a prefix often means *up*, a sense which it does not carry as a preposition: ἀναβαίνω *go up*; ἀναβλέπω *look up*; ἀναπηδάω *leap up*.

3. *Intensifying prefixes*

A number of prepositions are used as prefixes to intensify the meaning of the verb. Thus:

> ἀναιρέω *destroy utterly*; ἀποφεύγω *flee away, escape* (cf. φεύγω *flee*); ἐκθαυμάζω *wonder greatly*; διαπορέω *be in great perplexity* (but διά *through* = *thorough*, is naturally used in this intensive sense); ἐπιτελέω *accomplish*; ἐγκαταλείπω *abandon* (an intensifying of καταλείπω); παροξύνω *provoke to anger*; συναρπάζω *seize and carry away*.

Thus almost every Greek preposition can be employed with this intensifying effect. Compare the wide range of English prepositions which may be used with a verb with the same effect; "clear off", "eat up", "swallow down", "fall over".

4. *Prepositions which acquire special meanings as prefixes*

Certain prepositions are also used with special meanings in compounds:

ἀνά *again*, *afresh*, and sometimes *back*: ἀναβλέπω *recover sight*; ἀνασταυρόω *crucify afresh*; ἀναχωρέω *retire*; ἀνακυλίω *roll back*.

ἀπό *back*; ἀπαιτέω *demand back*; ἀποτίνω *pay back*; ἀποκρίνομαι literally: *decide for oneself back*, i.e. *answer*.

ἀπό also has a negative force, *un-*: ἀποκαλύπτω *uncover, reveal*; ἀποστεγάζω *unroof*.

μετά conveys the idea of *change* (Latin and English *trans-*): μεθερμηνεύω *translate*; μετοικίζω *cause to migrate*. Its prepositional meaning of *with* introduces into a compound verb the idea of *sharing*: μετέχω *partake of*; μεταδίδωμι *give a share of*.

παρά sometimes has the meaning of *mis-*, perhaps arising from its meaning of *beside*, since that which falls beside misses the mark: παρακούω *mis-hear, take no heed*; παραλογίζομαι *miscalculate*.

περί is sometimes the equivalent of *super-* or *sur-*: περιλείπομαι *survive*; περιποιέομαι *make to survive for oneself, gain*; περιεργάζομαι *be a busybody* (περί here conveys the idea of overdoing it).

πρός, normally meaning *to*, as a prefix sometimes has the closely allied meaning *in addition, besides*: προσδαπανάω *spend besides*; προσοφείλω *owe besides*.

ὑπό curiously means *away* in certain verbs: ὑπάγω *depart*; ὑποχωρέω *retire*; ὑποστρέφω *turn back, return*.

The Augment in Compound Verbs

The augment in historic past tenses of the indicative mood, as has already been stated (p. 66), is to be placed before the stem of the simplex, and not before the prefix. Thus ἐκβάλλω: aor. indic. act., ἐξέβαλον.

A few peculiarities are, however, to be noticed in the augmenting of the following verbs:

ἀνοίγω *open*, has a variety of augments. In the aor. act. ἀνέῳξα, ἤνοιξα, and ἠνέῳξα are variously found, the two last showing an anomalous augment of the prefix. So also in ἠνοίχθην and ἠνεῴχθην, which are found for the aor. pass. as well as ἀνεῴχθην (see also note on p. 123).

ἀποκαθίστημι *restore*: this double compound has a double augment, and we find ἀπεκατεστάθη for the 3rd pers. sing. aor. indic. pass.

(Mark 3.5), and ἀπεκατέστη for the 3rd pers. sing. str. aor. act. (Mark 8.25).

ἀφίημι *forgive*: the imperfect of this verb (or more strictly of its thematized form ἀφίω) appears with an augment before the prefix; thus, 3rd pers. sing. ἤφιεν (Mark 1.34 and 11.16).

διακονέω *serve*; this verb is not a compound with διά, but is derived from διάκονος *servant*. Its augment should, therefore, be placed at the beginning; e.g., in the imperfect, ἐδιακόνουν; but διηκόνουν is found, placing the augment as though δια- were a prefix.

εὐαγγελίζομαι *proclaim good tidings*: this verb also is not strictly a compound verb, being derived from εὐαγγέλιον. But an imperfect εὐηγγελιζόμην is found, the augment being placed as though εὐ- were a prefix.

καθέζομαι *sit*: this verb is a genuine compound of κατά and ἕζομαι. But here, probably because the simplex had become obsolete, and it was forgotten that the verb was a compound, the augment is placed at the beginning; imperfect: ἐκαθεζόμην.

καθεύδω *sleep*: like καθέζομαι this is a compound with simplex obsolete. Imperfect: ἐκάθευδον.

καθίζω *sit*: like the two preceding verbs this also is a compound with simplex obsolete. Aorist: ἐκάθισα.

προφητεύω *prophesy*: this is not a compound with πρό, but is derived from προφήτης. N.T. has augment at the beginning correctly, ἐπροφητεύσαμεν (Matt. 7.22). But inferior readings have προεφητεύσαμεν. (Classical Greek augments this verb anomalously in this way.)

THE USE OF PREPOSITIONS AFTER COMPOUND VERBS

1. The preposition used in the prefix may be repeated with a noun or pronoun in the appropriate case. Thus:

ἐκβάλλει ἐκ τοῦ θησαυροῦ (Matt. 13.52) *he bringeth out of his treasure*.
εἰσελθὼν εἰς Καφαρναούμ (Mark 2.1) *when he entered into Capernaum*.
τὸν κονιορτὸν ἀπὸ τῶν ποδῶν ὑμῶν ἀποτινάσσετε (Luke 9.5) *shake off the dust from your feet*.

2. The construction may follow without repetition of the prefix. Thus:

ἀπολέλυσαι τῆς ἀσθενείας σου (Luke 13.12) *thou art loosed from thine infirmity*.
κατηγορήσω ὑμῶν (John 5.45) *I will accuse you*.

3. The compound verb may have as its object a noun or pronoun without the preposition, and in a case different from that which the preposition in the prefix takes. Thus, verbs compounded with ἀντί are frequently followed by the dative rather than by the genitive:

ἀντιλέγει τῷ Καίσαρι (John 19.12) *he speaketh against Caesar.*
διήρχετο τὴν ʿΙερειχώ (Luke 19.1) *he was passing through Jericho.*
διεπορεύοντο τὰς πόλεις (Acts 16.4) *they went on their way through the cities.*

4. Sometimes the compound verb is followed by a preposition other than that in the prefix. This is often demanded by the sense of the sentence:

ἀνέβη ἀπὸ τοῦ ὕδατος (Matt. 3.16) *he went up from the water.*
ὕπαγε εἰς τὸν οἶκόν σου (Mark 2.11) *go unto thy house.*

But not infrequently another preposition is used where a repetition of the prefix would be expected:

ἀπέστειλαν ... ἐξ ʿΙεροσολύμων ἱερεῖς (John 1.19) *they sent from Jerusalem priests.*
ἐξερχόμενοι ἀπὸ τῆς πόλεως ἐκείνης (Luke 9.5) *when ye depart from that city.*

VERBS IN –μι (6) –εἶμι, κάθημαι

εἶμι *I shall go*, has ι- as its verb stem. The verb is common in classical Greek, but only has a few forms surviving in the N.T.

3rd pers. plur. pres. indic.: -ἴασι(ν) *they will go.*
3rd pers. sing. imperf. indic.: -ῄει *he was going.*
3rd pers. plur. imperf. indic.: -ῄεσαν *they were going.*
Infinitive: -ἰέναι.
Participle: -ἰών (gen. -ἰόντος) -ἰοῦσα, -ἰόν.

In the N.T. it is only found compounded:

ἄπειμι *depart.*
σύνειμι *come together.*

κάθημαι *I sit*, has ἡς- as its stem. Strictly this verb is a compound with prefix κατα-, but it is not treated as a compound, and the augment is placed before the prefix.
The 2nd pers. sing. pres. indic., κάθη, and the imperative, κάθου, are

from κάθομαι, another instance of -μι verbs being assimilated to -ω verb forms.

	Pres. indic.	Imperfect indic.	Pres. imperative	Pres. subjunctive
Sing. 1.	κάθημαι	ἐκαθήμην		καθῶμαι
2.	κάθη	ἐκάθησο	κάθου	καθῇ
3.	κάθηται	ἐκάθητο		καθῆται
Plur. 1.	καθήμεθα	ἐκαθήμεθα		καθώμεθα
2.	κάθησθε	ἐκάθησθε		καθῆσθε
3.	κάθηνται	ἐκάθηντο		καθῶνται

Present infinitive: καθῆσθαι.
Present participle: καθήμενος, καθημένη, καθήμενον.

The fut. indic. is καθήσομαι.

There is a compound, συνκάθημαι *sit together with*.

SEMITISMS

Languages can greatly influence one another, as is clearly illustrated by the impact which Norman French made upon Old English in the eleventh and twelfth centuries. Throughout its history English has borrowed from other languages; Greek, Latin, French, Dutch, even Turkish and Chinese.

Koine Greek from the end of the fourth century B.C. established itself as the language of trade and literature in the Semitic world. It is to be expected, therefore, that words, phrases, and idioms of Semitic languages would have made their way into Greek.

Semitisms abound in the New Testament. Their occurrence may be attributed to any of four causes:

1. That by the time the New Testament authors wrote, *Koine* Greek had already absorbed many Semitisms which were in use even by Greek writers who knew no Hebrew or any other Semitic language. Recently discovered papyrus fragments show many Semitic words and idioms used in the Greek of ordinary secular affairs.

2. That most of the New Testament writers were Jews, whose usual language of speech was Aramaic, and who had a considerable knowledge of the Hebrew Bible; and that consequently they introduced Semitisms, more or less unconsciously, into the Greek they wrote.

3. That the writers consciously modelled themselves on the style of the

Greek version of the Old Testament (the Septuagint) which often gives a very literal translation of the Hebrew.

4. That the New Testament writers incorporated documents which were translations of Aramaic into Greek, and which contained Semitisms in proportion to the literalness of the translation.

To decide which of these four causes operated in the case of the Semitisms of any one author, or any one passage, is a work of scholarship. Indeed the whole subject of the Semitisms in the New Testament is one for advanced study, for which, also, some knowledge of Hebrew and Aramaic is necessary. Here we can only draw the attention of the student to some of the more frequent Semitic points of style and idioms which are found in the New Testament, and to those which are most likely to cause puzzlement; and we must do so briefly, and without the full explanation which would demand some knowledge at least of Hebrew in the reader. The more advanced student may be referred to the Appendix "Semitisms in the New Testament" in J. H. Moulton's *Grammar of New Testament Greek*, vol. ii, and to the short account in chapter xxv, "Semitisms", in C. F. D. Moule's *An Idiom-Book of New Testament Greek*. Both of these give many references for further study.

1. *Parataxis*

A good Greek style welds the sentences together closely, and each sentence has one main verb, all other verbs being subordinated in adverbial clauses of one kind or another. On the other hand Hebrew tended to place main verbs side by side, joining them together with a conjunction. This is known as *parataxis*, from παρατάσσω *to set side by side*. St Mark has many examples of this. Two which occur early in the gospel are:

καὶ ἐξεπορεύετο πρὸς αὐτὸν πᾶσα ἡ ᾿Ιουδαία χώρα καὶ οἱ ῾Ιεροσολυμεῖται πάντες, καὶ ἐβαπτίζοντο ὑπ᾿ αὐτοῦ (Mark 1.5) *and there went out unto him all the country of Judaea, and all they of Jerusalem; and they were baptized of him.*

Here a more typical Greek style would have subordinated the ἐξεπορεύετο by means of a temporal clause or a participial clause.

καὶ ἦν ἐν τῇ ἐρήμῳ . . . καὶ ἦν μετὰ τῶν θηρίων, καὶ οἱ ἄγγελοι διηκόνουν αὐτῷ (Mark 1.13) *and he was in the wilderness . . . and he was with the wild beasts; and the angels ministered unto him.*

Here a more typical Greek style would, perhaps, have subordinated the first two clauses by employing the participle ὄντι in agreement with αὐτῷ.

2. Καὶ ἐγένετο

Hebrew frequently uses a verb meaning "it was so" or "it came to pass" preceded by the conjunction *and*, and followed by another main verb. This Semitism appears in the N.T. in three variations:

(*a*) καὶ ἐγένετο followed by another main verb in a past indicative tense:

καὶ ἐγένετο ἐν ἐκείναις ταῖς ἡμέραις ἦλθεν Ἰησοῦς (Mark 1.9) *and it came to pass in those days, (that) Jesus came.*

(*b*) καὶ ἐγένετο followed by καί and another main verb:

καὶ ἐγένετο ἐν μιᾷ τῶν ἡμερῶν καὶ αὐτὸς ἦν διδάσκων (Luke 5.17) *and it came to pass on one of (those) days, that he was teaching.*

(*c*) καὶ ἐγένετο followed by the accusative and infinitive:

καὶ ἐγένετο αὐτὸν ἐν τοῖς σάββασιν παραπορεύεσθαι (Mark 2.23) *and it came to pass, that he was going on the Sabbath day.*

This idiom is far more frequent in the Lucan writings than elsewhere: St Mark has only four examples of it.

3. Ἐν *with Articular Infinitive expressing Time*

In Hebrew the preposition most closely represented by ἐν is commonly used with an infinitive to express time; thus, "in the going", i.e. in the period of the going, and so "while he (she, etc.) went". This idiom is found in the N.T.:

ἐν τῷ σπείρειν (Mark 4.4) *as he sowed.*
βασανιζομένους ἐν τῷ ἐλαύνειν (Mark 6.48) *distressed in rowing.*

Again it is most frequent in the Lucan writings, and is often used together with the καὶ ἐγένετο construction:

καὶ ἐγένετο ἐν τῷ πορεύεσθαι ... καὶ αὐτὸς διήρχετο (Luke 17.11) *and it came to pass, as they were on the way ... that he was passing through. ...*

4. Προστίθημι *expressing Repetition*

In Hebrew the verb "to add" is used with an infinitive following to express the repetition of an action; thus, "he added to go", i.e. he went again. St Luke is evidently influenced by this in the following:

προσθεὶς εἶπεν παραβολήν (Luke 19.11) *he spake another parable* (R.V. translates literally, "he added and spake a parable").

καὶ προσέθετο ἕτερον πέμψαι δοῦλον (Luke 20.11) *and he sent yet another servant.*

5. *The Future Indicative expressing Command*

The part of the Hebrew verb most closely representing the Greek future indicative is used to express commands. This construction probably influences passages like the following:

> ἔσται ὑμῶν διάκονος (Mark 10.43) *let him be your minister* (rather than the "he shall be" of R.V.). Compare also in the next verse ἔσται πάντων δοῦλος *let him be servant of all.*
>
> εἴ τις θέλει πρῶτος εἶναι, ἔσται πάντων ἔσχατος (Mark 9.35) *if any man would be first, let him be last of all.*

In all these passages greater point is given by taking the futures as imperatives.

6. *Verb and Cognate Noun expressing Emphasis*

In Hebrew the form of the verb known as the infinitive absolute may be closely associated with another part of the same verb to express emphasis. This idiom appears in the N.T. in three main variations:

(*a*) A verb, together with a cognate noun in the dative:

> ἐπιθυμίᾳ ἐπεθύμησα (Luke 22.15) *I have greatly desired* (R.V. translates literally "with desire I have desired").
>
> ἀναθέματι ἀνεθεματίσαμεν ἑαυτούς (Acts 23.14) *we have made a solemn vow* (R.V. "We have bound ourselves under a great curse").

Mark 7.10 is similar: θανάτῳ τελευτάτω *let him surely die*, though here the noun is cognate in meaning with the verb, but not in root.

(*b*) A verb, together with a participle of the same verb:

> ἵνα βλέποντες βλέπωσιν καὶ μὴ ἴδωσιν, καὶ ἀκούοντες ἀκούωσιν καὶ μὴ συνιῶσιν (Mark 4.12). Here, perhaps the emphasis is best reproduced in English by a repetition of the verb: "that they may look and look, and yet not see, and listen and listen, and yet not understand".
>
> εὐλογῶν εὐλογήσω σε καὶ πληθύνων πληθυνῶ σε (Heb. 6.14) *I will abundantly bless thee, and will greatly multiply thee.*

(*c*) A verb, together with a cognate noun in the accusative:

> ἐφοβήθησαν φόβον μέγαν (Mark 4.41) *they feared exceedingly.*

ἐχάρησαν χαρὰν μεγάλην σφόδρα (Matt. 2.10) *they rejoiced exceedingly*. (Here R.V. translates more literally "they rejoiced with exceeding great joy").

Though Semitic influence is probable here, the cognate accusative is perfectly natural in classical Greek; see pp. 180f.

7. *Adjectival Genitive after* υἱός

The use of a genitive noun after υἱός or υἱοί often reproduces a closely corresponding Semitic idiom. The genitive noun has the force of an adjective:

οἱ υἱοὶ τοῦ νυμφῶνος (Mark 2.19) *the sons of the bridechamber*, i.e. the bridegroom's attendants.

So οἱ υἱοὶ τοῦ φωτός in Luke 16.8, *sons of light*, means enlightened men, those in whom the light of God shines; cf. τέκνα φωτός (Eph. 5.8).

8. *Redundant Pronoun after a Relative Pronoun*

The Hebrew relative is indeclinable and without gender, and therefore necessitates a personal pronoun in the clause which follows. This has influenced a few N.T. passages in which an unnecessary pronoun appears after a relative:

ἔρχεται ὁ ἰσχυρότερός μου . . . οὗ οὐκ εἰμὶ ἱκανὸς κύψας λῦσαι τὸν ἱμάντα τῶν ὑποδημάτων αὐτοῦ (Mark 1.7) *there cometh . . . he that is mightier than I, the latchet of whose shoes I am not worthy to stoop down and unloose.*

γυνὴ . . . ἧς εἶχεν τὸ θυγάτριον αὐτῆς πνεῦμα ἀκάθαρτον (Mark 7.25) *a woman, whose little daughter had an unclean spirit.*

In these two sentences αὐτοῦ and αὐτῆς are clearly redundant.

9. Εἰ *introducing a clause expressing Strong Denial*

The Hebrew word most closely corresponding to the Greek εἰ can introduce a clause expressing strong denial. This idiom appears to have influenced the following:

ἀμὴν λέγω ὑμῖν, εἰ δοθήσεται τῇ γενεᾷ ταύτῃ σημεῖον (Mark 8.12) *verily I say unto you, a sign shall by no means be given to this generation.*

εἰ εἰσελεύσονται εἰς τὴν κατάπαυσίν μου (Heb. 4.3) *they shall by no means enter into my rest.*

10. *Distributives expressed by Repetition*

The repetition of a numeral to denote distribution is a Semitic idiom:

αὐτοὺς ἀποστέλλειν δύο δύο (Mark 6.7) *to send them forth two by two*.

Similar, although not actually employing numerals, are:

ἀνακλιθῆναι πάντας συμπόσια συμπόσια (Mark 6.39) *all to sit down by companies*.

ἀνέπεσαν πρασιαὶ πρασιαί (Mark 6.40) *they sat down in ranks*.

Εἷς κατὰ εἷς in Mark 14.19 seems to be a mixture of this Semitism with the *Koine* Greek καθεῖς or καθ'εἷς *one by one, severally* (Rom. 12.5).

Vocabulary

ἀνατολή (1) *rising (of sun), east*
ἀγορά (1) *market-place*
παραγγελία (1) *command, charge*
νῖκος, ους, τό (3) *victory*
'Ιερειχώ, ἡ (indecl.) *Jericho*
ἄνομος, ον *lawless, wicked*
κενός, ή, όν *empty*

σωματικός, ή, όν *bodily*
χρηστός, ή, όν *kind*
ἀδικέω *injure, hurt*
ἐξαποστέλλω *send forth*
ἐπαιτέω *beg*
κατάγω *bring down*
καταπίνω *drink down, swallow up*

EXERCISE 85

Translate into English:

1. ἐγένετο δὲ ἐν τῷ ἐγγίζειν αὐτὸν εἰς 'Ιερειχὼ τυφλός τις ἐκάθητο παρὰ τὴν ὁδὸν ἐπαιτῶν. 2. καὶ εἶδον ἄλλον ἄγγελον ἀναβαίνοντα ἀπὸ ἀνατολῆς ἡλίου, ἔχοντα σφραγῖδα Θεοῦ ζῶντος, καὶ ἔκραξεν φωνῇ μεγάλῃ τοῖς τέσσαρσιν ἀγγέλοις οἷς ἐδόθη αὐτοῖς ἀδικῆσαι τὴν γῆν καὶ τὴν θάλασσαν. 3. παραγγελίᾳ παρηγγείλαμεν ὑμῖν μὴ διδάσκειν ἐπὶ τῷ ὀνόματι τούτῳ, καὶ ἰδοὺ πεπληρώκατε τὴν 'Ιερουσαλὴμ τῆς διδαχῆς ὑμῶν. 4. καὶ συνάγεται πρὸς αὐτὸν ὄχλος πλεῖστος, ὥστε αὐτὸν εἰς πλοῖον ἐμβάντα καθῆσθαι ἐν τῇ θαλάσσῃ, καὶ πᾶς ὁ ὄχλος πρὸς τὴν θάλασσαν ἐπὶ τῆς γῆς ἦσαν. 5. καὶ ἐπηρώτων αὐτὸν οἱ ὄχλοι λέγοντες Τί οὖν ποιήσωμεν; ἀποκριθεὶς δὲ ἔλεγεν αὐτοῖς Ὁ ἔχων δύο χιτῶνας μεταδότω τῷ μὴ ἔχοντι, καὶ ὁ ἔχων βρώματα ὁμοίως ποιείτω. 6. οἱ δὲ γεωργοὶ ἐξαπέστειλαν αὐτὸν δείραντες κενόν. καὶ προσέθετο ἕτερον πέμψαι δοῦλον· οἱ δὲ κἀκεῖνον δείραντες καὶ ἀτιμάσαντες ἐξαπέστειλαν κενόν. καὶ προσέθετο τρίτον πέμψαι.

EXERCISE 86

Translate into Greek:

1. Let us place these children beside that table.   2.   He distributes the bread to those who sit at meat.   3. By our sins the Son of man is crucified afresh.   4. When he stretched out his hands they were restored immediately.   5. The women who went up with Jesus to Jerusalem ministered to him and to the apostles.   6. Those who speak against Caesar will be utterly destroyed.   7. While the crowd was coming together, the robber escaped into the street.   8. Bethlehem, being interpreted, means House of Bread.   9. Sit together with us here, for we want to talk with you.   10. And it came to pass that as they were wondering greatly at these things the angel departed from them.

EXERCISE 87

Translate into Greek:

1. If you had said that you were coming, the doors would have been opened.   2. The kind master gave yet another penny to the workmen. 3. The children walked in fours from the synagogue to the market-place. 4. For this reason they were exceedingly sorrowful, and returned to their houses.   5. When Christ rose death was swallowed up in victory.   6. The Lord will utterly destroy the lawless one who shall be revealed in the last days.   7. Even if you do not like him you ought to pay back what you owe.   8. We went up the mountain to bring the sheep down.   9. The stone being rolled back, the women entered and saw a young man sitting clothed with a white garment.   10. And it came to pass that when Jesus was baptized the heavens were opened, and John, who stood by, saw the Holy Spirit descending in a bodily shape upon him.

APPENDIX 1

ACCENTS (See also p. 11)

THE MAIN RULES

1. *The acute accent* may be written over long or short syllables; the circumflex only over a syllable which contains a vowel long by nature (viz. η, ω) or a diphthong.

2. *The acute accent* may stand on any one of the last three syllables of a word:

(*a*) It may stand on the last syllable, when it is called *oxytone* (i.e. *sharp-toned*). The acute on the last syllable is, however, turned into a grave, except when an enclitic (see below) or a stop follows. The accent on the interrogative τίς, τί does not turn into a grave. A grave accent on the last syllable (it can stand nowhere else) is called *barytone*.

(*b*) It may stand on the last syllable but one. This is called *paroxytone*. If the last syllable is long the acute accent can be placed no further back than the penultimate syllable.

(*c*) It may stand on the last syllable but two, provided that the last syllable of the word is short. It is then called *proparoxytone*.

3. *The circumflex accent* may stand on either of the last two syllables of a word:

(*a*) It may stand on the last syllable. If the last syllable is long a circumflex can stand in no other position than over it. This is called *perispomenon*.

(*b*) It may stand on the last syllable but one, provided that the last syllable is short. It is then called *properispomenon*.

The following are examples (all from Mark 1):

oxytone: ὁδόν, ἀγαπητός, αὐτούς.
paroxytone: εὐαγγελίου, προφήτῃ, ἀποστέλλω, προσώπου.
proparoxytone: γέγραπται, ἑτοιμάσατε, ἐγένετο, βάπτισμα.
perispomenon: Ἰησοῦ, Χριστοῦ, ποταμῷ, ἁλεεῖς.
properispomenon: βοῶντος, ἦλθεν, πνεῦμα, ἦσαν.
barytone: ἀρχὴ, in verse 1 (no stop or enclitic following).

ENCLITICS

There are some words which throw their accents back on to the preceding word if it is able to receive it. Enclitics are words which, for the most part, have no full meaning of their own, but depend on a preceding word to make the significance clear (enclitic from ἐγκλίνω *lean on*).

The enclitics are the 1st and 2nd personal pronouns singular in the oblique cases, με, μου, μοι and σε, σου, σοι; the indefinite pronoun τις, τι, in all cases; the indefinite adverbs που *anywhere*, ποθεν *from anywhere*, ποτε *ever*, πω *yet*, πως *anyhow*; the particles γε *at least*, τε *and*, and τοι *truly*; the pres. indic. of εἰμί *I am*, except for εἶ, the 2nd pers. sing.; the pres. indic. of φημί *I say*, except for φῆς, the 2nd pers. sing.

The rules of accentuation when enclitics are involved are as follows:

1. When the last syllable of the word preceding an enclitic has an accent, the accent of the enclitic disappears. If the accent on the preceding syllable is a grave it is changed back to the original acute:

oἱ ἀδελφοί μου (Mark 3.34) *my brethren.*
oὗτοι δέ εἰσιν (Mark 4.15) *and these are.*

2. When the last syllable of the word preceding the enclitic has no accent of its own, it receives the accent of the enclitic as an acute accent, but not when this would give two acute accents on successive syllables:

ὁ ἰσχυρότερός μου (Mark 1.7) *he that is stronger than I.*
ὀπίσω μου (not ὀπίσώ μου: Mark 1.7) *after me.*
oὗτοί εἰσιν (Mark 4.16) *these are.*

3. An enclitic of two syllables retains its accent on the second syllable if the preceding word has an acute on the last syllable but one:

ἐν οἴκῳ ἐστίν (Mark 2.1) *he is in the house* (acute because full stop follows).
ἄλλοι εἰσὶν ... (Mark 4.18) *others are....*
λόγοι τινές *certain words.*
λόγων τινῶν *of certain words.*

PROCLITICS

Certain words also possess no accent having no full meaning of their own and deriving their full significance from what *follows*. They are sometimes called *atona* (without accent). The most important proclitics are four forms of the article, ὁ, ἡ, οἱ, αἱ; the prepositions εἰς, ἐκ or ἐξ, and ἐν; the conjunctions εἰ *if*, and ὡς *as, when, that*; and the negative οὐ, unless it is followed by a stop, when it is accented, οὔ.

A proclitic, however, receives an accent if it is followed by an enclitic: εἴ τις θέλει (Mark 9.35) *if any man will*.

THE ACCENTUATION OF VERBS

The general rule is that the acute accent is placed as far back as the usual rules permit, i.e. on the third syllable from the end when the last syllable is short, and on the second syllable from the end when the last syllable is long:

λύομεν, ἐλύετε.
πρεσβεύω, ἐλύθην.

It is to be noted that final -αι and -οι are regarded as short syllables except in the optative (which is rare in the N.T.). Thus: λύομαι; but the 3rd pers. sing. pres. opt. of πρεσβεύω would be πρεσβεύοι, not πρέσβευοι.

Where the last syllable is short and the preceding syllable long, the latter takes a circumflex if it has an accent at all. Thus εἶπον, εὗρε. But ἤκουον not ἠκοῦον, because the last syllable being short, the accent must be on the third syllable from the end if there is a third syllable.

THE ACCENT ON COMPOUND VERBS

1. Where there is an augment or reduplication, the accent cannot precede the augment or reduplication. Thus in ἄπαγε (2nd pers. sing. pres. imper. of ἀπάγω) the accent goes three syllables back, there being no augment. But in ἀπῆγον the accent is halted at the augment, and is a circumflex since the last syllable is short and the second from last long.

2. Where the prefix has two syllables (e.g. ἐπί, ἀπό, παρά) the accent may go as far back as the syllable immediately preceding the verbal part. Thus ἀπόδος (2nd pers. sing. aor. imper. act. of ἀποδίδωμι), not ἄποδος.

3. In the infinitives and participles of compound verbs in -μι the accent may not go back to the prefix at all. Thus ἀποδοῦναι, not ἀπόδουναι; ἀποδούς, not ἄποδους. But in the indicative of -μι verbs the accent goes back as far as possible, e.g. ἔξεστιν; so also in the imperative, ἄφες.

Irregular Accents on certain Verb Forms

Imperatives

The 2nd pers. sing. str. aor. mid. imper. (which is not common) is peri-spomenon, πιθοῦ.

The following five common 2nd pers. sing. str. aor. imperatives act. are oxytone; ἐλθέ, εὑρέ, ἰδέ, λαβέ, εἰπέ. But when compounded the accent follows the normal rule, ἄπελθε.

Subjunctives

The aorist passive subjunctive of all verbs, and the pres. and str. aor. act. and mid. subjunctive of verbs in -μι (except those in -νυμι) are accented as though they were contracted forms (see below). Thus, λυθῶ, λυθῇς, λυθῇ, λυθῶμεν, λυθῆτε, λυθῶσι (as though from λυθέω, λυθέῃς, etc.). So also ἱστῶ, ἱστῇς, etc., στῶ, στῇς, etc., τιθῶμαι, τιθῇ, τιθῆται, τιθώμεθα, τιθῆσθε, τιθῶνται (pres. subj. mid.), and θῶμαι, θῇ, θῆται, θώμεθα, θῆσθε, θῶνται (str. aor. subj. mid.).

Infinitives

The last syllable but one is accented (circumflex if long, acute if short) in the following infinitives:

Wk. aor. act. infin.: τιμῆσαι, βαπτίσαι.
Str. aor. mid. infin. (not common): βαλέσθαι.
Perf. mid. and pass. infin.: λελύσθαι, τετιμῆσθαι.
All infinitives ending in -ναι: λελυκέναι, ἱστάναι, λυθῆναι, στῆναι.

The str. aor. act. infin. is perispomenon: βαλεῖν, λιπεῖν.

Participles

The following are oxytone:

str. aor. act. part.: βαλών.

All participles ending in -ς declined in the 3rd declension, with the exception of the wk. aor. act. participle: thus, λελυκώς, λυθείς, ἱστάς, διδούς, but λύσας, στήσας.

The perfect middle and passive participle is paroxytone, λελυμένος, βεβλημένος. The verbal adjective in -τεος is also paroxytone: λυτέος.

Participles are like adjectives in that in the fem. and neut. and in the oblique cases the accent remains on the same syllable as in the nom. sing. masc. so far as the general laws allow. It will become circumflex when the last syllable is short and the accent is on the penultimate syllable being

long. Thus, λελυκώς, λελυκυῖα, λελυκός; genitive: λελυκότος, λελυ-
κυίας, λελυκότος; λυθείς, λυθεῖσα, λυθέν; genitive: λυθέντος, λυθείσης,
λυθέντος.

Accents on Contracted Verbs

The accentuation of contracted verbs will hold no difficulty if it is
remembered that the ordinary rules of accentuation are applied to the
uncontracted form. The effect of contraction on the accent is described
on p. 111.

THE ACCENTUATION OF NOUNS AND ADJECTIVES

The student is advised not to spend much time in learning the accents
of nouns and adjectives at any early stage in his study of the Greek
language. But the following points can be quickly learned and applied:

1. The position of the accent on the nom. sing. of nouns and adjectives
cannot be found by the application of any rules. Compare ἄγγελος and
ἀριθμός; μόνος and σοφός. Generally speaking, the accent is placed as
near to the beginning of the word as the ordinary rules allow; but there
are very many exceptions.

2. In the oblique cases the accent falls on the same syllable as in the nom.
sing. so far as the ordinary rules permit. Note that final -αι and -οι are
regarded as short syllables. Thus ἄνθρωπος: plur. ἄνθρωποι. But

    (*a*) The gen. plur. of all 1st decl. nouns is perispomenon: κεφαλή,
    κεφαλῶν; ἐπαγγελία, ἐπαγγελιῶν; ὥρα, ὡρῶν.
    (*b*) The gen. and dat. sing. and plur. of words which in the nom. sing.
    are accented on the last syllable are perispomenon. Thus, ἐντολή:
    gen., ἐντολῆς, ἐντολῶν; dat., ἐντολῇ, ἐντολαῖς; ἀδελφός: gen.,
    ἀδελφοῦ, ἀδελφῶν; dat., ἀδελφῷ, ἀδελφοῖς.

3. Nouns of two syllables ending in -αρα, -ευρα, -ουρα, and all nouns
ending in -ορα are oxytone. Thus χαρά *joy*; πλευρά *side*; οὐρά *tail*;
διασπορά *dispersion*.

4. All nouns ending in -ευς are oxytone. Thus γραμματεύς *scribe*;
βασιλεύς *king*.

5. Adjectives ending in -ικος and -υς are also oxytone. Thus προ-
βατικός, ή, όν *of sheep*; βαρύς, εῖα, ύ *heavy*.

6. Nouns ending in -εια, derived from verbs ending in -ευω, are
paroxytone. Thus βασιλεία *kingdom* (from βασιλεύω); πρεσβεία *em-
bassy* (from πρεσβεύω).

7. 3rd decl. words of one syllable have the accent on the last syllable of the gen. and dat. sing. and plur. Thus νύξ *night*; gen., νυκτός, νυκτῶν; dat., νυκτί, νυξί; ποῦς *foot*: gen., ποδός, ποδῶν; dat. ποδί, ποσί. Exceptions are πᾶς *all*, which accents the *stem* in the gen. and dat. plur. πάντων, πᾶσι; but sing., παντός, παντί; παῖς *servant*, and οὖς *ear*, accent the stem in the gen. plur. παίδων, ὤτων; but παισί(ν), ὠσί(ν).

8. In 3rd decl. nouns like πόλις the -εως and -εων of the gen. were apparently regarded as one syllable. Thus πόλεως, πόλεων.

9. Contracted adjectives of the 2nd and 1st decl. have a circumflex on the last syllable throughout. Thus ἀργυροῦς, ἀργυρᾶ, ἀργυροῦν, and so in the oblique cases.

APPENDIX 2

WORDS DIFFERING IN ACCENTS OR BREATHINGS

The following is a list of words with similar spelling but differing in accent or breathing which the student should note:

ἁγία nom. sing. fem. of ἅγιος *holy*.

ἅγια nom., voc., acc., plur. neut. of ἅγιος.

ἄγων nom. masc. sing. pres. participle of ἄγω *I lead*.

ἀγών (ὁ) 3rd decl. noun, *contest*. Gen. ἀγῶνος.

ἀλλά conjunction, *but*.

ἄλλα nom. and acc. neut. plur. of adj. ἄλλος *other*.

ἄρα particle, *then*.

ἆρα interrogative particle.

ἀρά (ἡ) 1st decl. noun, *curse*. Gen. ἀρᾶς.

αὐτή nom. fem. sing. of pronoun αὐτός *he, self*.

αὕτη nom. fem. sing. of pronoun οὗτος *this*.

αὐταί nom. fem. plur. of αὐτός.

αὗται nom. fem. plur. of οὗτος.

δώῃ 3rd pers. sing. aor. subj. of δίδωμι *I give*.

δῴη 3rd pers. sing. aor. opt. of δίδωμι.

εἰ conjunction, *if*.

εἶ 2nd pers. sing. pres. indic. of εἰμί *I am*.

εἶπε 3rd pers. sing. aor. indic. act., *he said*

εἰπέ 2nd pers. sing. aor. imper., *say*.

εἰς preposition, *to*.

εἷς masc. of adjective, *one*.

ἔξω adverb, *outside*.

ἕξω 1st pers. sing. fut. indic. of ἔχω *I have*.

ἐν preposition, *in*.

ἕν nom. and acc. neut. of εἷς *one*.

ἡ nom. fem. sing. of the article, ὁ, *the*.

ἥ nom. fem. sing. of the relative pronoun ὅς *who*.

ἤ conjunction, *or, than*.

ᾖ 3rd pers. sing. subjunctive of εἰμί *I am*.

ᾗ dat. fem. sing. of relative pronoun ὅς *who*.

ἦν 1st pers. sing. imperf. indic. of εἰμί *I am*.

ἤν another form of the conditional conjunction ἐάν *if.*
ἥν acc. fem. sing. of the relative pronoun ὅς *who.*
ὁ nom. masc. sing. of the article, *the.*
ὅ nom. and acc. neut. sing. of the relative pronoun ὅς *who.*
ὄν nom. and acc. neut. sing., pres. participle of εἰμί *I am.*
ὅν acc. masc. sing. of the relative pronoun ὅς *who.*
οὐ negative particle, *not.*
οὗ gen. masc. and neut. sing. of the relative pronoun ὅς *who.*
πόσιν acc. sing. of 3rd decl. noun πόσις (ἡ) (gen. πόσεως), *drinking.*
ποσίν dat. plur. of 3rd decl. noun ποῦς (ὁ) (gen. ποδός), *foot.*
πότε interrogative adverb, *when?*
ποτέ indefinite particle, *at some time.*
ποῦ interrogative adverb, *where?*
που indefinite particle, *somewhere, anywhere.*
πῶς interrogative adverb, *how?*
πως indefinite particle, *somehow.*
ταῦτα nom. and acc. neut. plur. of the pronoun οὗτος *this.*
ταὐτά contracted form of τὰ αὐτά (neut. plur. of αὐτός) *the same things.*
τίς, τί interrogative pronoun, *who? what?*
τις, τι indefinite pronoun, *a certain man, a certain thing, something.*
φίλων gen. masc. fem. and neut. plur. of φίλος *friendly.*
φιλῶν nom. masc. sing. present participle of φιλῶ *I love.*
χειρῶν gen. plur. of χείρ *hand.*
χείρων comparative adjective, *worse.*
ὦ ⎱ with vocative, o.
ὦ ⎰ 1st pers. sing. subjunctive of εἰμί *I am.*
ᾧ dat. masc. and neut. sing. of relative pronoun ὅς *who.*
ὤν nom. masc. sing., present participle of εἰμί *I am.*
ὧν gen. masc. fem., and neut. plur. of the relative pronoun ὅς *who.*

The distinction between verbal forms with similar spelling but different accents should be studied in the conjugations given in Appendix 4; e.g. λῦσαι, λύσαι; τιμᾷ, τίμα; φιλεῖ, φίλει; δήλου, δηλοῦ.

DECLENSION OF NOUNS, ADJECTIVES AND PRONOUNS

NOUNS

FIRST DECLENSION

Feminine

	ἡ ἀρχή *beginning*	ἡ ἡμέρα *day* α pure	ἡ δόξα *glory* α impure
Sing. N.	ἀρχή	ἡμέρα	δόξα
V.	ἀρχή	ἡμέρα	δόξα
A.	ἀρχήν	ἡμέραν	δόξαν
G.	ἀρχῆς	ἡμέρας	δόξης
D.	ἀρχῇ	ἡμέρᾳ	δόξῃ
Plur. N., V.	ἀρχαί	ἡμέραι	δόξαι
A.	ἀρχάς	ἡμέρας	δόξας
G.	ἀρχῶν	ἡμερῶν	δοξῶν
D.	ἀρχαῖς	ἡμέραις	δόξαις

Masculine

	ὁ προφήτης *prophet*	ὁ νεανίας *a youth*	ὁ βορρᾶς *north wind*
Sing. N.	προφήτης	νεανίας	βορρᾶς
V.	προφῆτα	νεανία	βορρᾶ
A.	προφήτην	νεανίαν	βορρᾶν
G.	προφήτου	νεανίου	βορρᾶ
D.	προφήτῃ	νεανίᾳ	βορρᾷ
Plur. N., V.	προφῆται	νεανίαι	
A.	προφήτας	νεανίας	
G.	προφητῶν	νεανιῶν	
D.	προφήταις	νεανίαις	

SECOND DECLENSION

	Masculine (a few nouns in -ος are fem.) ὁ λόγος *word*	Neuter τὸ ἔργον *work*
Sing. N.	λόγος	ἔργον
V.	λόγε	ἔργον
A.	λόγον	ἔργον
G.	λόγου	ἔργου
D.	λόγῳ	ἔργῳ
Plur. N., V.	λόγοι	ἔργα
A.	λόγους	ἔργα
G.	λόγων	ἔργων
D.	λόγοις	ἔργοις

THIRD DECLENSION
Masculine or Feminine

	ὁ φύλαξ *guard* Stem ending in guttural, φυλακ-	ἡ σάλπιγξ *trumpet* Stem ending in guttural, σαλπιγγ-	ὁ Ἄραψ *Arab* Stem ending in labial, Ἀραβ-	ὁ (ἡ) παῖς *boy, girl* Stem ending in dental, παιδ-
Sing. N.	φύλαξ	σάλπιγξ	Ἄραψ	παῖς
V.	φύλαξ	σάλπιγξ	Ἄραψ	παῖς
A.	φύλακα	σάλπιγγα	Ἄραβα	παῖδα
G.	φύλακος	σάλπιγγος	Ἄραβος	παιδός
D.	φύλακι	σάλπιγγι	Ἄραβι	παιδί
Plur. N., V.	φύλακες	σάλπιγγες	Ἄραβες	παῖδες
A.	φύλακας	σάλπιγγας	Ἄραβας	παῖδας
G.	φυλάκων	σαλπίγγων	Ἀράβων	παίδων
D.	φύλαξι(ν)	σάλπιγξι(ν)	Ἄραψι(ν)	παισί(ν)

	ὁ ἄρχων *ruler* Stem ending in dental, ἀρχοντ-	ὁ ὀδούς *tooth* Stem ending in dental, ὀδοντ-	ὁ ἱμάς *strap* Stem ending in dental, ἱμαντ-	ὁ ποιμήν *shepherd* Stem ending in nasal, strong flexion, ποιμεν-
Sing. N.	ἄρχων	ὀδούς	ἱμάς	ποιμήν
V.	ἄρχων	ὀδούς	ἱμάς	ποιμήν
A.	ἄρχοντα	ὀδόντα	ἱμάντα	ποιμένα
G.	ἄρχοντος	ὀδόντος	ἱμάντος	ποιμένος
D.	ἄρχοντι	ὀδόντι	ἱμάντι	ποιμένι
Plur. N., V.	ἄρχοντες	ὀδόντες	ἱμάντες	ποιμένες
A.	ἄρχοντας	ὀδόντας	ἱμάντας	ποιμένας
G.	ἀρχόντων	ὀδόντων	ἱμάντων	ποιμένων
D.	ἄρχουσι(ν)	ὀδοῦσι(ν)	ἱμᾶσι(ν)	ποιμέσι(ν)

	ὁ ἡγεμών *leader* Stem ending in nasal, strong flexion, ἡγεμον-	ὁ Ἕλλην *Greek* Stem ending in nasal, no strong flexion, Ἑλλην-	ὁ αἰών *an age* Stem ending in nasal, no strong flexion, αἰων-	ὁ πατήρ *father* Stem ending in liquid, strong flexion, πατερ-
Sing. N.	ἡγεμών	Ἕλλην	αἰών	πατήρ
V.	ἡγεμών	Ἕλλην	αἰών	πάτερ
A.	ἡγεμόνα	Ἕλληνα	αἰῶνα	πατέρα
G.	ἡγεμόνος	Ἕλληνος	αἰῶνος	πατρός
D.	ἡγεμόνι	Ἕλληνι	αἰῶνι	πατρί
Plur. N., V.	ἡγεμόνες	Ἕλληνες	αἰῶνες	πατέρες
A.	ἡγεμόνας	Ἕλληνας	αἰῶνας	πατέρας
G.	ἡγεμόνων	Ἑλλήνων	αἰώνων	πατέρων
D.	ἡγεμόσι(ν)	Ἕλλησι(ν)	αἰῶσι(ν)	πατράσι(ν)

	ὁ ἀνήρ *man* Stem ending in liquid, strong flexion, ἀνερ-	ὁ ῥήτωρ *orator* Stem ending in liquid, partial strong flexion, ῥητορ-	ὁ σωτήρ *saviour* Stem ending in liquid, no strong flexion, σωτηρ-	ἡ χείρ *hand* Stem ending in liquid, no strong flexion, χειρ-
Sing. N.	ἀνήρ	ῥήτωρ	σωτήρ	χείρ
V.	ἄνερ	ῥῆτορ	σωτήρ	χείρ
A.	ἄνδρα	ῥήτορα	σωτῆρα	χεῖρα
G.	ἀνδρός	ῥήτορος	σωτῆρος	χειρός
D.	ἀνδρί	ῥήτορι	σωτῆρι	χειρί
Plur. N., V.	ἄνδρες	ῥήτορες	σωτῆρες	χεῖρες
A.	ἄνδρας	ῥήτορας	σωτῆρας	χεῖρας
G.	ἀνδρῶν	ῥητόρων	σωτήρων	χειρῶν
D.	ἀνδράσι(ν)	ῥήτορσι(ν)	σωτήρσι(ν)	χερσί(ν)

	ὁ μάρτυς *witness* Stem ending in liquid, no strong flexion, μαρτυρ-	ἡ πόλις *city* Stem ending in ι, changing to ε before vowels, and in dat. plur., πολι- πολε-	ὁ πῆχυς *cubit* Stem ending in υ, changing to ε before vowels, and in dat. plur., πηχυ-, πηχε-	ὁ ἰχθύς *fish* Stem ending in υ (unchanging) ἰχθυ-
Sing. N.	μάρτυς	πόλις	πῆχυς	ἰχθύς
V.	μάρτυς	πόλι		ἰχθύ
A.	μάρτυρα	πόλιν	πῆχυν	ἰχθύν
G.	μάρτυρος	πόλεως	πήχεως or πήχεος	ἰχθύος
D.	μάρτυρι	πόλει	πήχει	ἰχθύι
Plur. N., V.	μάρτυρες	πόλεις	πήχεις	ἰχθύες
A.	μάρτυρας	πόλεις	πήχεις	ἰχθύας
G.	μαρτύρων	πόλεων	πηχῶν or πήχεων	ἰχθύων
D.	μάρτυσι(ν)	πόλεσι(ν)	πήχεσι(ν)	ἰχθύσι(ν)

	ὁ βασιλεύς *king* Stem ending in ευ, βασιλευ-	ὁ βοῦς *ox* Stem ending in ου, βου-
Sing. N.	βασιλεύς	βοῦς
V.	βασιλεῦ	βοῦ
A.	βασιλέα	βοῦν
G.	βασιλέως	βοός
D.	βασιλεῖ	βοΐ
Plur. N., V.	βασιλεῖς	βόες
A.	βασιλεῖς	βόας
G.	βασιλέων	βοῶν
D.	βασιλεῦσι(ν)	βουσί(ν)

Neuter

	τὸ γράμμα *letter* Nom. in -α, stem ending in ατ- γραμματ-	τὸ κέρας *horn* Nom. in -ας, stem ending in ατ- κερατ-	τὸ γένος *race* Nom. in -ος, stem ending in εσ- γενεσ-
Sing. N.	γράμμα	κέρας	γένος
V.	γράμμα	κέρας	γένος
A.	γράμμα	κέρας	γένος
G.	γράμματος	κέρατος	γένους
D.	γράμματι	κέρατι	γένει
Plur. N., V.	γράμματα	κέρατα	γένη
A.	γράμματα	κέρατα	γένη
G.	γραμμάτων	κεράτων	γενῶν
D.	γράμμασι(ν)	κέρασι(ν)	γένεσι(ν)

The Article

	Masculine	*Feminine*	*Neuter*
Sing. N.	ὁ	ἡ	τό
A.	τόν	τήν	τό
G.	τοῦ	τῆς	τοῦ
D.	τῷ	τῇ	τῷ
Plur. N.	οἱ	αἱ	τά
A.	τούς	τάς	τά
G.	τῶν	τῶν	τῶν
D.	τοῖς	ταῖς	τοῖς

ADJECTIVES

Types are illustrated by participles where necessary.

SECOND AND FIRST DECLENSIONS:

THREE TERMINATIONS

	ἀγαθός, ή, όν *good*			ἅγιος, ᾱ, ον *holy* Feminine has α pure		
	Masculine	*Feminine*	*Neuter*	*Masculine*	*Feminine*	*Neuter*
Sing. N.	ἀγαθός	ἀγαθή	ἀγαθόν	ἅγιος	ἁγία	ἅγιον
V.	ἀγαθέ	ἀγαθή	ἀγαθόν	ἅγιε	ἁγία	ἅγιον
A.	ἀγαθόν	ἀγαθήν	ἀγαθόν	ἅγιον	ἁγίαν	ἅγιον
G.	ἀγαθοῦ	ἀγαθῆς	ἀγαθοῦ	ἁγίου	ἁγίας	ἁγίου
D.	ἀγαθῷ	ἀγαθῇ	ἀγαθῷ	ἁγίῳ	ἁγίᾳ	ἁγίῳ
Plur. N., V.	ἀγαθοί	ἀγαθαί	ἀγαθά	ἅγιοι	ἅγιαι	ἅγια
A.	ἀγαθούς	ἀγαθάς	ἀγαθά	ἁγίους	ἁγίας	ἅγια
G.	ἀγαθῶν	ἀγαθῶν	ἀγαθῶν	ἁγίων	ἁγίων	ἁγίων
D.	ἀγαθοῖς	ἀγαθαῖς	ἀγαθοῖς	ἁγίοις	ἁγίαις	ἁγίοις

CONTRACTED ADJECTIVES

	διπλοῦς, ῆ, οῦν *double* Stem διπλο-			ἀργυροῦς, ᾶ, οῦν *of silver* Stem ἀργυρε-; fem. has α pure		
	Masculine	*Feminine*	*Neuter*	*Masculine*	*Feminine*	*Neuter*
Sing. N.	διπλοῦς	διπλῆ	διπλοῦν	ἀργυροῦς	ἀργυρᾶ	ἀργυροῦν
V.	διπλοῦς	διπλῆ	διπλοῦν	ἀργυροῦς	ἀργυρᾶ	ἀργυροῦν
A.	διπλοῦν	διπλῆν	διπλοῦν	ἀργυροῦν	ἀργυρᾶν	ἀργυροῦν
G.	διπλοῦ	διπλῆς	διπλοῦ	ἀργυροῦ	ἀργυρᾶς	ἀργυροῦ
D.	διπλῷ	διπλῇ	διπλῷ	ἀργυρῷ	ἀργυρᾷ	ἀργυρῷ
Plur. N., V.	διπλοῖ	διπλαῖ	διπλᾶ	ἀργυροῖ	ἀργυραῖ	ἀργυρᾶ
A.	διπλοῦς	διπλᾶς	διπλᾶ	ἀργυροῦς	ἀργυρᾶς	ἀργυρᾶ
G.	διπλῶν	διπλῶν	διπλῶν	ἀργυρῶν	ἀργυρῶν	ἀργυρῶν
D.	διπλοῖς	διπλαῖς	διπλοῖς	ἀργυροῖς	ἀργυραῖς	ἀργυροῖς

Here the termination follows ο as in διπλο-ος, or ε as in ἀργυρε-ος and χρυσε-ος *golden*, and contraction takes place. Note the regular rules of contraction (see p. 10) do not apply throughout. If they did we should have διπλῶ for διπλῆ, διπλῶ for διπλᾶ, and ἀργυρῆ for ἀργυρᾶ. But assimilation to the ἀγαθός type takes place.

Second Declension:

TWO TERMINATIONS

	ἀδύνατος, ον *impossible*	
	Masc. and fem.	*Neut.*
Sing. N.	ἀδύνατος	ἀδύνατον
V.	ἀδύνατε	ἀδύνατον
A.	ἀδύνατον	ἀδύνατον
G.	ἀδυνάτου	ἀδυνάτου
D.	ἀδυνάτῳ	ἀδυνάτῳ
Plur. N., V.	ἀδύνατοι, etc.	ἀδύνατα, etc.

Third and First Declensions:

THREE TERMINATIONS

	πᾶς, πᾶσα, πᾶν *all, every* Stem ending in αντ-			λυθείς, εῖσα, έν *loosed* (aor. part. pass.). Stem ending in εντ-		
	Masculine	*Feminine*	*Neuter*	*Masculine*	*Feminine*	*Neuter*
Sing. N., V.	πᾶς	πᾶσα	πᾶν	λυθείς	λυθεῖσα	λυθέν
A.	πάντα	πᾶσαν	πᾶν	λυθέντα	λυθεῖσαν	λυθέν
G.	παντός	πάσης	παντός	λυθέντος	λυθείσης	λυθέντος
D.	παντί	πάσῃ	παντί	λυθέντι	λυθείσῃ	λυθέντι
Plur. N., V.	πάντες	πᾶσαι	πάντα	λυθέντες	λυθεῖσαι	λυθέντα
A.	πάντας	πάσας	πάντα	λυθέντας	λυθείσας	λυθέντα
G.	πάντων	πασῶν	πάντων	λυθέντων	λυθεισῶν	λυθέντων
D.	πᾶσι(ν)	πάσαις	πᾶσι(ν)	λυθεῖσι(ν)	λυθείσαις	λυθεῖσι(ν)

	ἑκών, οῦσα, όν *willing* Stem ending in οντ-			δεικνύς, ῦσα, ύν *showing* (pres. part. act.). Stem ending in υντ-		
	Masculine	*Feminine*	*Neuter*	*Masculine*	*Feminine*	*Neuter*
Sing. N., V.	ἑκών	ἑκοῦσα	ἑκόν	δεικνύς	δεικνῦσα	δεικνύν
A.	ἑκόντα	ἑκοῦσαν	ἑκόν	δεικνύντα	δεικνῦσαν	δεικνύν
G.	ἑκόντος	ἑκούσης	ἑκόντος	δεικνύντος	δεικνύσης	δεικνύντος
D.	ἑκόντι	ἑκούσῃ	ἑκόντι	δεικνύντι	δεικνύσῃ	δεικνύντι
Plur. N., V.	ἑκόντες	ἑκοῦσαι	ἑκόντα	δεικνύντες	δεικνῦσαι	δεικνύντα
A.	ἑκόντας	ἑκούσας	ἑκόντα	δεικνύντας	δεικνύσας	δεικνύντα
G.	ἑκόντων	ἑκουσῶν	ἑκόντων	δεικνύντων	δεικνυσῶν	δεικνύντων
D.	ἑκοῦσι(ν)	ἑκούσαις	ἑκοῦσι(ν)	δεικνῦσι(ν)	δεικνύσαις	δεικνῦσι(ν)

	τιμῶν, ῶσα, ῶν *honouring* (pres. part. act.). Stem ending in ωντ- (contracted from αοντ-)			φιλῶν, οῦσα, οῦν *loving* (pres. part. act.). Stem ending in ουντ- (contracted from εοντ-)		
	Masculine	Feminine	Neuter	Masculine	Feminine	Neuter
Sing. N., V.	τιμῶν	τιμῶσα	τιμῶν	φιλῶν	φιλοῦσα	φιλοῦν
A.	τιμῶντα	τιμῶσαν	τιμῶν	φιλοῦντα	φιλοῦσαν	φιλοῦν
G.	τιμῶντος	τιμώσης	τιμῶντος	φιλοῦντος	φιλούσης	φιλοῦντος
D.	τιμῶντι	τιμώσῃ	τιμῶντι	φιλοῦντι	φιλούσῃ	φιλοῦντι
Plur. N., V.	τιμῶντες	τιμῶσαι	τιμῶντα	φιλοῦντες	φιλοῦσαι	φιλοῦντα
A.	τιμῶντας	τιμώσας	τιμῶντα	φιλοῦντας	φιλούσας	φιλοῦντα
G.	τιμώντων	τιμωσῶν	τιμώντων	φιλούντων	φιλουσῶν	φιλούντων
D.	τιμῶσι(ν)	τιμώσαις	τιμῶσι(ν)	φιλοῦσι(ν)	φιλούσαις	φιλοῦσι(ν)

	φανερῶν, οῦσα, οῦν *showing* (pres. part. act.). Stem ending in ουντ- (contracted from οοντ-)			ταχύς, εῖα, ύ *swift*. Stem ending in υ, changing to ε before a vowel, and in dat. plur.		
	Masculine	Feminine	Neuter	Masculine	Feminine	Neuter
Sing. N., V.	φανερῶν	φανεροῦσα	φανεροῦν	ταχύς	ταχεῖα	ταχύ
A.	φανεροῦντα	φανεροῦσαν	φανεροῦν	ταχύν	ταχεῖαν	ταχύ
G.	φανεροῦντος	φανερούσης	φανεροῦντος	ταχέως	ταχείας	ταχέως
D.	φανεροῦντι	φανερούσῃ	φανεροῦντι	ταχεῖ	ταχείᾳ	ταχεῖ
Plur. N., V.	φανεροῦντες	φανεροῦσαι	φανεροῦντα	ταχεῖς	ταχεῖαι	ταχέα
A.	φανεροῦντας	φανερούσας	φανεροῦντα	ταχεῖς	ταχείας	ταχέα
G.	φανερούντων	φανερουσῶν	φανερούντων	ταχέων	ταχειῶν	ταχέων
D.	φανεροῦσι(ν)	φανερούσαις	φανεροῦσι(ν)	ταχέσι(ν)	ταχείαις	ταχέσι(ν)

	λελυκώς, υῖα, ός *having loosed* (perf. part. act.). Stem ending in οτ-			ἑστώς, ῶσα, ός *standing* (str. perf. part. of ἵστημι). Stem ending in ωτ-. Fem. on analogy of pres. part		
	Masculine	Feminine	Neuter	Masculine	Feminine	Neuter
Sing. N., V.	λελυκώς	λελυκυῖα	λελυκός	ἑστώς	ἑστῶσα	ἑστός
A.	λελυκότα	λελυκυῖαν	λελυκός	ἑστῶτα	ἑστῶσαν	ἑστός
G.	λελυκότος	λελυκυίας	λελυκότος	ἑστῶτος	ἑστώσης	ἑστῶτος
D.	λελυκότι	λελυκυίᾳ	λελυκότι	ἑστῶτι	ἑστώσῃ	ἑστῶτι
Plur. N., V.	λελυκότες	λελυκυῖαι	λελυκότα	ἑστῶτες	ἑστῶσαι	ἑστῶτα
A.	λελυκότας	λελυκυίας	λελυκότα	ἑστῶτας	ἑστώσας	ἑστῶτα
G.	λελυκότων	λελυκυιῶν	λελυκότων	ἑστώτων	ἑστωσῶν	ἑστώτων
D.	λελυκόσι(ν)	λελυκυίαις	λελυκόσι(ν)	ἑστῶσι(ν)	ἑστώσαις	ἑστῶσι(ν)

	μέλας, αινα, αν *black* Stem ending in ν		
Sing. N.	μέλας	μέλαινα	μέλαν
V.	μέλαν	μέλαινα	μέλαν
A.	μέλανα	μέλαιναν	μέλαν
G.	μέλανος	μελαίνης	μέλανος
D.	μέλανι	μελαίνη	μέλανι
Plur. N., V.	μέλανες	μέλαιναι	μέλανα
A.	μέλανας	μελαίνας	μέλανα
G.	μελάνων	μελαινῶν	μελάνων
D.	μέλασι(ν)	μελαίναις	μέλασι(ν)

IRREGULAR:

THREE TERMINATIONS

	μέγας, άλη, α *great*			πολύς, πολλή, πολύ *much* (plur., *many*)		
	Masculine	*Feminine*	*Neuter*	*Masculine*	*Feminine*	*Neuter*
Sing. N., V.	μέγας	μεγάλη	μέγα	πολύς	πολλή	πολύ
A.	μέγαν	μεγάλην	μέγα	πολύν	πολλήν	πολύ
G.	μεγάλου	μεγάλης	μεγάλου	πολλοῦ	πολλῆς	πολλοῦ
D.	μεγάλῳ	μεγάλη	μεγάλῳ	πολλῷ	πολλῇ	πολλῷ
Plur. N., V.	μεγάλοι	μεγάλαι	μεγάλα	πολλοί	πολλαί	πολλά
A.	μεγάλους	μεγάλας	μεγάλα	πολλούς	πολλάς	πολλά
G.	μεγάλων	μεγάλων	μεγάλων	πολλῶν	πολλῶν	πολλῶν
D.	μεγάλοις	μεγάλαις	μεγάλοις	πολλοῖς	πολλαῖς	πολλοῖς

THIRD DECLENSION:

TWO TERMINATIONS

	ἄφρων, ον *foolish* Stem ending in ν		ἄρσην, εν *male* Stem ending in ν	
	Masc. and *fem.*	*Neut.*	*Masc.* and *fem.*	*Neut.*
Sing. N., V.	ἄφρων	ἄφρον	ἄρσην	ἄρσεν
A.	ἄφρονα	ἄφρον	ἄρσενα	ἄρσεν
G.	ἄφρονος	ἄφρονος	ἄρσενος	ἄρσενος
D.	ἄφρονι	ἄφρονι	ἄρσενι	ἄρσενι
Plur. N., V.	ἄφρονες	ἄφρονα	ἄρσενες	ἄρσενα
A.	ἄφρονας	ἄφρονα	ἄρσενας	ἄρσενα
G.	ἀφρόνων	ἀφρόνων	ἀρσένων	ἀρσένων
D.	ἄφροσι(ν)	ἄφροσι(ν)	ἄρσεσι(ν)	ἄρσεσι(ν)

	μείζων, ον *greater* (so other comparatives in -ων). Stem ending in ν		ἀληθής, ές *true.* Stem ending in ες-	
	Masc. and *fem.*	*Neut.*	*Masc.* and *fem.*	*Neut.*
Sing. N.	μείζων	μεῖζον	ἀληθής	ἀληθές
V.	μείζων	μεῖζον	ἀληθές	ἀληθές
A.	μείζονα and μείζω	μεῖζον	ἀληθῆ	ἀληθές
G.	μείζονος	μείζονος	ἀληθοῦς	ἀληθοῦς
D.	μείζονι	μείζονι	ἀληθεῖ	ἀληθεῖ
Plur. N., V.	μείζονες and μείζους	μείζονα and μείζω	ἀληθεῖς	ἀληθῆ
A.	μείζονας and μείζους	μείζονα and μείζω	ἀληθεῖς	ἀληφῆ
G.	μειζόνων	μειζόνων	ἀληθῶν	ἀληθῶν
D.	μείζοσι(ν)	μείζοσι(ν)	ἀληθέσι(ν)	ἀληθέσι(ν)

PRONOUNS

PERSONAL PRONOUNS

		1st person, ἐγώ *I*	2nd person, σύ *thou*	3rd person, αὐτός, ή, όν *he, she, it* (also as adjective, *same, self*)		
				Masc.	*Fem.*	*Neut.*
Sing.	N.	ἐγώ	σύ	αὐτός	αὐτή	αὐτό
	A.	ἐμέ, με	σέ, σε	αὐτόν	αὐτήν	αὐτό
	G.	ἐμοῦ, μου	σοῦ, σου	αὐτοῦ	αὐτῆς	αὐτοῦ
	D.	ἐμοί, μοι	σοί, σοι	αὐτῷ	αὐτῇ	αὐτῷ
Plur.	N.	ἡμεῖς	ὑμεῖς	αὐτοί	αὐταί	αὐτά
	A.	ἡμᾶς	ὑμᾶς	αὐτούς	αὐτάς	αὐτά
	G.	ἡμῶν	ὑμῶν	αὐτῶν	αὐτῶν	αὐτῶν
	D.	ἡμῖν	ὑμῖν	αὐτοῖς	αὐταῖς	αὐτοῖς

POSSESSIVE PRONOUNS

These are adjectival, and are declined as second and first declension adjectives:

ἐμός, ἐμή, ἐμόν *my*
ἡμέτερος, α, ον *our*
σός, σή, σόν *thy, your* (singular)
ὑμέτερος, α, ον *your* (plural)

Demonstrative Pronouns

	οὗτος *this*			ἐκεῖνος *that*		
	Masculine	Feminine	Neuter	Masculine	Feminine	Neuter
Sing. N.	οὗτος	αὕτη	τοῦτο	ἐκεῖνος	ἐκείνη	ἐκεῖνο
A.	τοῦτον	ταύτην	τοῦτο	ἐκεῖνον	ἐκείνην	ἐκεῖνο
G.	τούτου	ταύτης	τούτου	ἐκείνου	ἐκείνης	ἐκείνου
D.	τούτῳ	ταύτῃ	τούτῳ	ἐκείνῳ	ἐκείνῃ	ἐκείνῳ
Plur. N.	οὗτοι	αὗται	ταῦτα	ἐκεῖνοι	ἐκεῖναι	ἐκεῖνα
A.	τούτους	ταύτας	ταῦτα	ἐκείνους	ἐκείνας	ἐκεῖνα
G.	τούτων	τούτων	τούτων	ἐκείνων	ἐκείνων	ἐκείνων
D.	τούτοις	ταύταις	τούτοις	ἐκείνοις	ἐκείναις	ἐκείνοις

Declined like οὗτος, but without its initial τ are:

τοσοῦτος, τοσαύτη, τοσοῦτο or τοσοῦτον *so great, so much* (plur. *so many*).

τοιοῦτος, τοιαύτη, τοιοῦτο or τοιοῦτον *such.*

τηλικοῦτος, τηλικαύτη, τηλικοῦτο *so great.*

	ὅδε *this*		
	Masculine	Feminine	Neuter
Sing. N.	ὅδε	ἥδε	τόδε
A.	τόνδε	τήνδε	τόδε
G.	τοῦδε	τῆσδε	τοῦδε
D.	τῷδε	τῇδε	τῷδε
Plur. N.	οἵδε	αἵδε	τάδε
A.	τούσδε	τάσδε	τάδε
G.	τῶνδε	τῶνδε	τῶνδε
D.	τοῖσδε	ταῖσδε	τοῖσδε

Reflexive Pronouns

	1st person ἐμαυτόν *myself*		2nd person σεαυτόν *thyself*		3rd person ἑαυτόν, ήν, ό[1] *himself, herself, itself*		
	Masc.	Fem.	Masc.	Fem.	Masc.	Fem.	Neut.
Sing. A.	ἐμαυτόν	ἐμαυτήν	σεαυτόν	σεαυτήν	ἑαυτόν	ἑαυτήν	ἑαυτό
G.	ἐμαυτοῦ	ἐμαυτῆς	σεαυτοῦ	σεαυτῆς	ἑαυτοῦ	ἑαυτῆς	ἑαυτοῦ
D.	ἐμαυτῷ	ἐμαυτῇ	σεαυτῷ	σεαυτῇ	ἑαυτῷ	ἑαυτῇ	ἑαυτῷ

[1] αὑτόν, ήν, ὁ is also found.

The plural of all three persons is the same:

	ἑαυτούς, άς, ά *ourselves, yourselves, themselves*		
	Masculine	*Feminine*	*Neuter*
Plur. A.	ἑαυτούς	ἑαυτάς	ἑαυτά
G.	ἑαυτῶν	ἑαυτῶν	ἑαυτῶν
D.	ἑαυτοῖς	ἑαυταῖς	ἑαυτοῖς

INTERROGATIVE PRONOUN

	τίς *who?*	
	Masc. and fem.	*Neut.*
Sing. N.	τίς	τί
A.	τίνα	τί
G.	τίνος	τίνος
D.	τίνι	τίνι
Plur. N.	τίνες	τίνα
A.	τίνας	τίνα
G.	τίνων	τίνων
D.	τίσι(ν)	τίσι(ν)

INDEFINITE PRONOUN

	τις *any*	
	Masc. and fem.	*Neut.*
Sing. N.	τις	τι
A.	τινά	τι
G.	τινός	τινός
D.	τινί	τινί
Plur. N.	τινές	τινά
A.	τινάς	τινά
G.	τινῶν	τινῶν
D.	τισί(ν)	τισί(ν)

Relative Pronouns

	ὅς *who*			ὅστις *who, whoever.* Only the parts below are found in the N.T.		
	Masculine	*Feminine*	*Neuter*	*Masculine*	*Feminine*	*Neuter*
Sing. N.	ὅς	ἥ	ὅ	ὅστις	ἥτις	ὅτι
A.	ὅν	ἥν	ὅ			
G.	οὗ	ἧς	οὗ	ὅτου		ὅτου
D.	ᾧ	ᾗ	ᾧ			
Plur. N.	οἵ	αἵ	ἅ	οἵτινες	αἵτινες	ἅτινα
A.	οὕς	ἅς	ἅ			
G.	ὧν	ὧν	ὧν			
D.	οἷς	αἷς	οἷς			

Reciprocal Pronoun

	ἀλλήλους *one, another*
	Masculine
A.	ἀλλήλους
G.	ἀλλήλων
D.	ἀλλήλοις

No fem. or neut. forms are found in the N.T.

NUMERALS

	εἷς *one*		
	Masc.	*Fem.*	*Neut.*
N.	εἷς	μία	ἕν
A.	ἕνα	μίαν	ἕν
G.	ἑνός	μιᾶς	ἑνός
D.	ἑνί	μιᾷ	ἑνί

οὐδείς, οὐδεμία, οὐδέν, and μηδείς, μηδεμία, μηδέν *nobody*, are declined similarly.

	δύο *two*	τρεῖς *three*		τέσσαρες *four*	
	Mas., fem. and neut.	*Masc. and fem.*	*Neut.*	*Masc. and fem.*	*Neut.*
N.	δύο	τρεῖς	τρία	τέσσαρες	τέσσαρα
A.	δύο	τρεῖς	τρία	τέσσαρας	τέσσαρα
G.	δύο	τριῶν	τριῶν	τεσσάρων	τεσσάρων
D.	δυσί(ν)	τρισί(ν)	τρισί(ν)	τέσσαρσι(ν)	τέσσαρσι(ν)

APPENDIX 4

PARADIGMS OF VERBS

Verbs in -ω (thematic); stems ending in υ and ι
λύω I loose; stem λυ-

ACTIVE VOICE

TENSE	NUMBER and PERSON	INDICATIVE MOOD			IMPERATIVE MOOD	CONJUNCTIVE MOOD		VERB INFINITE	
		PRIMARY	HISTORIC			PRIMARY	HISTORIC	Infinitive (noun)	Participle (adjective)
		Present	Imperfect			Subjunctive	Optative		
Present, *I loose* Imperfect, *I was loosing*	*Sing.* 1.	λύω	ἔλυον			λύω	λύοιμι	λύειν	*Masc.* λύων
	2.	λύεις	ἔλυες	λῦε		λύῃς	λύοις		*Fem.* λύουσα
	3.	λύει	ἔλυε(ν)	λυέτω		λύῃ	λύοι		*Neut* λῦον
	Plur. 1.	λύομεν	ἐλύομεν			λύωμεν	λύοιμεν		Stem λυοντ-
	2.	λύετε	ἐλύετε	λύετε		λύητε	λύοιτε		
	3.	λύουσι(ν)	ἔλυον	λυέτωσαν		λύωσι(ν)	λύοιεν		
Future, *I shall loose*	*Sing.* 1.	λύσω					Not found in N.T.	λύσειν (rare in N.T.)	*Masc.* λύσων
	2.	λύσεις							*Fem.* λύσουσα
	3.	λύσει							*Neut.* λῦσον
	Plur. 1.	λύσομεν							Stem λυσοντ- (rare in N T.)
	2.	λύσετε							
	3.	λύσουσι(ν)							

		Imperative	Subjunctive	Optative	Infinitive	Participle
Weak Aorist, *I loosed*	*Sing.* 1. ἔλυσα 2. ἔλυσας 3. ἔλυσε(ν)	λῦσον λυσάτω	λύσω λύσῃς λύσῃ	λύσαιμι λύσαις λύσαι or λύσειε(ν) (rare)	λῦσαι	*Masc.* λύσας *Fem.* λύσασα *Neut.* λῦσαν Stem λυσαντ-
	Plur. 1. ἐλύσαμεν 2. ἐλύσατε 3. ἔλυσαν	λύσατε λυσάτωσαν	λύσωμεν λύσητε λύσωσι(ν)	λύσαιμεν λύσαιτε λύσαιεν or λύσειαν		
Weak Perfect, *I have loosed*	Perfect λέλυκα *Sing.* 1. λέλυκα 2. λέλυκας 3. λέλυκε(ν) *Plur.* 1. λελύκαμεν 2. λελύκατε 3. λελύκασι(ν)	Not found in N.T.	Formed by perf. part. and subj. of εἰμί, λελυκὼς ὦ etc.	Not found in N.T.	λελυκέναι	*Masc.* λελυκώς *Fem.* λελυκυῖα *Neut.* λελυκός Stem λελυκοτ-
Pluperfect, *I had loosed*	Pluperfect (ἐ)λελύκειν *Sing.* 1. (ἐ)λελύκειν 2. (ἐ)λελύκεις 3. (ἐ)λελύκει *Plur.* 1. (ἐ)λελύκειμεν 2. (ἐ)λελύκειτε 3. (ἐ)λελύκεισαν					

Strong Aorist, Strong Perfect, Strong Pluperfect: not found in verbs with ν and 1 stems.

λύω: PASSIVE VOICE and
Present, Imperfect, Perfect, and Plupefect Tenses of MIDDLE VOICE

TENSE	NUMBER and PERSON	INDICATIVE MOOD PRIMARY	INDICATIVE MOOD HISTORIC	IMPERATIVE MOOD	CONJUNCTIVE MOOD PRIMARY	CONJUNCTIVE MOOD HISTORIC	VERB INFINITE Infinitive (noun)	VERB INFINITE Participle (adjective)
Present Mid., I loose for myself Pass., I am being loosed / Imperfect Mid., I was loosing for myself Pass., I was being loosed		Present	Imperfect		Subjunctive	Optative	λύεσθαι	M. λυόμενος F. λυομένη N. λυόμενον
	Sing. 1.	λύομαι	ἐλυόμην		λύωμαι	λυοίμην		
	2.	λύῃ	ἐλύου	λύου	λύῃ	λύοιο		
	3.	λύεται	ἐλύετο	λυέσθω	λύηται	λύοιτο		
	Plur. 1.	λυόμεθα	ἐλυόμεθα		λυώμεθα	λυοίμεθα		
	2.	λύεσθε	ἐλύεσθε	λύεσθε	λύησθε	λύοισθε		
	3.	λύονται	ἐλύοντο	λυέσθωσαν	λύωνται	λύοιντο		
Weak Future Pass., I shall be loosed	Sing. 1.	λυθήσομαι			Not found in N.T.		λυθήσεσθαι (but not found in N.T.)	M. λυθησόμενος F. λυθησομένη N. λυθησόμενον
	2.	λυθήσῃ						
	3.	λυθήσεται						
	Plur. 1.	λυθησόμεθα						
	2.	λυθήσεσθε						
	3.	λυθήσονται						
Weak Aorist Pass., I was loosed	Sing. 1.		ἐλύθην		λυθῶ	λυθείην	λυθῆναι	M. λυθείς F. λυθεῖσα N. λυθέν Stem λυθεντ–
	2.		ἐλύθης	λύθητι	λυθῇς	λυθείης		
	3.		ἐλύθη	λυθήτω	λυθῇ	λυθείη		
	Plur. 1.		ἐλύθημεν		λυθῶμεν	λυθείημεν		
	2.		ἐλύθητε	λύθητε	λυθῆτε	λυθείητε		
	3.		ἐλύθησαν	λυθήτωσαν	λυθῶσι(ν)	λυθείησαν		

		Perfect	Pluperfect					
Perfect Mid., *I have loosed for myself*	Sing. 1.	λέλυμαι	(ἐ)λελύμην		λελυμένος ὦ	λελυμένος εἴην		
	2.	λέλυσαι	(ἐ)λέλυσο	λέλυσο	,, ᾖς	,, εἴης		
Pass., *I have been loosed*	3.	λέλυται	(ἐ)λέλυτο	λελύσθω	,, ᾖ	,, εἴη	λελύσθαι	M. λελυμένος
	Plur. 1.	λελύμεθα	(ἐ)λελύμεθα		λελυμένοι ὦμεν	λελυμένοι εἴημεν		F. λελυμένη
Pluperfect Mid., *I had loosed for myself*	2.	λέλυσθε	(ἐ)λέλυσθε	λέλυσθε	,, ἦτε	,, εἴητε		N. λελυμένον
Pass., *I had been loosed*	3.	λέλυνται	(ἐ)λέλυντο	λελύσθωσαν	,, ὦσι(ν)	,, εἴησαν		

Strong Future Passive, Strong Aorist Passive: not found in verbs with υ and ι stems.

λύω: Tenses peculiar to the MIDDLE VOICE

TENSE	NUMBER and PERSON	INDICATIVE MOOD		IMPERATIVE MOOD	CONJUNCTIVE MOOD		VERB INFINITE	
		PRIMARY	HISTORIC		PRIMARY	HISTORIC	Infinitive (noun)	Participle (adjective)
					Subjunctive	Optative Not found in N.T.		
Future Middle, *I shall loose for myself*	*Sing.* 1.	λύσομαι					λύσεσθαι	M. λυσόμενος F. λυσομένη N. λυσόμενον
	2.	λύσῃ						
	3.	λύσεται						
	Plur. 1.	λυσόμεθα						
	2.	λύσεσθε						
	3.	λύσονται						
Weak Aorist Middle, *I loosed for myself*	*Sing.* 1.		ἐλυσάμην		λύσωμαι	λυσαίμην	λύσασθαι	M. λυσάμενος F. λυσαμένη N. λυσάμενον
	2.		ἐλύσω	λῦσαι	λύσῃ	λύσαιο		
	3.		ἐλύσατο	λυσάσθω	λύσηται	λύσαιτο		
	Plur. 1.		ἐλυσάμεθα		λυσώμεθα	λυσαίμεθα		
	2.		ἐλύσασθε	λύσασθε	λύσησθε	λύσαισθε		
	3.		ἐλύσαντο	λυσάσθωσαν	λύσωνται	λύσαιντο		

CONTRACTED VERBS:

Stems in α

τιμῶ (τιμά-ω), stem τιμα- *I honour*

ACTIVE

TENSE	NUMBER and PERSON	INDICATIVE MOOD		IMPERATIVE MOOD	CONJUNCTIVE MOOD		VERB INFINITE	
		PRIMARY	HISTORIC		PRIMARY	HISTORIC	Infinitive (noun)	Participle (adjective)
Present, *I honour*	*Sing.* 1.	Present τιμῶ	Imperfect ἐτίμων		Subjunctive τιμῶ	Optative Optatives of contracted verbs not found in the N.T.	τιμᾶν	M. τιμῶν
Imperfect, *I was honouring*	2.	τιμᾷς	ἐτίμας	τίμα	τιμᾷς			F. τιμῶσα
	3.	τιμᾷ	ἐτίμα	τιμάτω	τιμᾷ			N. τιμῶν
	Plur. 1.	τιμῶμεν	ἐτιμῶμεν		τιμῶμεν			Stem τιμωντ-
	2.	τιμᾶτε	ἐτιμᾶτε	τιμᾶτε	τιμᾶτε			
	3.	τιμῶσι(ν)	ἐτίμων	τιμάτωσαν	τιμῶσι(ν)			

MIDDLE AND PASSIVE

TENSE	NUMBER and PERSON	INDICATIVE MOOD		IMPERATIVE MOOD	CONJUNCTIVE MOOD		VERB INFINITE	
		PRIMARY	HISTORIC		PRIMARY	HISTORIC	Infinitive	Participle
Present Mid., *I honour for myself* Pass., *I am being honoured*	*Sing.* 1.	τιμῶμαι	ἐτιμώμην		τιμῶμαι		τιμᾶσθαι	M. τιμώμενος
	2.	τιμᾷ	ἐτιμῶ	τιμῶ	τιμᾷ			F. τιμωμένη
	3.	τιμᾶται	ἐτιμᾶτο	τιμάσθω	τιμᾶται			N. τιμώμενον
Imperfect Mid., *I was honouring for myself* Pass., *I was being honoured*	*Plur.* 1.	τιμώμεθα	ἐτιμώμεθα		τιμώμεθα			
	2.	τιμᾶσθε	ἐτιμᾶσθε	τιμᾶσθε	τιμᾶσθε			
	3.	τιμῶνται	ἐτιμῶντο	τιμάσθωσαν	τιμῶνται			

Other Tenses of τιμῶ

	ACTIVE	MIDDLE	PASSIVE
Future	τιμήσω	τιμήσομαι	τιμηθήσομαι
Weak Aorist	ἐτίμησα	ἐτιμησάμην	ἐτιμήθην
Perfect	τετίμηκα	τετίμημαι	τετίμημαι

Contracted Verbs:

Stems in ε

φιλῶ (φιλέ-ω), stem φιλε– *I love*

ACTIVE

TENSE	NUMBER and PERSON	INDICATIVE MOOD — PRIMARY	INDICATIVE MOOD — HISTORIC	IMPERATIVE MOOD	CONJUNCTIVE MOOD — PRIMARY	CONJUNCTIVE MOOD — HISTORIC	VERB INFINITE — Infinitive (noun)	VERB INFINITE — Participle (adjective)
Present, *I love*		Present	Imperfect		Subjunctive	Optative	φιλεῖν	M. φιλῶν
Imperfect, *I was loving*	*Sing.* 1.	φιλῶ	ἐφίλουν		φιλῶ	Optatives of contracted verbs not found in the N.T.		F. φιλοῦσα
	2.	φιλεῖς	ἐφίλεις	φίλει	φιλῇς			N. φιλοῦν
	3.	φιλεῖ	ἐφίλει	φιλείτω	φιλῇ			Stem φιλουντ–
	Plur. 1.	φιλοῦμεν	ἐφιλοῦμεν		φιλῶμεν			
	2.	φιλεῖτε	ἐφιλεῖτε	φιλεῖτε	φιλῆτε			
	3.	φιλοῦσι(ν)	ἐφίλουν	φιλείτωσαν	φιλῶσι(ν)			

MIDDLE AND PASSIVE

		Present	Imperfect			Infinitive	Participle
Present Mid., *I love for myself* Pass., *I am being loved*	*Sing.* 1. 2. 3.	φιλοῦμαι φιλῇ φιλεῖται	ἐφιλούμην ἐφιλοῦ ἐφιλεῖτο	φιλοῦ φιλείσθω	φιλῶμαι φιλῇ φιλῆται	φιλεῖσθαι	*M.* φιλούμενος *F.* φιλουμένη *N.* φιλούμενον
Imperfect Mid., *I was loving for myself* Pass., *I was being loved*	*Plur.* 1. 2. 3.	φιλούμεθα φιλεῖσθε φιλοῦνται	ἐφιλούμεθα ἐφιλεῖσθε ἐφιλοῦντο	φιλεῖσθε φιλείσθωσαν	φιλώμεθα φιλῆσθε φιλῶνται		

OTHER TENSES

	ACTIVE	MIDDLE	PASSIVE
Future	φιλήσω	φιλήσομαι	φιληθήσομαι
Weak Aorist	ἐφίλησα	ἐφιλησάμην	ἐφιλήθην
Perfect	πεφίληκα	πεφίλημαι	πεφίλημαι

Contracted Verbs:

Stems in o

δηλῶ (δηλό-ω), stem δηλο- *I show*

ACTIVE

TENSE	NUMBER and PERSON	INDICATIVE MOOD		IMPERATIVE MOOD	CONJUNCTIVE MOOD		VERB INFINITE	
		PRIMARY	HISTORIC		PRIMARY	HISTORIC	Infinitive (noun)	Participle (adjective)
Present, *I show* Imperfect, *I was showing*	*Sing.* 1.	Present δηλῶ	Imperfect ἐδήλουν		Subjunctive δηλῶ	Optative	δηλοῦν	M. δηλῶν F. δηλοῦσα N. δηλοῦν Stem δηλουντ-
	2.	δηλοῖς	ἐδήλους	δήλου	δηλοῖς	Optatives of contracted verbs not found in the N.T.		
	3.	δηλοῖ	ἐδήλου	δηλούτω	δηλοῖ			
	Plur. 1.	δηλοῦμεν	ἐδηλοῦμεν		δηλῶμεν			
	2.	δηλοῦτε	ἐδηλοῦτε	δηλοῦτε	δηλῶτε			
	3.	δηλοῦσι(ν)	ἐδήλουν	δηλούτωσαν	δηλῶσι(ν)			

MIDDLE AND PASSIVE

		Indicative	Imperfect	Imperative	Subjunctive	Infinitive	Participle
Present Mid., *I show for myself* Pass., *I am being shown*	*Sing.* 1.	δηλοῦμαι	ἐδηλούμην		δηλῶμαι	δηλοῦσθαι	M. δηλούμενος F. δηλουμένη N. δηλούμενον
	2.	δηλοῖ	ἐδηλοῦ	δηλοῦ	δηλοῖ		
	3.	δηλοῦται	ἐδηλοῦτο	δηλούσθω	δηλῶται		
Imperfect Mid., *I was showing for myself* Pass., *I was being shown*	*Plur.* 1.	δηλούμεθα	ἐδηλούμεθα		δηλώμεθα		
	2.	δηλοῦσθε	ἐδηλοῦσθε	δηλοῦσθε	δηλῶσθε		
	3.	δηλοῦνται	ἐδηλοῦντο	δηλούσθωσαν	δηλῶνται		

OTHER TENSES

	ACTIVE	MIDDLE	PASSIVE
Future	δηλώσω	δηλώσομαι	δηλωθήσομαι
Weak Aorist	ἐδήλωσα	ἐδηλωσάμην	ἐδηλώθην
Perfect	δεδήλωκα	δεδήλωμαι	δεδήλωμαι

ACTIVE

TENSE	NUMBER and PERSON	INDICATIVE MOOD		IMPERATIVE MOOD
		PRIMARY	HISTORIC	
Strong Aorist Active, *I cast*	*Sing.* 1.		ἔβαλον	
	2.		ἔβαλες	βάλε
	3.		ἔβαλε	βαλέτω
	Plur. 1.		ἐβάλομεν	
	2.		ἐβάλετε	βάλετε
	3.		ἔβαλον	βαλέτωσαν
Strong Perfect Active, *I have written* Strong Pluperfect Active, *I had got to know, I knew*	*Sing.* 1.	γέγραφα	Not found in the ordinary verb. The endings are shown by the past tense of οἶδα, *I know* { ἤδειν ἤδεις ἤδει ἤδειμεν ἤδειτε ἤδεισαν	Not found in the ordinary verb. The imperative of οἶδα is irregular { ἴσθι ἴστω ἴστε ἴστωσαν
	2.	γέγραφας		
	3.	γέγραφε(ν)		
	Plur. 1.	γεγράφαμεν		
	2.	γεγράφατε		
	3.	γεγράφασι(ν)		

MIDDLE

TENSE	NUMBER and PERSON	INDICATIVE MOOD		IMPERATIVE MOOD
		PRIMARY	HISTORIC	
Strong Aorist Middle, *I cast for myself*	*Sing.* 1.		ἐβαλόμην	
	2.		ἐβάλου	βαλοῦ
	3.		ἐβάλετο	βαλέσθω
	Plur. 1.		ἐβαλόμεθα	
	2.		ἐβάλεσθε	βάλεσθε
	3.		ἐβάλοντο	βαλέσθωσαν

PASSIVE

TENSE	NUMBER and PERSON	INDICATIVE MOOD		IMPERATIVE MOOD
		PRIMARY	HISTORIC	
Strong Future Passive, *I shall be changed*	*Sing.* 1.	ἀλλαγήσομαι		
	2.	ἀλλαγήσῃ		
	3.	ἀλλαγήσεται		
	Plur. 1.	ἀλλαγησόμεθα		
	2.	ἀλλαγήσεσθε		
	3.	ἀλλαγήσονται		
Strong Aorist Passive, *I was changed*	*Sing.* 1.		ἠλλάγην	
	2.		ἠλλάγης	ἀλλάγηθι
	3.		ἠλλάγη	ἀλλαγήτω
	Plur. 1.		ἠλλάγημεν	
	2.		ἠλλάγητε	ἀλλάγητε
	3.		ἠλλάγησαν	ἀλλαγήτωσαν

STRONG TENSES

STEMS ENDING IN CONSONANTS:

βάλλω, γράφω, οἶδα, and -ἀλλάσσω

ACTIVE

CONJUNCTIVE MOOD		VERB INFINITE	
PRIMARY	HISTORIC		
Subjunctive	Optative	Infinitive (noun)	Participle (adjective)
βάλω	βάλοιμι	βαλεῖν	M. βαλών
βάλῃς	βάλοις		F. βαλοῦσα
βάλῃ	βάλοι		N. βαλόν
βάλωμεν	βάλοιμεν		Stem βαλοντ-
βάλητε	βάλοιτε		
βάλωσι(ν)	βάλοιεν		
γεγραφὼς ὦ ⎱ εἰδῶ	Not found in N.T.	γεγραφέναι εἰδέναι	M. γεγραφώς, εἰδώς
„ ῇς εἰδῇς			F. γεγραφυῖα, εἰδυῖα
„ ῇ ⎰εἰδῇ			N. γεγραφός, εἰδός
γεγραφότες ὦμεν εἰδῶμεν			Stem γεγραφοτ-, εἰδοτ-
„ ῆτε εἰδῆτε			
„ ὦσι(ν) ⎰εἰδῶσι(ν)			

MIDDLE

βάλωμαι	βαλοίμην ⎱ The form	βαλέσθαι	M. βαλόμενος
βάλῃ	βάλοιο only ex-		F. βαλομένη
βάληται	βάλοιτο ⎰ists in		N. βαλόμενον
βαλώμεθα	βαλοίμεθα ⎰N.T. in		
βάλησθε	βάλοισθε ⎰γένοιτο		
βάλωνται	βάλοιντο		

PASSIVE

		ἀλλαγήσεσθαι (but not found in N.T.)	M. ἀλλαγησόμενος F. ἀλλαγησομένη N. ἀλλαγησόμενον (but not found in N.T.)
ἀλλαγῶ	ἀλλαγείην ⎱ but	ἀλλαγῆναι	M. ἀλλαγείς
ἀλλαγῇς	ἀλλαγείης not		F. ἀλλαγεῖσα
ἀλλαγῇ	ἀλλαγείη ⎰found		N. ἀλλαγέν
ἀλλαγῶμεν	ἀλλαγείημεν in the		Stem ἀλλαγεντ-
ἀλλαγῆτε	ἀλλαγείητε N.T.		
ἀλλαγῶσι(ν)	ἀλλαγείησαν		

PERFECT AND PLUPERFECT MIDDLE AND PASSIVE OF VERBS WITH STEMS ENDING IN A CONSONANT

	NUMBER and PERSON	PERFECT INDICATIVE	PLUPERFECT INDICATIVE	OTHER MOODS
GUTTURAL STEM E.g. τεταγ- (pres. indic. τάσσω)	Sing. 1. 2. 3. Plur. 1. 2. 3.	τέταγμαι τέταξαι τέτακται τετάγμεθα τέταχθε τεταγμένοι εἰσί(ν)	(ἐ)τετάγμην (ἐ)τέταξο (ἐ)τέτακτο (ἐ)τετάγμεθα (ἐ)τέταχθε τεταγμένοι ἦσαν	Imperative, Not found Subjunctive, peri- phrastic, τεταγμένος ὦ, etc. Infinitive, τετάχθαι Participle, τεταγμένος, η, ον
DENTAL STEM E.g. πεπειθ- (pres. indic. πείθω)	Sing. 1. 2. 3. Plur. 1. 2. 3.	πέπεισμαι πέπεισαι πέπεισται πεπείσμεθα πέπεισθε πεπεισμένοι εἰσί(ν)	(ἐ)πεπείσμην (ἐ)πέπεισο (ἐ)πέπειστο (ἐ)πεπείσμεθα (ἐ)πέπεισθε πεπεισμένοι ἦσαν	Imperative, Not found Subjunctive, peri- phrastic, πεπεισμένος ὦ, etc. Infinitive, πεπεῖσθαι Participle, πεπεισ- μένος, η, ον
LABIAL STEM E.g. γεγραφ- (pres. indic. γράφω)	Sing. 1. 2. 3. Plur. 1. 2. 3.	γέγραμμαι γέγραψαι γέγραπται γεγράμμεθα γέγραφθε γεγραμμένοι εἰσί(ν)	(ἐ)γεγράμμην (ἐ)γέγραψο (ἐ)γέγραπτο (ἐ)γεγράμμεθα (ἐ)γέγραφθε γεγραμμένοι ἦσαν	Imperative, Not found Subjunctive, peri- phrastic, γεγραμ- μένος ὦ, etc. Infinitive, γεγράφθαι. Participle, γεγραμ- μένος, η, ον
LIQUID STEM E.g. ἐσταλ- (pres. indic. στέλλω)	Sing. 1. 2. 3. Plur. 1. 2. 3.	ἔσταλμαι ἔσταλσαι ἔσταλται ἐστάλμεθα ἔσταλθε ἐσταλμένοι εἰσί(ν)	ἐστάλμην ἔσταλσο ἔσταλτο ἐστάλμεθα ἔσταλθε ἐσταλμένοι ἦσαν	Imperative, Not found Subjunctive, peri- phrastic, ἐσταλμένος ὦ, etc. Infinitive, ἐστάλθαι Participle, ἐσταλμένος, η, ον
NASAL STEM E.g. μεμιαν- (pres. indic. μιαίνω)	Sing. 1. 2. 3. Plur. 1. 2. 3.	μεμίαμμαι μεμίανσαι μεμίανται μεμιάμμεθα μεμίανθε μεμιαμμένοι εἰσί(ν)	The pluperfect of this type does not occur in the N.T.	Imperative, subjunctive, infinitive, do not occur in the N.T. Participle, μεμιαμ- μένος, η, ον

VERBS IN -μι (NON-THEMATIC)
ἵστημι *I place*, stem στα- and στη-
ACTIVE VOICE

TENSE	NUMBER and PERSON	INDICATIVE MOOD		IMPERATIVE MOOD	CONJUNCTIVE MOOD		VERB INFINITE	
		PRIMARY	HISTORIC		PRIMARY	HISTORIC	Infinitive (noun)	Participle (adjective)
		Present	Imperfect		Subjunctive	Optative		
Present, *I place* Imperfect, *I was placing*	Sing. 1.	ἵστημι	ἵστην	ἵστη	ἱστῶ	Not found in the N.T.	ἱστάναι	M. ἱστάς
	2.	ἵστης	ἵστης	ἱστάτω	ἱστῇς			F. ἱστᾶσα
	3.	ἵστησι(ν)	ἵστη		ἱστῇ			N. ἱστάν
	Plur. 1.	ἵσταμεν	ἵσταμεν	ἵστατε	ἱστῶμεν			Stem ἱσταντ-
	2.	ἵστατε	ἵστατε	ἱστάτωσαν	ἱστῆτε			
	3.	ἱστᾶσι(ν)	ἵστασαν		ἱστῶσι(ν)			
Strong Aorist, *I stood* (intransitive)	Sing. 1.		ἔστην	στῆθι or -στα	στῶ	Not found in the N.T.	στῆναι	M. στάς
	2.		ἔστης	στήτω	στῇς			F. στᾶσα
	3.		ἔστη		στῇ			N. στάν
[For weak aorist, see below]	Plur. 1.		ἔστημεν	στῆτε	στῶμεν			Stem σταντ-
	2.		ἔστητε	στήτωσαν	στῆτε			
	3.		ἔστησαν		στῶσι(ν)			
Perfect, present meaning, *I stand* (intransitive)	Sing. 1.	ἕστηκα	εἱστήκειν or ἑστήκειν, etc.	Not found in the N.T.	ἑστηκώς or ἑστώς } ὦ		ἑστάναι (strong) ἑστηκέναι, weak perf. infin. not found in the N.T.	Strong Weak M. ἑστώς ἑστηκώς
	2.	ἑστηκώς	εἱστήκεις		,, ,, ἦς / ᾖ			F. ἑστῶσα ἑστηκυῖα
Pluperfect, *I stood* (intransitive)	3.	ἕστηκε(ν)	εἱστήκει		ἑστηκότες or ἑστῶτες } ὦμεν, etc.			N. ἑστός ἑστηκός
	Plur. 1.	ἑστήκαμεν, etc.	εἱστήκειμεν, etc.					Stem ἑστωτ- ἑστηκοτ-

MIDDLE AND PASSIVE VOICES

Present Mid., *I place for myself* Pass., *I am being placed, I stand*	Sing. 1. 2. 3. Plur. 1. 2. 3.	ἵσταμαι ἵστασαι ἵσταται ἱστάμεθα ἵστασθε ἵστανται	ἱστάμην ἵστασο ἵστατο ἱστάμεθα ἵστασθε ἵσταντο	ἵστασο ἱστάσθω ἵστασθε ἱστάσθωσαν	ἱστῶμαι ἱστῇ ἱστῆται ἱστώμεθα ἱστῆσθε ἱστῶνται	Not found in the N.T.	ἵστασθαι	M. ἱστάμενος F. ἱσταμένη N. ἱστάμενον
Imperfect Mid., *I placed for myself* Pass., *I was being placed, I was standing*							

OTHER TENSES

	ACTIVE	MIDDLE	PASSIVE
Future	στήσω *I shall place*	στήσομαι *I shall stand*	σταθήσομαι *I shall be placed, I shall stand*
Weak Aorist	ἔστησα *I placed*		ἐστάθην *I was placed, I stood*

τίθημι *I place*; stem θε- and θη-
ACTIVE VOICE

TENSE	NUMBER and PERSON	INDICATIVE MOOD		IMPERATIVE MOOD	CONJUNCTIVE MOOD		VERB INFINITE	
		PRIMARY	HISTORIC		PRIMARY	HISTORIC	Infinitive (noun)	Participle (adjective)
Present, *I place*	*Sing.* 1.	Present τίθημι	Imperfect ἐτίθην		Subjunctive τιθῶ	Optative Not found in the N.T.	τιθέναι	*M.* τιθείς *F.* τιθεῖσα *N.* τιθέν Stem τιθεντ-
	2.	τίθης	ἐτίθεις	τίθει	τιθῇς			
Imperfect, *I was placing*	3.	τίθησι(ν)	ἐτίθει	τιθέτω	τιθῇ			
	Plur. 1.	τίθεμεν	ἐτίθεμεν		τιθῶμεν			
	2.	τίθετε	ἐτίθετε	τίθετε	τιθῆτε			
	3.	τιθέασι(ν) or τιθᾶσι(ν)	ἐτίθεσαν or ἐτίθουν	τιθέτωσαν	τιθῶσι(ν)			
Aorist, *I placed*	*Sing.* 1.		ἔθηκα (weak)		θῶ	Not found in the N.T.	θεῖναι	*M.* θείς *F.* θεῖσα *N.* θέν Stem θεντ-
	2.		ἔθηκας	θές	θῇς			
	3.		ἔθηκε(ν)	θέτω	θῇ			
	Plur. 1.		ἔθεμεν (strong)		θῶμεν			
	2.		ἔθετε	θέτε	θῆτε			
	3.		ἔθεσαν	θέτωσαν	θῶσι(ν)			

MIDDLE VOICE (The Passive of Pres. and Imperf. Tenses is supplied by κεῖμαι, see pp. 233f.)

		Imperfect indic.	Present indic.	Imperative	Subjunctive		Infinitive	Participle
Present, *I place for myself* / **Imperfect**, *I was placing for myself*	Sing. 1.	ἐτιθέμην	τίθεμαι		τιθῶμαι	Not found in the N.T.	τίθεσθαι	M. τιθέμενος
	2.	ἐτίθεσο	τίθεσαι	τίθεσο	τιθῇ			F. τιθεμένη
	3.	ἐτίθετο	τίθεται	τιθέσθω	τιθῆται			N. τιθέμενον
	Plur. 1.	ἐτιθέμεθα	τιθέμεθα		τιθώμεθα			
	2.	ἐτίθεσθε	τίθεσθε	τίθεσθε	τιθῆσθε			
	3.	ἐτίθεντο	τίθενται	τιθέσθωσαν	τιθῶνται			
Strong Aorist, *I placed for myself*	Sing. 1.	ἐθέμην			θῶμαι	Not found in the N.T.	θέσθαι	M. θέμενος
	2.	ἔθου	θοῦ		θῇ			F. θεμένη
	3.	ἔθετο	θέσθω		θῆται			N. θέμενον
	Plur. 1.	ἐθέμεθα			θώμεθα			
	2.	ἔθεσθε	θέσθε		θῆσθε			
	3.	ἔθεντο	θέσθωσαν		θῶνται			

OTHER TENSES

	ACTIVE	MIDDLE	PASSIVE
Future	θήσω	-θήσομαι	-τεθήσομαι
Aorist			ἐτέθην
Perfect	τέθεικα	τέθειμαι	τέθειμαι

δίδωμι I give; stem δο- and δω

ACTIVE VOICE

TENSE	NUMBER and PERSON	INDICATIVE MOOD		IMPERATIVE MOOD	CONJUNCTIVE MOOD		VERB INFINITE	
		PRIMARY	HISTORIC		PRIMARY	HISTORIC	Infinitive (noun)	Participle (adjective)
		Present	Imperfect		Subjunctive	Optative Not found in the N.T.	διδόναι	M. διδούς F. διδοῦσα N. διδόν Stem διδοντ-
Present, I give Imperfect, I was giving	Sing. 1.	δίδωμι	ἐδίδουν		διδῶ			
	2.	δίδως	ἐδίδους	δίδου	διδῷς or διδοῖς			
	3.	δίδωσι(ν)	ἐδίδου	διδότω	διδῷ or διδοῖ			
	Plur. 1.	δίδομεν	ἐδίδομεν		διδῶμεν			
	2.	δίδοτε	ἐδίδοτε	δίδοτε	διδῶτε			
	3.	διδόασι(ν)	ἐδίδοσαν	διδότωσαν	διδῶσι(ν)			
Aorist, I gave	Sing. 1.		ἔδωκα (weak)		δῶ		δοῦναι	M. δούς F. δοῦσα N. δόν Stem δοντ-
	2.		ἔδωκας	δός	δῷς or δοῖς			
	3.		ἔδωκε(ν)	δότω	δῷ, δοῖ, or δώη	δώη (only 3rd pers. sing, found in N.T.)		
	Plur. 1.		ἔδομεν (strong)		δῶμεν			
	2.		ἔδοτε	δότε	δῶτε			
	3.		ἔδοσαν	δότωσαν	δῶσι(ν)			

MIDDLE AND PASSIVE VOICES

		Indicative	Subjunctive	Imperative	Infinitive	Participle
Present Mid., *I am giving for myself* Pass., *I am being given*	*Sing.* 1. 2. 3. *Plur.* 1. 2. 3.	δίδομαι δίδοσαι δίδοται διδόμεθα δίδοσθε δίδονται	Not found in the N.T.	δίδοσο διδόσθω δίδοσθε διδόσθωσαν	δίδοσθαι	M. διδόμενος F. διδομένη N. διδόμενον
Imperfect Mid., *I was giving for myself* Pass., *I was being given*	*Sing.* 1. 2. 3. *Plur.* 1. 2. 3.	ἐδιδόμην ἐδίδοσο ἐδίδοτο ἐδιδόμεθα ἐδίδοσθε ἐδίδοντο				
Strong Aorist Mid. only, *I gave for myself*	*Sing.* 1. 2. 3. *Plur.* 1. 2. 3.	ἐδόμην ἔδου ἔδοτο ἐδόμεθα ἔδοσθε ἔδοντο ⎱ But not found in the N.T.	Not found in the N.T.	δοῦ δόσθω δόσθε δόσθωσαν	δόμαι δῷ, etc., as present endings but not found in the N.T.	δόσθαι but not found in the N.T.
					δόμενος, η, ον, but not found in the N.T.	

OTHER TENSES

	ACTIVE	MIDDLE	PASSIVE
Future	δώσω	-δώσομαι	δοθήσομαι
Aorist			ἐδόθην
Perfect	δέδωκα	δέδομαι	δέδομαι

δείκνυμι *I show*; stem δεικ-

ACTIVE VOICE

| TENSE | NUMBER and PERSON | INDICATIVE MOOD | | IMPERATIVE MOOD | CONJUNCTIVE MOOD | | VERB INFINITE | |
		PRIMARY	HISTORIC		PRIMARY	HISTORIC	Infinitive (noun)	Participle (adjective)
Present, *I show* Imperfect, *I was showing*	*Sing.* 1. 2. 3. *Plur.* 1. 2. 3.	Present δείκνυμι δείκνυς δείκνυσι(ν) δείκνυμεν δείκνυτε δείκνύασι(ν)	Imperfect ἐδείκνυν ἐδείκνυς ἐδείκνυ ἐδείκνυμεν ἐδείκνυτε ἐδείκνυσαν	δείκνυ δεικνύτω δείκνυτε δεικνύτωσαν	Subjunctive δεικνύω δεικνύῃς δεικνύῃ δεικνύωμεν δεικνύητε δεικνύωσι(ν)	Optative Not found in the N.T.	δεικνύναι	M. δεικνύς F. δεικνῦσα N. δεικνύν Stem. δεικνυντ-

MIDDLE AND PASSIVE VOICES

					Not found in the N.T.	δείκνυσθαι	M. δεικνύμενος F. δεικνυμένη N. δεικνύμενον
Present Mid., I show for myself Pass., I am being shown	Sing. 1. 2. 3. Plur. 1. 2. 3.	δείκνυμαι δείκνυσαι δείκνυται δεικνύμεθα δείκνυσθε δείκνυνται	ἐδεικνύμην ἐδείκνυσο ἐδείκνυτο ἐδεικνύμεθα ἐδείκνυσθε ἐδείκνυντο	δείκνυσο δεικνύσθω δείκνυσθε δεικνύσθωσαν	δεικνύωμαι δεικνύῃ δεικνύηται δεικνυώμεθα δεικνύησθε δεικνύωνται		
Imperfect Mid., I was showing for myself Pass., I was being shown							

OTHER TENSES

	ACTIVE	MIDDLE	PASSIVE
Future	δείξω		
Aorist	ἔδειξα	ἐδειξάμην	ἐδείχθην
Perfect		δέδειγμαι	δέδειγμαι

εἰμί *I am*; stem ἐσ-

TENSE	NUMBER and PERSON	INDICATIVE MOOD — Present (PRIMARY)	INDICATIVE MOOD — Imperfect (HISTORIC)	IMPERATIVE MOOD	CONJUNCTIVE MOOD — Subjunctive (PRIMARY)	CONJUNCTIVE MOOD — Optative (HISTORIC)	VERB INFINITE — Infinitive (noun)	VERB INFINITE — Participle (adjective)
Present, *I am* / Imperfect, *I was*	*Sing.* 1.	εἰμί	ἤμην		ὦ	εἴην	εἶναι	M. ὤν
	2.	εἶ	ἦς or ἦσθα	ἴσθι	ᾖς	εἴης		F. οὖσα
	3.	ἐστί(ν)	ἦν	ἔστω or ἤτω	ᾖ	εἴη		N. ὄν
	Plur. 1.	ἐσμέν	ἦμεν or ἤμεθα		ὦμεν	εἴημεν or εἶμεν		Stem ὀντ-
	2.	ἐστέ	ἦτε	ἔστε	ἦτε	εἴητε or εἶτε		
	3.	εἰσί(ν)	ἦσαν	ἔστωσαν or ἤτωσαν	ὦσι(ν)	εἴησαν or εἶεν		
Future, *I shall be*	*Sing.* 1.	ἔσομαι					ἔσεσθαι	M. ἐσόμενος
	2.	ἔσῃ						F. ἐσομένη
	3.	ἔσται						N. ἐσόμενον
	Plur. 1.	ἐσόμεθα						
	2.	ἔσεσθε						
	3.	ἔσονται						

οἶδα *I know*; the Perfect of the stem ἰδ-

TENSE	NUMBER and PERSON	INDICATIVE MOOD — Perfect (PRIMARY)	INDICATIVE MOOD — Pluperfect (HISTORIC)	IMPERATIVE MOOD	CONJUNCTIVE MOOD — Subjunctive (PRIMARY)	CONJUNCTIVE MOOD — Optative (HISTORIC)	VERB INFINITE — Infinitive (noun)	VERB INFINITE — Participle (adjective)
Perfect (present meaning), *I know* / Pluperfect, *I knew*	*Sing.* 1.	οἶδα	ᾔδειν		εἰδῶ	Not found in the N.T.	εἰδέναι	M. εἰδώς
	2.	οἶδας	ᾔδεις	ἴσθι	εἰδῇς			F. εἰδυῖα
	3.	οἶδε(ν)	ᾔδει	ἴστω	εἰδῇ			N. εἰδός
	Plur. 1.	οἴδαμεν	ᾔδειμεν		εἰδῶμεν			Stem εἰδοτ-
	2.	οἴδατε	ᾔδειτε	ἴστε	εἰδῆτε			
	3.	οἴδασι(ν)	ᾔδεισαν	ἴστωσαν	εἰδῶσι(ν)			

Appendix 5

LIST OF VERBS

Lists of verbs have been given in the course of the book as follows:

The list given below includes verbs which employ suppletives,[1] the -μι verbs apart from those like δείκνυμι, a few verbs which are similar in spelling, some of the most common New Testament verbs, and also a few which have not been included in the lists in the body of the book.

This list is given for the purpose of quick reference by the student. If he fails to find the verb he wants in it he should consult the appropriate list among those mentioned above.

---

[1] I.e. verbs which are defective in some tenses, and have these tenses supplied (Latin *suppleo*) from different roots: e.g. ϝειδ and ὀπ are suppletives of ὁράω *see*.

	English	Future	Aorist	Perf. Act.	Perf. Pass.	Aor. Pass.	Remarks
ἄγω	lead	ἄξω	ἤγαγον	ἦρκα	ἦγμαι	ἤχθην	Aor. infin. ἐλεῖν
αἱρέω	take	ἑλῶ	-εἶλον		-ἥρημαι	-ἡρέθην	Aor. infin. ἆραι
αἴρω	raise	ἀρῶ	ἦρα		ἦρμαι	ἤρθην	
ἀκούω	hear	ἀκούσω ἀκούσομαι	ἤκουσα	ἀκήκοα		ἠκούσθην	
ἀναλίσκω ἀναλόω	destroy	ἀναλώσω	ἀνήλωσα			ἀνηλώθην	Aor. infin. ἀναλῶσαι
-βαίνω βάλλω	go throw	-βήσομαι βαλῶ	-ἔβην ἔβαλον	-βέβηκα βέβληκα	βέβλημαι	ἐβλήθην	
γίγνομαι	become	γενήσομαι	ἐγενόμην ἐγενήθην	γέγονα γεγένημαι	γεγένημαι		For γί-γνομαι
γιγνώσκω	come to know	γνώσομαι	ἔγνων	ἔγνωκα	ἔγνωσμαι	ἐγνώσθην	For γι-γνώσκω See p. 117
δέομαι	need, beseech						See p. 117. 3rd pers. sing. pres. indic., δεῖ, used impersonally, *it is necessary*
δέω	bind	δήσω	ἔδησα	δέδεκα	δέδεμαι	ἐδέθην	
δίδωμι	give	δώσω	ἔδωκα but 1st p.p. ἔδομεν	δέδωκα	δέδομαι·	ἐδόθην	
δύναμαι	am able	δυνήσομαι	ἐδυνήθην ἠδυνήθην		-δέδυμαι		ἐδυνάσθην (and ἠδ-) also found for aor.
δύνω, -δύω	set (of sun)		ἔδυσα, ἔδυν				Both str. and wk. aor. are found
ἐθίζω [ἔθω]	accustom *be accustomed*			εἴωθα	εἴθισμαι		Pres. obsolete. Perf. has pres. meaning
-είκω [εἴκω]	yield *am like*		εἶξα	ἔοικα			Pres. obsolete. Perf. has pres. meaning See p. 316
εἰμί	am	ἔσομαι					

-εἶμι	*go*						See p. 264
ἑλκόω	*make sore*						
ἕλκω	*drag*	ἑλκύσω	εἵλκυσα		εἵλκομαι		
ἐλπίζω	*hope*	ἐλπιῶ	ἤλπισα	ἤλπικα			See p. 225
ἐπίσταμαι	*understand*						
ἐραυνάω ἐρευνάω	*search*		ἠραύνησα				
ἐργάζομαι	*work*		ἠργασάμην εἰργασάμην	εἴργασμαι		-ειργάσθην	
ἔρχομαι	*come*	ἐλεύσομαι	ἦλθον	ἐλήλυθα			Aor. infin. ἐλθεῖν τρώγω also used as suppletive
ἐσθίω, ἔσθω	*eat*	φάγομαι	ἔφαγον				
ἔχω	*have*	ἕξω	ἔσχον	ἔσχηκα			Imperf., εἶχον
ζάω	*live*	ζήσω, ζήσομαι	ἔζησα				For ζάω. See p. 114
θάπτω	*bury*		ἔθαψα			ἐτάφην	
θιγγάνω	*touch*		ἔθιγον				
-θνῄσκω	*die*	-θανοῦμαι	-ἔθανον	τέθνηκα			Perf. not compounded
-ἵημι	*send*	-ἥσω	-ἧκα	-εἷκα	-ἕωμαι part. εἱμένος	-ἕθην	See pp. 243ff.
ἱλάσκομαι	*propitiate*		ἱλάσθην				
-ἵστημι	*stand*	στήσω	ἔστην (str.) ἔστησα (wk.)	ἕστηκα (intrans.) -ἕστακα (trans.)		ἐστάθην	See pp. 220ff.
καθαίρω	*purify*		ἐκάθαρα		κεκάθαρμαι		Not a κατά compound. Distinguish καθαίρω, compound of αἱρέω
καθαρίζω κάθημαι	*purify* *sit*	καθαριῶ καθήσομαι	ἐκαθέρισα		κεκαθέρισμαι	ἐκαθερίσθην	ἐκαθαρ- also found Imperf. ἐκαθήμην. See pp. 264f.
καίω	*burn*	καύσω	ἔκαυσα		κέκαυμαι	-ἐκάην (str.) ἐκαύθην (wk.)	

	English	Future	Aorist	Perf. Act.	Perf. Pass.	Aor. Pass.	Remarks
καλέω	call	καλέσω	ἐκάλεσα	κέκληκα	κέκλημαι	ἐκλήθην	
καυχάομαι	boast	καυχήσομαι	ἐκαυχησάμην	κεκαύχημαι			
κεῖμαι	lie						Imperf. ἐκείμην. See pp. 233f.
κίχρημι	lend		ἔχρησα				
κλαίω	weep	κλαύσω / κλαύσομαι	ἔκλαυσα				
κλάω	break		ἔκλασα			ἐκλάσθην	
κλείω	shut	κλείσω	-ἔκλεισα		κέκλεισμαι	-ἐκλείσθην	
κρέμαμαι	hang		ἐκρέμασα			ἐκρεμάσθην	
λαμβάνω	take	λήμψομαι	ἔλαβον	εἴληφα	εἴλημμαι	ἐλήμφθην / ἐρρέθην	Fut. λήψομαι in some texts
λέγω	say	ἐρῶ	εἶπον	εἴρηκα	εἴρημαι	-ἐλέχθην	
-λέγω	gather	-λέξω	-έλεξα		λέλεγμαι		
λούω	wash		ἔλουσα		λέλουμαι / λέλουσμαι		
μέλλω	intend	μελλήσω					Imperf. ἤμελλον, ἔμελλον
μέλω	be a care	-μελήσω				-ἐμελήθην	3rd pers. sing. pres. used impersonally, it matters
μνηστεύω / -νέμω	betroth / distribute				ἐμνήστευμαι	ἐμνηστεύθην / -ἐνεμήθην	
οἰκτείρω / οἰκτίρω	pity	οἰκτειρήσω					
οἴομαι	think						1st pers. sing. pres. shortened to οἶμαι See p. 259
-όλλυμι	destroy	-ὀλέσω, -ὀλῶ	-ὤλεσα	-όλωλα			
ὀνίνημι	profit		ὠνάμην				Aor. opt., ὀναίμην

Present	Meaning	Future	Aorist	Perfect	Perfect mid./pass.	Aorist passive	Notes
ὁράω	see	ὄψομαι	εἶδον	ἑόρακα, ἑόρακα		ὤφθην	Imperf. ἑώρων. Aor. infin. ἰδεῖν
ὀφείλω	owe		ὠψάμην (mid.), ὤφελον				Str. aor. ὤφελον has become a particle, used sometimes to express a wish; e.g. I Cor. 4.8
πάσχω	suffer		ἔπαθον	πέπονθα			
πείθω	persuade	πείσω	ἔπεισα	πέποιθα	πέπεισμαι	ἐπείσθην	
πίμπρημι	burn		-ἔπρησα				
πίνω	drink	πίομαι	ἔπιον	πέπωκα		ἐπόθην	Aor. infin. πεῖν. Classical Greek, πίμπλημι. See p. 132
-πίμπλάω	fill		ἔπλησα		πέπλησμαι	ἐπλήσθην	See p. 263
πίπτω	fall	πεσοῦμαι	ἔπεσον	πέπτωκα			
προφητεύω	prophesy	προφητεύσω	ἐπροφήτευσα				
-πτύω	spit	-πτύσω	ἔπτυσα			-ἐπτύσθην	
ῥύομαι	deliver	ῥύσομαι	ἐρυσάμην, ἐρρυσάμην			ἐρύσθην	
-σείω	shake	σείσω	ἔσεισα			ἐσείσθην	
-στέλλω	send	-στελῶ	-ἔστειλα	-ἔσταλκα	-ἔσταλμαι	-ἐστάλην	
στρέφω	turn	-στρέψω	ἔστρεψα		ἔστραμμαι, ἔστρεμμαι	ἐστράφην	
τελέω	accomplish	-τελέσω, -τελοῦμαι	ἐτέλεσα	τετέλεκα	τετέλεσμαι	ἐτελέσθην	Meaning of simplex doubtful; ἀνατέλλω means "raise"
-τέλλω			-ἔτειλα	-τέταλκα	-τέταλμαι		
-τέμνω	cut		-ἔτεμον		-τέτμημαι	ἐτμήθην	
τίθημι	place	θήσω	ἔθηκα	τέθεικα	τέθειμαι	ἐτέθην	Imperfect, ἐτίθουν. Aor. infin. θεῖναι: see pp. 230ff.
τίκτω	bring forth	τέξομαι	ἔτεκον			ἐτέχθην	
-τρέπω	turn		-ἔτρεψα			-ἐτράπην	
τρέφω	nourish		ἔθρεψα		τέθραμμαι	-ἐτράφην	Root, θρεφ
τρέχω	run		ἔδραμον				Root, θρεχ

	English	Future	Aorist	Perf. Act.	Perf. Pass.	Aor. Pass.	Remarks
τύπτω	strike	πατάξω	ἔπαισα, ἐπάταξα			ἐπλήγην	παίω, πατάσσω, and πλήσσω are suppletives
φέρω	bear	οἴσω	ἤνεγκα	-ἐνήνοχα		ἠνέχθην	Aor. infin., ἐνεγκεῖν
φημί	say						Imperf, ἔφην. See p. 225
φθείρω	destroy	φθερῶ	ἔφθειρα		-ἔφθαρμαι	ἐφθάρην	
-χέω } -χύννω	pour	-χεῶ	-ἔχεα		-κέχυμαι	-ἐχύθην	
χράομαι	use		ἐχρησάμην	κέχρημαι			For χράομαι: see p. 114
χρίω	anoint		ἔχρισα				Verbal adj., χριστός
-ὠθέω	thrust		-ὦσα, -ἔωσα				

ENGLISH–GREEK VOCABULARY

No details of the Greek words (e.g. declension of nouns and principal parts of verbs) are given here. The student should refer to the Greek Index and turn to the pages there indicated.

The following list is not in any sense complete. It contains, however, all the words which the student will need in doing the English–Greek exercises in the book.

abst. abstract; *adj.* adjective; *adv.* adverb; *caus.* causal; *con.* conjunction; *consec.* consecutive; *dem.* demonstrative; *fin.* final; *interrog.* interrogative; *intr.* intransitive; *n.* noun; *pl.* plural; *prep.* preposition; *rel.* relative; *s.* singular; *tr.* transitive; *v.* verb.

abase ταπεινόω
able, be δύναμαι
abound περισσεύω
about περί
about to, be μέλλω
above ὑπέρ
Abraham Ἀβραάμ
according to κατά
accuse κατηγορέω, αἰτιάομαι
accustomed, be εἴωθα
add προστίθημι
affairs (*see* p. 171)
afflict κακόω
affliction θλίψις
afraid, be φοβέομαι
after (prep.) μετά, ὀπίσω
again πάλιν
all πᾶς
allow ἐπιτρέπω
already ἤδη
also καί
although καίπερ
amazed, be *use* ἐξίστημι
among ἐν
and καί
Andrew Ἀνδρέας
angel ἄγγελος
anger ὀργή
anoint ἀλείφω
answer ἀποκρίνομαι
apostle ἀπόστολος
appear φαίνομαι
appoint καθίστημι
Arab Ἄραψ
arise ἐγείρομαι, *or use* ἀνίστημι

arm (n.) βραχίων
arouse ἐγείρω
arrange τάσσω
as καθώς
ascend ἀναβαίνω
ascertain πυνθάνομαι
ask ἐρωτάω (question); αἰτέω (request)
asleep, fall, *use* κοιμάω
assembly ἐκκλησία
at hand, be *use* ἐγγίζω
authority ἐξουσία
away from ἀπό

bad κακός
baptize βαπτίζω
be εἰμί
beam δοκός
bear φέρω (carry); τίκτω (bring forth)
beat (v.) δέρω
because ὅτι
because of διά
become γίνομαι
bed κλινή, κράββατος
before (prep.) ἔμπροσθεν
begin ἄρχομαι
behold βλέπω
believe πιστεύω
beloved ἀγαπητός
belt ζώνη
beseech δέομαι
beside παρά
best κράτιστος
Bethlehem Βηθλεέμ
betray προδίδωμι
better (adv.) βέλτιον, κρεῖσσον

beyond ὑπέρ
bid (v.) κελεύω
black μέλας
bless μακαρίζω, εὐλογέω
blessed μακάριος
blind (adj.) τυφλός
boat πλοῖον
bodily σωματικός
body σῶμα
bond δεσμός
book βιβλίον
both ... and τε ... καί
boy παῖς
branch κλῆμα
bread ἄρτος
break (v.) συντρίβω, κλάω; λύω (of Sabbath); διαρήσσω (of nets); b. in pieces συντρίβω
bride νύμφη
bridegroom νυμφίος
bring φέρω; b. down κατάγω; b. forth τίκτω; b. in εἰσάγω
brother ἀδελφός
brow ὀφρύς
build οἰκοδομέω
bunch (of grapes) βότρυς
bury θάπτω
but ἀλλά
buy ἀγοράζω
by ὑπό, παρά

Caesar Καῖσαρ
call (v.) καλέω; ὀνομάζω (name)
can (see be able)
care (n.) μέριμνα
care (v.) use μέλει
carefully ἀκριβῶς
carry φέρω
cast βάλλω; c. away ἀποβάλλω; c. out ἐκβάλλω
cause scandal, c. to stumble σκανδαλίζω
certain τις
charger (dish) πίναξ
chief priest ἀρχιερεύς
child τέκνον, παιδίον
choke πνίγω
Christ Χριστός
church ἐκκλησία
circumcize περιτέμνω
city πόλις
cleanse καθαρίζω
cliff κρημνός
clothe ἐνδύω, περιβάλλω
clothing use ἱμάτιον

cloud νεφέλη
colt πῶλος
come ἔρχομαι; c. down κατέρχομαι; c. into εἰσέρχομαι; c. suddenly φθάνω; c. together συνέρχομαι; c. to pass γίνομαι
comfort (v.) παρακαλέω
comforted, be θαρρέω, θαρσέω
command (v.) κελεύω, ἐντέλλομαι
commandment ἐντολή
compel ἀναγκάζω
concerning περί
condemn κατακρίνω
confess ὁμολογέω
confound ἐξίστημι
conqueror, be a νικάω
consider βουλεύομαι
convict (v.) ἐλέγχω
cool (v.) καταψύχω
council συνέδριον
counsel, take βουλεύω
cover (v.) καλύπτω
cross σταυρός
crowd ὄχλος
crown (n.) στέφανος
crucify afresh ἀνασταυρόω
cry out ἀνακράζω
cubit πῆχυς
custom use εἴωθα
cut τέμνω; c. down ἐκκόπτω

dare τολμάω
darken σκοτίζω
darkness σκότος
daughter θυγάτηρ
day ἡμέρα; in the days of ἐπί
dead (adj.) νεκρός
dead, be use -θνήσκω
death θάνατος
death, put to θανατόω
deceive ψεύδομαι
declare γνωρίζω
deed ἔργον; δύναμις (mighty d.)
deny ἀπαρνέομαι
depart ὑπάγω
descend καταβαίνω
desert (n.) ἔρημος
destroy καταλύω, φθείρω
devil δαιμόνιον
die ἀποθνήσκω
diligently ἐπιμελῶς
dip βάπτω
disciple μαθητής
dishonour (v.) ἀτιμάζω
disregard (v.) παρίημι

distribute διαδίδωμι
disturbance τάραχος
do ποιέω
dog κύων
door θύρα
draw nigh ἐγγίζω
drink πίνω
drive ἄγω; d. together συνάγω
dwell σκηνόω, κατοικέω

earth γῆ
earthly ἐπίγειος
earthquake σεισμός
easy εὔκοπος
eat ἐσθίω
Egypt Αἴγυπτος
elder πρεσβύτερος
element στοιχεῖον
elsewhere ἀλλαχοῦ
enemy ἐχθρός
enrolment ἀπογραφή
enter εἰσέρχομαι
entrusted, be πιστεύομαι
escape (v.) ἀποφεύγω
eternal αἰώνιος
even καί; not e. οὐδέ, μηδέ; e. so οὕτως
evening ὀψία
every one πᾶς
evil κακός, πονηρός
evil spirit δαιμόνιον
example ὑπόδειγμα
exalt ὑψόω
eye ὀφθαλμός

faith πίστις
faithful πιστός
fall πίπτω
fall asleep use κοιμάω
falsehood ψεῦδος
far, from πόρρωθεν
fast (v.) νηστεύω
father πατήρ
favour χάρις
fear (n.) φόβος
fear (v.) φοβέομαι
feast ἑορτή
feed (tr.) τρέφω
few ὀλίγοι
field ἀγρός
fifty πεντήκοντα
fig-tree συκῆ
fill πίμπλημι
find εὑρίσκω
finger δάκτυλος

firmly βεβαίως
first (adj.) πρῶτος; (adv.) πρῶτον
fish ἰχθύς
fisherman ἁλιεύς
five πέντε
flee φεύγω
flesh σάρξ
follow ἀκολουθέω
foolish μωρός
foot πούς
for (con.) γάρ; διά (because of); ὑπέρ (on behalf of)
forbid κωλύω
forget ἐπιλανθάνομαι
forgive χαρίζομαι, ἀφίημι
forsake ἐγκαταλείπω
forty τεσσαράκοντα
found θεμελιόω
foundation θεμέλιος
four τέσσαρες
fourteen δεκατέσσαρες
free (adj.) ἐλεύθερος
free (v.) ἐλευθερόω
friend φίλος
from ἀπό, ἐκ, ἐξ
fruit καρπός
fulfil πληρόω
furnish στρώννυμι

gain (v.) κερδαίνω
Galilaean Γαλιλαῖος
garden κῆπος
garment ἱμάτιον
gather, g. together συνάγω
gift δῶρον
girdle ζώνη
girl παῖς; young g. παιδίσκη
give δίδωμι; g. up ἀναδίδωμι; g. in marriage γαμίζω
glorify δοξάζω
glory (n.) δόξα
go βαίνω, πορεύομαι; g. down καταβαίνω; g. into εἰσέρχομαι; g. out ἐξέρχομαι; g. up ἀναβαίνω
God, god Θεός, θεός
gold (n.) χρυσός
good ἀγαθός, καλός
gospel εὐαγγέλιον
gospel, preach the εὐαγγελίζομαι
great μέγας; so g. τοσοῦτος
greater μείζων
Greek (n.) Ἕλλην; in G. Ἑλληνιστί
ground (n.) γῆ
guard (n.) φυλακή
guard (v.) φυλάσσω

hair θρίξ
hallow ἁγιάζω
hand (n.) χείρ; at the hands of ὑπό
hand, be at use ἐγγίζω
hand over παραδίδωμι
hang (intr.) κρέμαμαι
happen συμβαίνω; or use γίνομαι
hard σκληρός
harm (v.) βλάπτω
hasten σπεύδω
hate (v.) μισέω
have ἔχω
he αὐτός
head κεφαλή
heal θεραπεύω, ἰάομαι
hear ἀκούω
heart καρδία
heaven οὐρανός
Hebrew, in Ἑβραϊστί
heed, take βλέπω, σκοπέω
herald (n.) κῆρυξ
herd ἀγέλη
here ὧδε
Herod Ἡρῴδης
high ὑψηλός
high-priest ἀρχιερεύς
hill ὄρος
himself ἑαυτόν, αὐτόν
hireling μισθωτός
holy ἅγιος
honour (n.) τιμή
honour (v.) τιμάω
hope (n.) ἐλπίς
hope (v.) ἐλπίζω
hour ὥρα
house οἶκος, οἰκία; master of a h.
  οἰκοδεσπότης
housetop δῶμα
how (interrog.) πῶς; how many?
  πόσοι; how much? πόσος
humble (v.) ταπεινόω
hundred ἑκατόν
husbandman γεωργός

I ἐγώ
if εἰ, ἐάν
immediately εὐθύς
in ἐν; in order to ἵνα, ὅπως; in vain
  δωρεάν
inferior, be ὑστερέω
injustice ἀδικία
insult (v.) ὑβρίζω
interpret ἑρμηνεύω
into εἰς
Israel Ἰσραήλ

Jerusalem Ἱεροσόλυμα, Ἱερουσαλήμ
Jesus Ἰησοῦς
Jew Ἰουδαῖος
Jewish Ἰουδαῖος
John Ἰωάνης
Jordan Ἰορδάνης
Joseph Ἰωσήφ
journey (v.) πορεύομαι
joy χαρά
Judaea Ἰουδαία
Judas Ἰούδας
judge (n.) κριτής
judge (v.) κρίνω

keep τηρέω
key κλείς
kill ἀποκτείνω
kind (adj.) χρηστός
king βασιλεύς
kingdom βασιλεία
kinsman συγγενής
knock down ἐκκόπτω
know γινώσκω, οἶδα

lamb ἀμνός
lame (adj.) χωλός
lamp λαμπάς
land γῆ
last (adj.) ἔσχατος
law νόμος
lawful, be use ἔξεστι(ν)
lawless ἄνομος
lawyer νομικός
lay, lay down τίθημι; l. hold of ἅπτομαι
lead ἄγω; l. away ἀπάγω; l. out
  ἐξάγω; l. up ἀναφέρω
leader ἡγεμών
learn μανθάνω
least ἐλάχιστος (or use comparative)
leave λείπω, καταλείπω
lest ἵνα μή, μή
let down χαλάω, καθίημι
let out use ἐκδίδωμι
letter ἐπιστολή
Levi Λευεί
Levite Λευείτης
lie κεῖμαι; ψεύδομαι (deceive)
life ζωή
lift αἴρω; l. up ὑψόω
lightning ἀστραπή
like (adj.) ὅμοιος
like (v.) φιλέω; be l. ὁμοιόω (in passive)
liken ὁμοιόω
likewise ὁμοίως
listen to ἀκούω

little μικρός
loaf ἄρτος
look at βλέπω
loose λύω
Lord, lord Κύριος, κύριος
love (n.) ἀγάπη
love (v.) ἀγαπάω
Lydda Λύδδα

mad, be μαίνομαι
maid, maiden κοράσιον, παιδίσκη
man ἄνθρωπος; old m. γέρων
manifest φανερός
many πολλοί
Mark Μάρκος
mark (n.) σκοπός
market-place ἀγορά
marriage, give in γαμίζω
marry γαμέω
Mary Μαρία, Μαριάμ
master of a house οἰκοδεσπότης
meat βρῶμα
messenger ἄγγελος
Messiah Μεσσίας
mighty deed δύναμις
minister to διακονέω
money ἀργύριον
month μήν
morning, in the πρωΐ
Moses Μωυσῆς
mother μητήρ
mountain ὄρος
mouth στόμα
move to revolt ἀφίστημι
much πολύς
multitude ὄχλος
murder (v.) φονεύω
murderer ἀνθρωποκτόνος
must use δεῖ
myself ἐμαυτόν

name (n.) ὄνομα
name (v.) ὀνομάζω
nation ἔθνος
native-place πατρίς
near (adv.) ἐγγύς
necessary, be use δεῖ
neither ... nor οὔτε ... οὔτε, μήτε ...
    μήτε
net δίκτυον
never οὐδέποτε, μηδέποτε
new νέος
next day use ἐπαύριον
night νύξ
nine ἐννέα

ninety ἐνενήκοντα
ninth ἔνατος
no longer οὐκέτι, μήκετι
nobody οὐδείς, μηδείς
nor οὔτε, μήτε
not οὐ, οὐκ, οὐχ, μή; not even οὐδέ,
    μηδέ; not only οὐ μόνον
nothing οὐδέν, μηδέν
now νῦν
number (v.) ἀριθμέω

O ὦ
obey ὑπακούω
of (concerning) περί
offer (v.) προσφέρω
oil ἔλαιον
old παλαιός; old man γέρων; of o.
    time ἀρχαῖος
on ἐν, ἐπί
one εἷς
one another ἀλλήλους
open (v.) ἀνοίγω
or ἤ
orator ῥήτωρ
order (v.) κελεύω, παραγγέλλω, ἐντέλ-
    λομαι
order to, in ἵνα, ὅπως
other ἄλλος
ought ὀφείλω or use δεῖ
our ἡμέτερος or use ἡμεῖς
out of ἐκ, ἐξ
outside ἔξω
over ὑπέρ
owe ὀφείλω
own, one's ἴδιος
ox βοῦς

palm-tree φοῖνιξ
parable παραβολή
parent γονεύς
pass by παράγω; p. through διέρχο-
    μαι; come to p. γίνομαι
Passover πάσχα
Paul Παῦλος
pay back ἀποτίνω
peace εἰρήνη
penny δηνάριον
people λαός
perceive γινώσκω
perish use ἀπόλλυμι
persecute διώκω
persuade πείθω
Peter Πέτρος
Pharisee Φαρισαῖος
piece of silver ἀργύριον

Pilate Πειλᾶτος
pinnacle πτερύγιον
place (n.) τόπος; native-place πατρίς
place (v.) τίθημι; p. beside παρατίθημι;
    p. under ὑποτίθημι; p. upon ἐπιτί-
    θημι
place, have χωρέω
plait (v.) πλέκω
plant (v.) φυτεύω
please, be pleasing ἀρέσκω
pleasure ἡδονή
poor πτωχός
possess ἔχω
possible δυνατός
power δύναμις
pray εὔχομαι, προσεύχομαι
prayer προσευχή
preach κηρύσσω
preach the gospel εὐαγγελίζομαι
precious ἔντιμος
prepare, make preparation ἑτοιμάζω
present (v.) παρίστημι
prevent κωλύω
price τιμή
pride ὑπερηφανία
prison φυλακή
prisoner δέσμιος
promise (n.) ἐπαγγελία
promise (v.) ἐπαγγέλλομαι
prophet προφήτης
publican τελώνης
publicly δημοσίᾳ
purify ἁγνίζω
purpose (v.) use προτίθημι
pursue διώκω
put βάλλω; p. forth προβάλλω;
    p. on, of clothes ἐνδύω, περιβάλλω;
    p. to death θανατόω

raise ἐγείρω; r. up ἀνίστημι
reach ἀφικνέομαι
read ἀναγινώσκω
receive λαμβάνω, δέχομαι; receive
    sight ἀναβλέπω
rejoice χαίρω
release ἀπολύω
remain μένω
remember μιμνήσκομαι, μνημονεύω
rend ῥήγνυμι, ῥήσσω
render ἀποδίδωμι
repent μετανοέω
require ζητέω
resist use ἀνθίστημι
restore (give back) ἀποδίδωμι; r. to
    health ἀποκαθίστημι

resurrection ἀνάστασις
return (v. intr.) ὑποστρέφω
reveal ἀποκαλύπτω
revolt, move to ἀφίστημι
reward (n.) μισθός
rich πλούσιος
riches πλοῦτος
right δεξιός
righteous δίκαιος
righteousness δικαιοσύνη
rise use ἀνίστημι; r. up against, use
    ἐπανίστημι
river ποταμός
robber λῃστής
rock πέτρα
roll back ἀνακυλίω
room, upper ἀνάγαιον
root (n.) ῥίζα
rule (v.) ἄρχω
ruler ἄρχων
run τρέχω

Sabbath σάββατον
sacrifice (v.) θύω
saint ἅγιος
sake of, for the διά
Samaria Σαμαρία
same αὐτός
sanctify ἁγιάζω
save σώζω, σῴζω
saviour σωτήρ
say λέγω, φημί
saying (n.) ῥῆμα
scandal, cause σκανδαλίζω
scourge (n.) μάστιξ
scribe γραμματεύς
scripture γραφή
sea θάλασσα
sea-shore αἰγιαλός
seal (n.) σφραγίς
seal (v.) σφραγίζω
second δεύτερος
secretly λάθρᾳ
see βλέπω, ὁράω
seed σπέρμα
seek ζητέω
self αὐτός
sell πωλέω
send πέμπω, ἀποστέλλω
separate (v.) χωρίζω
serpent ὄφις
servant δοῦλος
serve δουλεύω
set ἵστημι; s. before παρατίθημι
seven ἑπτά

shape εἶδος
sheep πρόβατον
shepherd ποιμήν
shining λαμπρός
shout (v.) κράζω
show (v.) δείκνυμι
shut κλείω
sick ἀσθενής
sign σημεῖον
sight, receive ἀναβλέπω
silver (n.) ἄργυρος; piece of s. ἀργύριον
sin (n.) ἁμαρτία
sin (v.) ἁμαρτάνω
sinful ἁμαρτωλός
sinner ἁμαρτωλός
sister ἀδελφή
sit καθίζω; s. at table, at meat
    ἀνάκειμαι; s. down καθίζω, ἀνα-
    πίπτω; s. together συνκάθημαι
six ἕξ
slave δοῦλος
slow βραδύς
small μικρός
so οὕτως; so great τοσοῦτος; so many
    τοσοῦτοι; so that ὥστε
sober σώφρων
soft μαλακός
soldier στρατιώτης
some τις
some... others οἱ μὲν... οἱ δέ (or
    ἄλλοι)
son υἱός
sorrowful, be λυπέω
soul ψυχή
sow σπείρω
speak λέγω; s. against ἀντιλέγω
spear λόγχη
spirit πνεῦμα; evil s. δαιμόνιον
spread abroad διαφημίζω
spring up ἐξανατέλλω
stand (tr.) ἵστημι; intr. ἕστηκα, ἵστα-
    μαι, στήκω; stand around use περι-
    ίστημι; stand by use παρίστημι
stature ἡλικία
stay μένω
steadfast ἑδραῖος
steal κλέπτω
still (adv.) ἔτι
stone (n.) λίθος
straight εὐθύς
strap ἱμάς
stray πλανάω
street ῥύμη
stretch -τείνω; s. forth ἐκτείνω
strike παίω

strong, be ἰσχύω
stumble, cause to σκανδαλίζω
such τοιοῦτος
suffer πάσχω
sun ἥλιος
supper δεῖπνον
swallow up καταπίνω
swear ὄμνυμι
sword μάχαιρα
synagogue συναγωγή

table τράπεζα
take λαμβάνω; t. away ἀπαίρω;
    t. heed βλέπω, σκοπέω; t. up αἴρω,
    ἀναλαμβάνω
talk with συνλαλέω, συλλαλέω
taste γεύομαι
teach διδάσκω
teacher διδάσκαλος
teaching (n.) διδαχή
tell λέγω
temple ἱερόν
tempt πειράζω
temptation πειρασμός
ten δέκα
tent σκηνή
test (v.) πειράζω
that (dem.) ἐκεῖνος; fin. con. ἵνα; caus.
    con. ὅτι; consec. con. ὥστε
themselves ἑαυτούς
then (adv.) τότε; (con.) οὖν
thence ἐκεῖθεν
there ἐκεῖ
therefore οὖν, διὰ ταῦτα
thief κλέπτης
think νομίζω
third τρίτος; the t. time τρίτον
thirty τριάκοντα
this οὗτος
thorn ἄκανθα
thousand χίλιοι, χιλιάς
three τρεῖς; t. times τρίς
through διά
throw βάλλω
thrust out ἐκβάλλω
thunder βροντή
thy σός
time, in due κατὰ καιρόν
Timothy Τιμόθεος
Titus Τίτος
to εἰς, πρός
today σήμερον
tomb μνημεῖον
tongue γλῶσσα
tooth ὀδούς

torch λαμπάς
touch ἅπτομαι
towards πρός
translate ἑρμηνεύω
treasure (n.) θησαυρός
tree δένδρον
tribe φυλή
trouble (v.) ταράσσω
true ἀληθινός
trumpet σάλπιγξ
trust (v.) πέποιθα (see πείθω)
truth, truthfulness ἀλήθεια
tunic χιτών
turn (v. tr.) στρέφω; turn back, intr.
   ὑποστρέφω
twelfth δωδέκατος
twelve δώδεκα
two δύο

unclean ἀκάθαρτος; u. spirit δαιμόνιον
under ὑπό
understand γινώσκω, ἐπίσταμαι
until (con.) ἕως, ἕως οὗ, ἄχρι οὗ, πρίν,
   πρὶν ἤ
until (prep.) ἕως, ἄχρι(ς), μέχρι(ς)
unto εἰς, πρός
upon ἐπί
upper room ἀνάγαιον
use (v.) χράομαι
utterly see p. 268

vain, in δωρεάν
value (v.) τιμάω
vein φλέψ
victory νῖκος
victorious, be νικάω
village κώμη
vineyard ἀμπελών
virgin παρθένος
voice φωνή

wait for προσδέχομαι
walk πορεύομαι, περιπατέω
wall τεῖχος
wander πλανάω
want (v.) βούλομαι, θέλω
wash (v.) λούω
watch (v.) παρατηρέω
water ὕδωρ
water-pot ὑδρία
way ὁδός
we ἡμεῖς
weep κλαίω
weigh down καταβαρύνω
well (adv.) καλῶς, εὖ

well (n.) φρέαρ
what (interrog.) τί
wheat σῖτος
when (interrog.) πότε
when (rel. con.) ὅτε
whensoever ὅταν
where (interrog.) ποῦ
where (rel. con.) οὗ, ὅπου
wherever ὅπου ἄν
which (rel.) ὅς
while ἐν ᾧ, ἕως or use articular infinitive
while, a little μικρόν
white λευκός
who (interrog.) τίς
who (rel.) ὅς
whole ὅλος; ὑγιής (healthy)
whosoever ὅς ἄν
why (interrog.) τί, διὰ τί
wicked πονηρός
wickedness ἀδικία
wilderness ἔρημος
will (n.) θέλημα
will (v.), be willing βούλομαι, θέλω
wind (n.) ἄνεμος
wine οἶνος
wineskin ἀσκός
wipe ἐκμάσσω
wise σοφός
wish (v.) βούλομαι
with σύν, μετά
wither ξηραίνω
without χωρίς
witness (n.) μάρτυς; μαρτυρία (abst.)
woman γυνή
wonder at θαυμάζω; w. greatly
   ἐκθαυμάζω
word λόγος
work (n.) ἔργον
work (v.) ἐργάζομαι
worker, workman ἐργάτης
world κόσμος
worse ἥσσων, ἥττων, χείρων
worship (v.) προσκυνέω
write γράφω
writing (n.) γραφή

year ἔτος
you (s.) σύ; (pl.) ὑμεῖς
young child παιδίον
young girl παιδίσκη
young man νεανίσκος, νεανίας
your (s.) σός; (pl.) ὑμέτερος
yourselves ἑαυτούς

Zebedee Ζεβεδαῖος

GREEK INDEX

It is of no lasting help to the student of a language if the Vocabularies at the end of a Grammar book appear to relieve him of the need to look a word up in order to make certain of its meaning and use. The page references given below indicate where the meaning of the word may be found, and, in many cases, where its use may be studied.

INDEX